AF541113

Microwave and RF Design
Basic Principles

Microwave and RF Design
Basic Principles

– *Author* –

Oppilan Bondalapati

Jaithra Desiraju

BIO-GREEN BOOKS

New Delhi - 110 002

ISBN: 9788196302566 (Hardback)

Published by : BIO-GREEN BOOKS
4736/23, Ansari Road, Darya Ganj,
New Delhi – 110 002
Phone: 011-23245578, 35004336
E-mail: biogreenbooks@gmail.com

Printed at : Replika Press Pvt. Ltd.

Preface

Microwave and RF design is a branch of electrical engineering that deals with the design and development of electronic circuits and systems that operate in the microwave and radio frequency (RF) range. This field is critical in the development of modern communication systems, including wireless networks, satellite communication, radar systems, and microwave ovens. The basic principles of microwave and RF design include transmission line theory, microwave network analysis, impedance matching, and passive components. Transmission line theory is critical in the design of high-frequency circuits because the behavior of circuits in the microwave range is significantly different from that of low-frequency circuits. Network analysis is used to analyze the behavior of circuits, and impedance matching is used to optimize the performance of circuits by matching the impedance of components to the transmission line.

Passive components, such as filters, couplers, and baluns, are widely used in microwave and RF circuits. These components are designed to manipulate the flow of electromagnetic energy in circuits, allowing designers to control the behavior of the circuit and optimize its performance. Antennas are also a critical component in microwave and RF design. Antennas are used to transmit and receive electromagnetic waves, allowing communication systems to function. Designers must consider factors such as antenna size, shape, and orientation when designing communication systems. The field of microwave and RF design has advanced significantly in recent years due to the development of new materials and fabrication techniques. Planar transmission lines, which are widely used in microwave circuits, can now be fabricated using printed circuit board (PCB) technology. This has significantly reduced the cost of manufacturing microwave circuits and has made it easier to design and develop microwave and RF circuits. Microwave and RF design is a critical field in the development of modern communication systems. The principles of microwave and RF design include transmission line theory, network analysis, impedance matching, passive components, and antenna design. With advancements in material science and fabrication techniques, microwave and RF design is becoming increasingly accessible to designers and engineers, leading to the development of more efficient and effective communication systems.

The book covers all the basic principles and concepts of microwave engineering, including transmission lines, network analysis, and passive components. The book provides an in-depth coverage of transmission lines, including planar transmission lines and extraordinary transmission line effects and covers both analytical and graphical network analysis techniques. The book includes clear explanations of complex concepts and equations, making it easy for readers to understand and apply the principles of microwave and RF design. The book includes relevant illustrations and diagrams to help readers visualize the concepts and applications of microwave and RF design. The book is suitable for beginners who have a basic understanding of electrical engineering and are interested in learning about microwave and RF design.

We are grateful to all those persons as well as various books, manuals, periodicals, magazines, journals etc. that helped in the preparation of this book. In spite of the best efforts, it is possible that some errors may have occurred into the compilation and editing of the book. Further queries, constructive suggestions and criticisms for the improvement of the book are always welcomed and shall be thankfully acknowledged.

Oppilan Bondalapati
Jaithra Desiraju

Contents

CHAPTER 1

Introduction to Microwave Engineering

1.1 RF and Microwave Engineering

Radio communications is the main driver of RF system development, leading to RF technology evolution at an unprecedented pace. A radio signal is a signal that is coherently generated, radiated by a transmit antenna, propagated through the air, collected by a receive antenna, and then amplified and information extracted. The radio spectrum is part of the electromagnetic (EM) spectrum exploited by humans for communications. A broad categorization of the EM spectrum is shown in Table 1-1. Today radios operate from 3 Hz (for submarine communications) to 300 GHz (proposed for 6G cellular communications).

Table 1-1: Broad electromagnetic spectrum divisions.

Name or band	Frequency	Wavelength
Radio frequency	3 Hz – 300 GHz	100 000 km – 1 mm
Microwave	300 MHz – 300 GHz	1 m – 1 mm
Millimeter (mm) band	110 – 300 GHz	2.7 mm – 1.0 mm
Infrared	300 GHz – 400 THz	1 mm – 750 nm
Far infrared	300 GHz – 20 THz	1 mm – 15 μm
Long-wavelength infrared	20 THz – 37.5 THz	15–8μm
Mid-wavelength infrared	37.5 – 100 THz	8–3 μm
Short-wavelength infrared	100 THz – 214 THz	3–1.4 μm
Near infrared	214 THz – 400 THz	1.4 μm – 750 nm
Visible	400 THz – 750 THz	750 – 400 mm
Ultraviolet	750 THz – 30 PHz	400 – 10 nm
X-Ray	30 PHz – 30 EHz	10 – 0.01 nm
Gamma Ray	> 15 EHz	< 0.02 nm

Gigahertz, GHz = 10^9 Hz; terahertz, THz = 10^{12} Hz; pentahertz, PHz = 10^{15} Hz; exahertz, EHz = 10^{18} Hz.

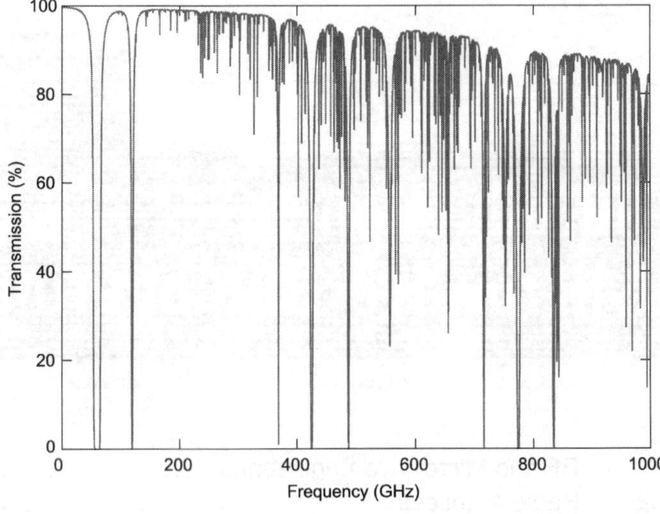

Figure 1-1: Atmospheric transmission at Mauna Kea, with a height of 4.2 km, on the Island of Hawaii where the atmospheric pressure is 60% of that at sea level and the air is dry with a precipitable water vapor level of 0.001 mm. After [1].

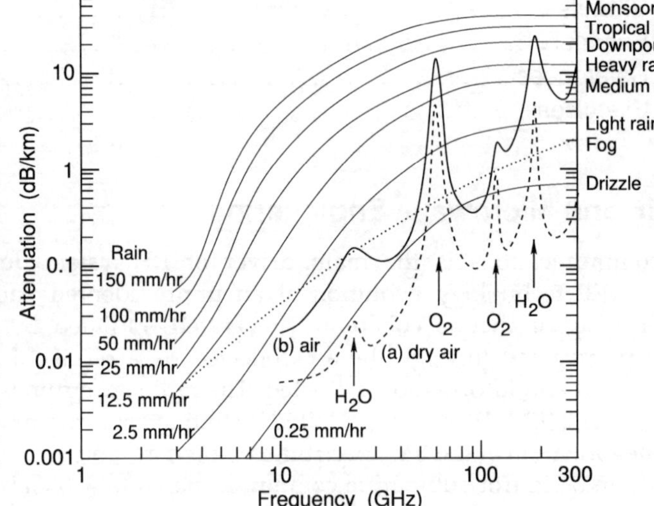

Figure 1-2: Excess attenuation due to atmospheric conditions showing the effect of rain on RF transmission at sea level. Curve (a) is atmospheric attenuation, due to excitation of molecular resonances, of very dry air at 0 °C, curve (b) is for typical air (i.e. less dry) at 20 °C. The attenuation shown for fog and rain is additional (in dB) to the atmospheric absorbtion shown as curve (b).

Propagating RF signals in air are absorbed by molecules in the atmosphere primarily by molecular resonances such as the bending and stretching of bonds which converts EM energy into heat. The transmittance of radio signals versus frequency in dry air at an altitude of 4.2 km is shown in Figure 1-1 and there are many transmission holes due to molecular resonances. The lowest frequency molecular resonance in dry air is the oxygen resonance centered at 60 GHz, but below that the absorption in dry air is very small. Attenuation increases with higher water vapor pressure peakin at 22 GHz and broadening due to the close packing of molecules in air. The effect of water is seen in Figure 1-2 and it is seen that rain and water vapor have little effect on cellular communications which are below 5 GHz except for millimeter-wave 5G.

RF signals diffract and so can bend around structures and penetrate into valleys. The ability to diffract reduces with increasing frequency. However, as frequency increases the size of antennas decreases and the capacity to carry information increases. A very good compromise for mobile communications is at UHF, 300 MHz to 4 GHz, where antennas are of convenient size and there is a good ability to diffract around objects and even penetrate walls.

1.1.1 Electromagnetic Fields

Communicating using EM signals built from an understanding of magnetic induction based on the experiments of Faraday in 1831 [2] in which he investigated the relationship of magnetic fields and currents, and now known as Faraday's law. This and the other static field laws are not enough to describe radio waves. The required description is embodied in Maxwell's equations and after these were developed it took little time before radio was invented.

1.1.2 Static Field Laws

There are two components of the EM field, the **electric field**, E, with units of volts per meter (V/m), and the **magnetic field**, H, with units of amperes per meter (A/m). There are also two flux quantities with the first being D, the **electric flux** density, with units of coulombs per square meter (C/m^2), and the other is B, the **magnetic flux** density, with units of teslas (T). B and H, and D and E, are related to each other by the properties of the medium, which are embodied in the quantities μ and ε (with the caligraphic letter, e.g. $\mathcal{B}$, denoting a time domain quantity):

$$\bar{\mathcal{B}} = \mu\bar{\mathcal{H}} \qquad (1.1) \qquad\qquad \bar{\mathcal{D}} = \varepsilon\bar{\mathcal{E}}, \qquad (1.2)$$

where the over bar denotes a vector quantity, and μ is the **permeability** of the medium and describes the ability to store **magnetic energy** in a region. The permeability in free space (or vacuum) is $\mu_0 = 4\pi \times 10^{-7}$ H/m and then

$$\bar{\mathcal{B}} = \mu_0\bar{\mathcal{H}}. \qquad (1.3)$$

The other material quantity is the **permittivity**, ε, and in a vacuum

$$\bar{\mathcal{D}} = \varepsilon_0\bar{\mathcal{E}}, \qquad (1.4)$$

where $\varepsilon_0 = 8.854 \times 10^{-12}$ F/m is the permittivity of a vacuum. The **relative permittivity**, ε_r, the **relative permeability**, μ_r, are defined as

$$\varepsilon_r = \varepsilon/\varepsilon_0 \quad \text{and} \quad \mu_r = \mu/\mu_0. \qquad (1.5)$$

Biot-Savart Law

The Biot–Savart law relates current to magnetic field as, see Figure 1-3,

$$d\bar{H} = \frac{Id\bar{\ell} \times \hat{a}_R}{4\pi R^2}, \qquad (1.6)$$

with units of amperes per meter in the SI system. In Equation (1.6) $d\bar{H}$ is the incremental static H field, I is current, $d\bar{\ell}$ is the vector of the length of a filament of current I, $\hat{a}_R$ is the unit vector in the direction from the current filament to the magnetic field, and R is the distance between the filament and the magnetic field. The $d\bar{H}$ field is directed at right angles to $\hat{a}_R$ and the current filament. So Equation (1.6) says that a filament of current produces a

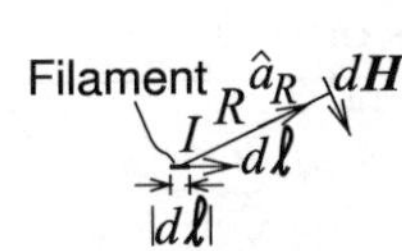

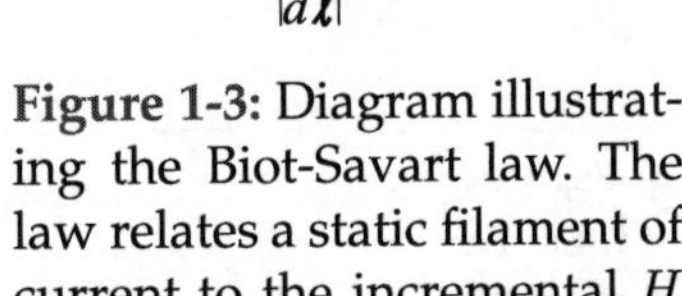

Figure 1-3: Diagram illustrating the Biot-Savart law. The law relates a static filament of current to the incremental H field at a distance.

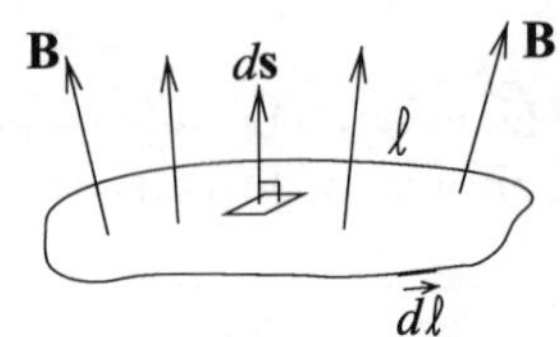

Figure 1-4: Diagram illustrating Faraday's law. The contour ℓ encloses the surface.

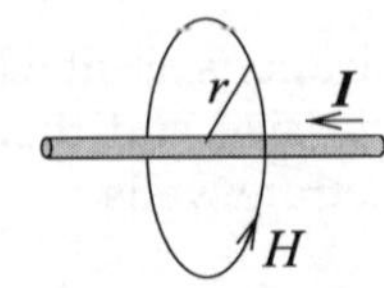

Figure 1-5: Diagram illustrating Ampere's law. Ampere's law relates the current, I, on a wire to the magnetic field around it, H.

magnetic field at a point. The total magnetic field from a current on a wire or surface can be found by modeling the wire or surface as a number of current filaments, and the total magnetic field at a point is obtained by integrating the contributions from each filament.

Faraday's Law of Induction

Faraday's law relates a time-varying magnetic field to an induced voltage drop, V, around a closed path, which is now understood to be $\oint_\ell \bar{\mathcal{E}} \cdot d\ell$, that is, the closed contour integral of the electric field,

$$V = \oint_\ell \bar{\mathcal{E}} \cdot d\ell = -\oint_s \frac{\partial \bar{\mathcal{B}}}{\partial t} \cdot ds, \tag{1.7}$$

and this has the units of volts in the SI unit system. The operation described in Equation (1.7) is illustrated in Figure 1-4.

Ampere's Circuital Law

Ampere's circuital law, often called just Ampere's law, relates direct current and the static magnetic field $\bar{\mathcal{H}}$, see Figure 1-5:

$$\oint_\ell \bar{H} \cdot d\ell = I_{\text{enclosed}}. \tag{1.8}$$

That is, the integral of the magnetic field around a loop is equal to the current enclosed by the loop. Using symmetry, the magnitude of the magnetic field at a distance r from the center of the wire shown in Figure 1-5 is

$$H = |I|/(2\pi r). \tag{1.9}$$

Gauss's Law

The final static EM law is Gauss's law, which relates the static electric flux density vector, $\bar{D}$, to charge. With reference to Figure 1-6, Gauss's law in integral form is

$$\oint_s \bar{D} \cdot ds = \int_v \rho_v \cdot dv = Q_{\text{enclosed}}. \tag{1.10}$$

This states that the integral of the constant electric flux vector, $\bar{D}$, over a closed surface is equal to the total charge enclosed by the surface, Q_{enclosed}.

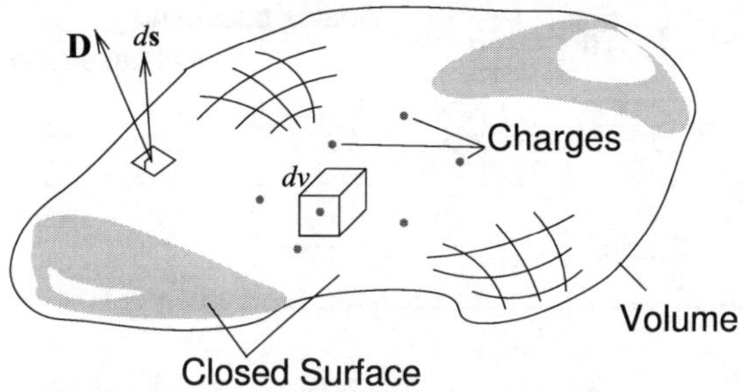

Figure 1-6: Diagram illustrating Gauss's law. Charges are distributed in the volume enclosed by the closed surface. An incremental area is described by the vector $d\mathbf{S}$, which is normal to the surface and whose magnitude is the area of the incremental area.

Gauss's Law of Magnetism

Gauss's law of magnetism parallels Gauss's law which applies to electric fields and charges. In integral form the law is

$$\oint_s \bar{B} \cdot ds = 0. \tag{1.11}$$

This states that the integral of the constant magnetic flux vector, $\bar{D}$, over a closed surface is zero reflecting the fact that magnetic charges do not exist.

1.1.3 Maxwell's Equations

The essential step in the invention of radio was the development of Maxwell's equations in 1861. Before Maxwell's equations were postulated, several static EM laws were known. Taken together they cannot describe the propagation of EM signals, but they can be derived from Maxwell's equations. Maxwell's equations cannot be derived from the static electric and magnetic field laws. Maxwell's equations embody additional insight relating spatial derivatives to time derivatives, which leads to a description of propagating fields. Maxwell's equations are

$$\nabla \times \bar{\mathcal{E}} = -\frac{\partial \bar{\mathcal{B}}}{\partial t} - \bar{\mathcal{M}} \tag{1.12}$$

$$\nabla \cdot \bar{\mathcal{D}} = \rho_V \tag{1.13}$$

$$\nabla \times \bar{\mathcal{H}} = \frac{\partial \bar{\mathcal{D}}}{\partial t} + \bar{\mathcal{J}} \tag{1.14}$$

$$\nabla \cdot \bar{\mathcal{B}} = \rho_{mV}. \tag{1.15}$$

The additional quantities in Equations (1.12)–(1.15) are

- $\bar{\mathcal{J}}$, the **electric current** density, with units of amperes per square meter (A/m^2);
- ρ_V, the **electric charge** density, with units of coulombs per cubic meter (C/m^3);
- ρ_{mV}, the magnetic charge density, with units of webers per cubic meter (Wb/m^3); and
- $\bar{\mathcal{M}}$, the magnetic current density, with units of volts per square meter (V/m^2).

Magnetic charges do not exist, but their introduction through the **magnetic charge** density, ρ_{mV}, and the **magnetic current** density, $\bar{\mathcal{M}}$, introduce an aesthetically appealing symmetry to Maxwell's equations. Maxwell's equations are differential equations, and as with most differential equations, their solution is obtained with particular boundary conditions, which in radio engineering are imposed by conductors.

Maxwell's equations have three types of derivatives. First, there is the time derivative, $\partial/\partial t$. Then there are two spatial derivatives, $\nabla\times$, called **curl**, capturing the way a field circulates spatially (or the amount that it curls up on itself), and $\nabla\cdot$, called the **div** operator, describing the spreading-out of a

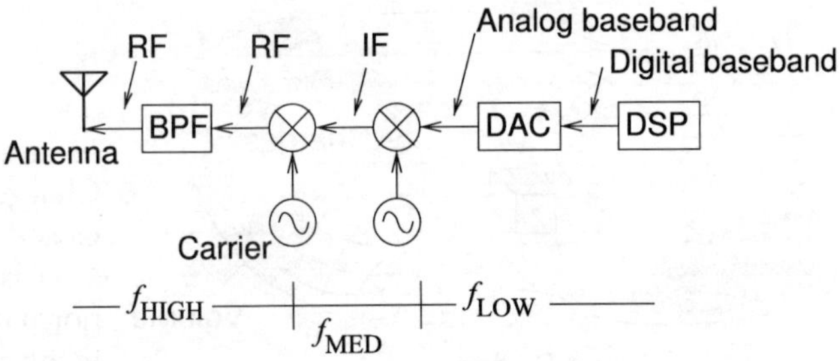

Figure 1-7: A simple transmitter with low, f_{LOW}, medium, f_{MED}, and high frequency, f_{HIGH}, sections. The mixers can be idealized as multipliers that boost the frequency of the input baseband or IF signal by the frequency of the carrier.

field. In rectangular coordinates, curl, $\nabla\times$, describes how much a field circles around the x, y, and z axes. That is, the curl describes how a field circulates on itself. So Equation (1.12) relates the amount an electric field circulates on itself to changes of the B field in time. So a spatial derivative of electric fields is related to a time derivative of the magnetic field. Also in Equation (1.14) the spatial derivative of the magnetic field is related to the time derivative of the electric field. These are the key elements that result in self-sustaining propagation.

Div, $\nabla\cdot$, describes how a field spreads out from a point. How fast a field varies with time, $\partial\bar{\mathcal{B}}/\partial t$ and $\partial\bar{\mathcal{D}}/\partial t$, depends on frequency. The more interesting derivatives are $\nabla\times\bar{\mathcal{E}}$ and $\nabla\times\bar{\mathcal{H}}$ which describe how fast a field can change spatially—this depends on wavelength relative to geometry. If the cross-sectional dimensions of a transmission line are less than a wavelength ($\lambda/2$ or $\lambda/4$ in different circumstances when there are conductors), then it will be impossible for the fields to curl up on themselves and so there will be only one solution (with no or minimal variation of the E and H fields) or, in some cases, no solution to Maxwell's equations.

1.2 Radio Architecture

A radio device is comprised of reasonably well-defined units, see Figure 1-7. The analog baseband signal can have frequency components that range from DC to many megahertz.

Up until the mid-1970s most wireless communications were based on centralized high-power transmitters and reception was expected until the signal level fell below a noise threshold. These systems were particularly sensitive to interference, therefore systems transmitting at the same frequency were geographically separated so that signals fell below the background noise threshold before there was a chance of interfering with a neighboring system operating at the same frequency. This situation is illustrated in Figure 1-8.

Cellular communications is based on the concept of cells in which a terminal unit communicates with a basestation at the center of a cell. For communication in closely spaced cells to work, interference from other radios must be managed. This is facilitated using the ability to recover from errors available with error correction schemes.

In 1981 the U.S. FCC defined cellular radio as "a high capacity land mobile system in which assigned spectrum is divided into discrete channels which are assigned in groups to geographic cells covering a cellular geographic area. The discrete channels are capable of being reused within the service area." The key attributes here are (a) the concept of cells arranged in clusters and the total number of channels available is divided among the cells in

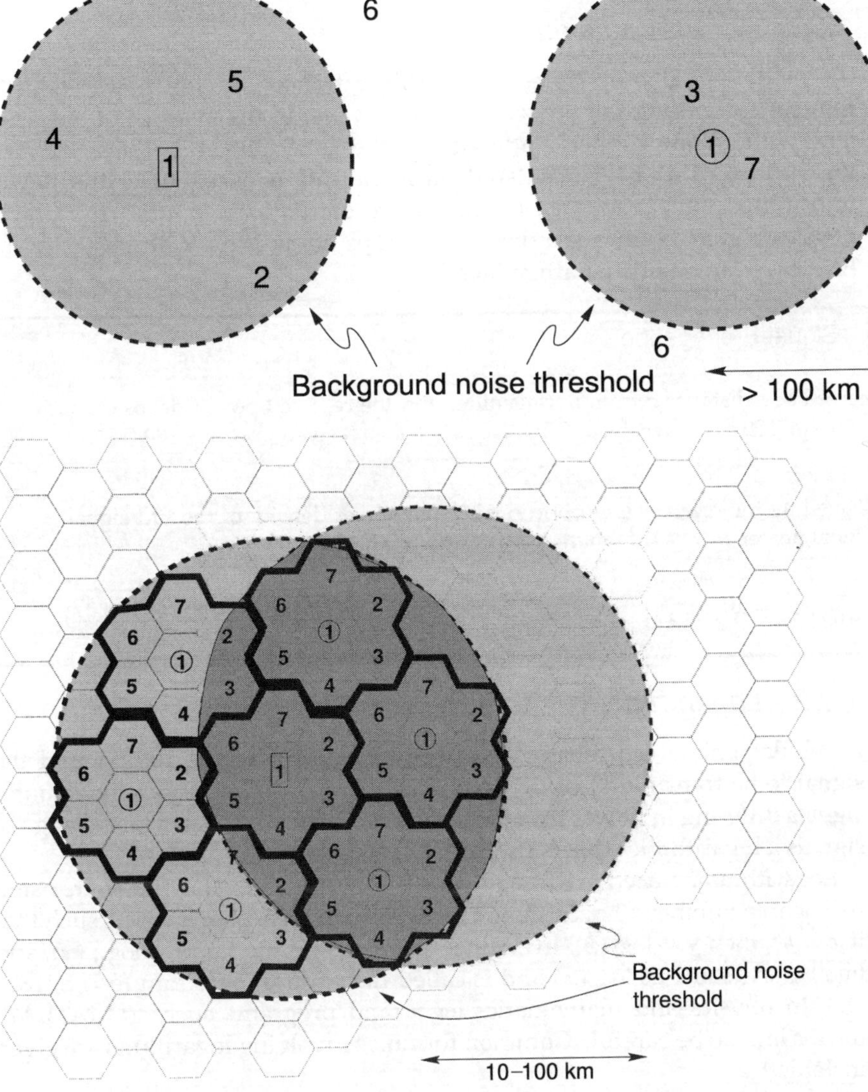

Figure 1-8: Interference in a conventional radio system. The two transmitters, 1, are at the centers of the coverage circles defined by the background noise threshold.

Figure 1-9: Interference in a cellular radio system.

a cluster and the full set is repeated in each cluster, e.g. in Figure 1-9 a 7-cell cluster is shown; and (b) frequency reuse. with frequencies used in one cell are reused in the corresponding cell in another cluster. As the cells are relatively close, it is important to dynamically control the power radiated by each radio, as radios in one cell will produce interference in other clusters.

Achieving maximum frequency reuse is essential. In a cellular system, there is a radical departure in concept from this. Consider the interference in a cellular system as shown in Figure 1-9. The signals in corresponding cells in different clusters interfere with each other and the interference is much larger than that of the background noise. Error correction coding to correct errors resulting from interference.

1.3 RF Power Calculations

1.3.1 RF Propagation

As an RF signal propagates away from a transmitter the power density reduces conserving the power in the EM wave. In the absence of obstacles and without atmospheric attenuation the total power passing through the surface of a sphere centered on a transmitter is equal to the power transmitted. Since the area of the sphere of radius r is $4\pi r^2$, the power density, e.g. in W/m^2, at a distance r drops off as $1/r^2$. With obstacles the EM wave can be further attenuated.

EXAMPLE 1.1 Signal Propagation

A signal is received at a distance r from a transmitter and the received power drops off as $1/r^2$. When $r = 1$ km, 100 nW is received. What is r when the received power is 100 fW?

Solution:

The signal collected by the receiver is proportional to the power density of the EM signal. The received signal power $P_r = k/r^2$ where k is a constant. This leads to

$$\frac{P_r(1\text{ km})}{P_r(r)} = \frac{100\text{ nW}}{100\text{ fW}} = 10^6 = \frac{kr^2}{k(1\text{ km})^2} = \frac{r^2}{(10^3\text{ m})^2}; \quad r = \sqrt{10^{12}\text{ m}^2} = 1000\text{ km} \quad (1.16)$$

1.3.2 Logarithm

A cellular phone can reliably receive a signal as small as 100 fW and the signal to be transmitted could be 1 W. So the same circuitry can encounter signals differing in power by a factor of 10^{13}. To handle such a large range of signals a logarithmic scale is used.

Logarithms are used in RF engineering to express the ratio of powers using reasonable numbers. Logarithms are taken with respect to a base b such that if $x = b^y$, then $y = \log_b(x)$. In engineering, $\log(x)$ is the same as $\log_{10}(x)$, and $\ln(x)$ is the same as $\log_e(x)$ and is called the natural logarithm (e = 2.71828 ...). In physics and mathematics $\log x$ (and programs such as MATLAB) means $\ln x$, so be careful. Common formulas involving logarithms are given in Table 1-2.

Table 1-2: Common logarithm formulas. In engineering $\log x \equiv \log_{10} x$ and $\ln x \equiv \log_2 x$.

Description	Formula	Example
Equivalence	$y = \log_b(x) \longleftrightarrow x = b^y$	$\log(1000) = 3$ and $10^3 = 1000$
Product	$\log_b(xy) = \log_b(x) + \log_b(y)$	$\log(0.13 \cdot 978) = \log(0.13) + \log(978)$ $= -0.8861 + 2.990 = 2.104$
Ratio	$\log_b(x/y) = \log_b(x) - \log_b(y)$	$\ln(8/2) = \ln(8) - \ln(2) = 3 - 1 = 2$
Power	$\log_b(x^p) = p\log_b(x)$	$\ln(3^2) = 2\ln(3) = 2 \cdot 1.0986 = 2.197$
Root	$\log_b(\sqrt[p]{x}) = \frac{1}{p}\log_b(x)$	$\log(\sqrt[3]{20}) = \frac{1}{3}\log(20) = 0.4337$
Change of base	$\log_b(x) = \dfrac{\log_k(x)}{\log_k(b)}$	$\ln(100) = \dfrac{\log(100)}{\log(2)} = \dfrac{2}{0.30103} = 6.644$

1.3.3 Decibels

RF signal levels are expressed in terms of the power of a signal. While power can be expressed in absolute terms, e.g. watts (W), it is more useful to use a logarithmic scale. The ratio of two power levels P and P_{REF} in bels (B) is

$$P(B) = \log\left(\frac{P}{P_{\text{REF}}}\right), \tag{1.17}$$

where P_{REF} is a reference power. Here $\log x$ is the same as $\log_{10} x$. Human senses have a logarithmic response and the minimum resolution tends to be about 0.1 B, so it is most common to use decibels (dB); 1 B = 10 dB. Common designations are shown in Table 1-3. Also, 1 mW = 0 dBm is a very common power level in RF and microwave power circuits where the m in dBm refers to the 1 mW reference. As well, dBW is used, and this is the power ratio with respect to 1 W with 1 W = 0 dBW = 30 dBm.

Table 1-3: Common power designations: (a) reference powers, P_{REF}; (b) power ratios in decibels(dB); and (c) powers in dBm and watts.

(a)

P_{REF}	Bell units	Decibel units
1 W	BW	dBW
1 mW = 10^{-3} W	Bm	dBm
1 fW = 10^{-15} W	Bf	dBf

(b)

Power ratio	in dB
10^{-6}	−60
0.001	−30
0.1	−20
1	0
10	10
1000	30
10^{6}	60

(c)

Power	Absolute power
−120 dBm	10^{-12} mW = 10^{-15} W = 1 fW
0 dBm	1 mW
10 dBm	10 mW
20 dBm	100 mW = 0.1 W
30 dBm	1000 mW = 1 W
40 dBm	10^{4} mW = 10 W
50 dBm	10^{5} mW = 100 W
−90 dBm	10^{-9} mW = 10^{-12} W = 1 pW
−60 dBm	10^{-6} mW = 10^{-9} W= 1 nW
−30 dBm	0.001 mW = 1 μW
−20 dBm	0.01 mW = 10 μW
−10 dBm	0.1 mW = 100 μW

EXAMPLE 1.2 Power Gain

An amplifier has a power gain of 1200. What is the power gain in decibels? If the input power is 5 dBm, what is the output power in dBm?

Solution:

Power gain in decibels, $G_{\text{dB}} = 10\log 1200 = 30.79$ dB.

The output power is $P_{\text{out}}|_{\text{dBm}} = P_{\text{dB}} + P_{\text{in}}|_{\text{dBm}} = 30.79 + 5 = 35.79$ dBm.

EXAMPLE 1.3 Gain Calculations

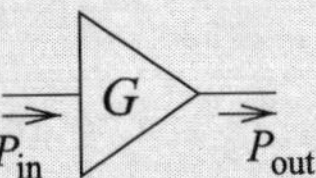

A signal with a power of 2 mW is applied to the input of an amplifier that increases the power of the signal by a factor of 20.

(a) What is the input power in dBm?

$$P_{\text{in}} = 2\text{ mW} = 10\cdot\log\left(\frac{2\text{ mW}}{1\text{ mW}}\right) = 10\cdot\log(2) = 3.010\text{ dBm} \approx 3.0\text{ dBm}. \tag{1.18}$$

(b) What is the gain, G, of the amplifier in dB?
The amplifier gain (by default this is power gain) is

$$G = 20 = 10 \cdot \log(20)\ \text{dB} = 10 \cdot 1.301\ \text{dB} = 13.0\ \text{dB}. \tag{1.19}$$

(c) What is the output power of the amplifier?

$$G = \frac{P_{\text{out}}}{P_{\text{in}}}, \quad \text{and in decibels } G|_{\text{dB}} = P_{\text{out}}|_{\text{dBm}} - P_{\text{in}}|_{\text{dBm}} \tag{1.20}$$

Thus the output power in dBm is

$$P_{\text{out}}|_{\text{dBm}} = G|_{\text{dB}} + P_{\text{in}}|_{\text{dBm}} = 13.0\ \text{dB} + 3.0\ \text{dBm} = 16.0\ \text{dBm}. \tag{1.21}$$

Note that dB and dBm are dimensionless but they do have meaning; dB indicates a power ratio but dBm refers to a power. Quantities in dB and one quantity in dBm can be added or subtracted to yield dBm, and the difference of two quantities in dBm yields a power ratio in dB.

In Examples 1.2 and 1.3 two digits following the decimal point were used for the output power expressed in dBm. This corresponds to an implied accuracy of about 0.01% or 4 significant digits of the absolute number. This level of precision is typical for the result of an engineering calculation.

EXAMPLE 1.4 Power Calculations

The output stage of an RF front end consists of an amplifier followed by a filter and then an antenna. The amplifier has a gain of 33 dB, the filter has a loss of 2.2 dB, and of the power input to the antenna, 45% is lost as heat due to resistive losses. If the power input to the amplifier is 1 W, then:

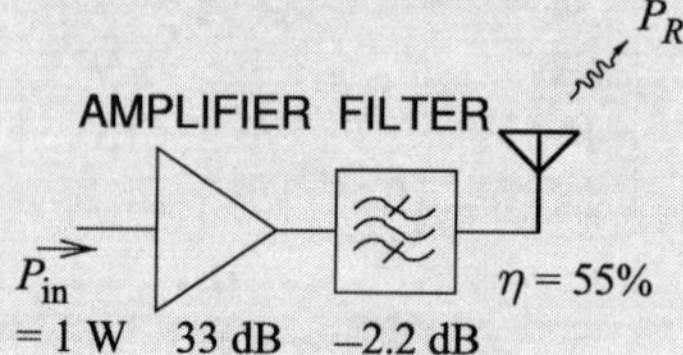

(a) What is the power input to the amplifier expressed in dBm?
$P_{\text{in}} = 1\ \text{W} = 1000\ \text{mW}, \quad P_{\text{dBm}} = 10\log(1000/1) = 30\ \text{dBm}.$

(b) Express the loss of the antenna in dB.
45% of the power input to the antenna is dissipated as heat.
The antenna has an efficiency, η, of 55% and so $P_2 = 0.55P_1$.
$\text{Loss} = P_1/P_2 = 1/0.55 = 1.818 = 2.60\ \text{dB}.$

(c) What is the total gain of the RF front end (amplifier + filter + antenna)?

$$\begin{aligned}\text{Total gain} &= (\text{amplifier gain})_{\text{dB}} + (\text{filter gain})_{\text{dB}} - (\text{loss of antenna})_{\text{dB}}\\ &= (33 - 2.2 - 2.6)\ \text{dB} = 28.2\ \text{dB}\end{aligned} \tag{1.22}$$

(d) What is the total power radiated by the antenna in dBm?

$$\begin{aligned}P_R &= P_{\text{in}}|_{\text{dBm}} + (\text{amplifier gain})_{\text{dB}} + (\text{filter gain})_{\text{dB}} - (\text{loss of antenna})_{\text{dB}}\\ &= 30\ \text{dBm} + (33 - 2.2 - 2.6)\ \text{dB} = 58.2\ \text{dBm}.\end{aligned} \tag{1.23}$$

(e) What is the total power radiated by the antenna?

$$P_R = 10^{58.2/10} = (661 \times 10^3)\ \text{mW} = 661\ \text{W}. \tag{1.24}$$

1.4 SI Units

The main SI units used in microwave engineering are given in Table 1-4.

- Symbols for units are written in upright roman font and are lowercase unless the symbol is derived from the name of a person. An exception is the use of L for liter to avoid possible confusion with l.

Table 1-4: Main SI units used in RF and microwave engineering.

SI unit	Name	Usage	In terms of fundamental units
A	ampere	current (abbreviated as amp)	Fundamental unit
cd	candela	luminous intensity	Fundamental unit
C	coulomb	charge	A·s
F	farad	capacitance	$\text{kg}^{-1}\cdot\text{m}^{-2}\cdot\text{A}^{-2}\cdot\text{s}^{4}$
g	gram	weight	=kg/1000
H	henry	inductance	$\text{kg}\cdot\text{m}^{2}\cdot\text{A}^{-2}\cdot\text{s}^{-2}$
J	joule	unit of energy	$\text{kg}\cdot\text{m}^{2}\cdot\text{s}^{-2}$
K	kelvin	thermodynamic temperature	Fundamental unit
kg	kilogram	SI fundamental unit	Fundamental unit
m	meter	length	Fundamental unit
mol	mole	amount of substance	Fundamental unit
N	newton	unit of force	$\text{kg}\cdot\text{m}\cdot\text{s}^{-2}$
Ω	ohm	resistance	$\text{kg}\cdot\text{m}^{2}\cdot\text{A}^{-2}\cdot\text{s}^{-3}$
Pa	pascal	pressure	$\text{kg}\cdot\text{m}^{-1}\cdot\text{s}^{-2}$
s	second	time	Fundamental unit
S	siemen	admittance	$\text{kg}^{-1}\cdot\text{m}^{-2}\cdot\text{A}^{2}\cdot\text{s}^{3}$
V	volt	voltage	$\text{kg}\cdot\text{m}^{2}\cdot\text{A}^{-1}\cdot\text{s}^{-3}$
W	watt	power	$J\cdot s^{-1}$

- A space separates a value from the symbol for the unit (e.g., 5.6 kg). There is an exception for degrees, with the symbol °, e.g. 45°.

When SI units are multiplied a center dot is used. For example, newton meters is written N·m. When a unit is derived from the ratio of symbols then either a solidus (/) or a negative exponent is used; the symbol for velocity (meters per second) is either m/s or $\text{m}\cdot\text{s}^{-1}$. The use of multiple solidi for a combination symbol is confusing and must be avoided. So the symbol for acceleration is $\text{m}\cdot\text{s}^{-2}$ or m/s^2 and not m/s/s.

Consider calculation of the thermal resistance of a rod of cross-sectional area A and length ℓ:

$$R_{\text{TH}} = \frac{\ell}{kA}. \tag{1.25}$$

If $A = 0.3\ \text{cm}^2$ and $\ell = 2$ mm, the thermal resistance is

$$\begin{aligned} R_{\text{TH}} &= \frac{(2\ \text{mm})}{(237\ \text{kW}\cdot\text{m}^{-1}\cdot\text{K}^{-1})\cdot(0.3\ \text{cm}^2)} \\ &= \frac{(2\cdot 10^{-3}\ \text{m})}{237\cdot(10^3\cdot\text{W}\cdot\text{m}^{-1}\cdot\text{K}^{-1})\cdot 0.3\cdot(10^{-2}\cdot\text{m})^2} \\ &= \frac{2\cdot 10^{-3}}{237\cdot 10^3\cdot 0.3\cdot 10^{-4}}\cdot\frac{\text{m}}{\text{W}\cdot\text{m}^{-1}\cdot\text{K}^{-1}\cdot\text{m}^2} \\ &= 2.813\cdot 10^{-4}\ \text{K}\cdot\text{W}^{-1} = 281.3\ \mu\text{K/W}. \end{aligned} \tag{1.26}$$

This would be an error-prone calculation if the thermal conductivity was taken as 237 kW/m/K.

SI prefixes are given in Table 1-5 and indicate the multiple of a unit (e.g., 1 pA is 10^{-12} amps). (Source: 2015 ISO/IEC 8000 [3].) In 2009 new definitions of the prefixes for bits and bytes were adopted [3] removing the confusion over the earlier use of quantities such as kilobit to represent either 1,000 bits or 1,024 bits. Now kilobit (kbit) always means 1,000 bits and a new term kibibit (Kibit) means 1,024 bit. Also the now obsolete usage of kbps is replaced by kbit/s (kilobit per second). The prefix k stands for kilo (i.e. 1,000) and Ki is the symbol for the binary prefix kibi- (i.e. 1,024). The symbol for byte (= 8 bits) is "B".

Table 1-5: SI prefixes.

SI Prefixes			SI Prefixes			Prefixes for bits and bytes		
Symbol	Factor	Name	Symbol	Factor	Name	Name		
10^{-24}	y	yocto	10^{1}	da	deca	kilobit	kbit	1000 bit
10^{-21}	z	zepto	10^{2}	h	hecto	megabit	Mbit	1000 kbit
10^{-18}	a	atto	10^{3}	k	kilo	gigabit	Gbit	1000 Mbit
10^{-15}	f	femto	10^{6}	M	mega	terabit	Tbit	1000 Gbit
10^{-12}	p	pico	10^{9}	G	giga	kibibit	Kibit	1024 bit
10^{-9}	n	nano	10^{12}	T	tera	mebibit	Mibit	1024 Kibit
10^{-6}	μ	micro	10^{15}	P	peta	gibibit	Gibit	1024 Mibit
10^{-3}	m	milli	10^{18}	E	exa	tebibit	Tibit	1024 Gibit
10^{-2}	c	centi	10^{21}	Z	zetta	kilobyte	kB	1000 B
10^{-1}	d	deci	10^{24}	Y	yotta	kibibyte	KiB	1024 B

1.5 Summary

The RF spectrum is used to support a tremendous range of applications, including voice and data communications, satellite-based navigation, radar, weather radar, mapping, environmental monitoring, air traffic control, police radar, perimeter surveillance, automobile collision avoidance, and many military applications.

In RF and microwave engineering there are always considerable approximations made in design, partly because of necessary simplifications that must be made in modeling, but also because many of the material properties required in a detailed design can only be approximately known. Most RF and microwave design deals with frequency-selective circuits often relying on line lengths that have a length that is a particular fraction of a wavelength. Many designs can require frequency tolerances of as little as 0.1%, and filters can require even tighter tolerances. It is therefore impossible to design exactly. Measurements are required to validate and iterate designs. Conceptual understanding is essential; the designer must be able to relate measurements, which themselves have errors, with computer simulations. The ability to design circuits with good tolerance to manufacturing variations and perhaps circuits that can be tuned by automatic equipment are skills developed by experienced designers.

1.6 References

[1] "Atmospheric microwave transmittance at mauna kea, wikipedia creative commons."

[2] "IEEE Virtual Museum," at http://www.ieee-virtual-museum.org Search term: 'Faraday'.

[3] "ISO/IEC 8000," intenational Standard Organization (ISO), International Electrotechnical Commission standard including standards on on letter symbols to be used in electrical technology. http://www.iso.org.

1.7 Exercises

1. What is the wavelength in free space of a signal at 4.5 GHz?
2. Consider a monopole antenna that is a quarter of a wavelength long. How long is the antenna if it operates at 3 kHz?
3. Consider a monopole antenna that is a quarter of a wavelength long. How long is the antenna if it operates at 500 MHz?
4. Consider a monopole antenna that is a quarter of a wavelength long. How long is the antenna if it operates at 2 GHz?
5. A dipole antenna is half of a wavelength long. How long is the antenna at 2 GHz?
6. A dipole antenna is half of a wavelength long. How long is the antenna at 1 THz?
7. A transmitter transmits an FM signal with a bandwidth of 100 kHz and the signal is received by a receiver at a distance r from the transmitter. When $r = 1$ km the signal power received by the receiver is 100 nW. When the receiver moves further away from the transmitter the power received drops off as $1/r^2$. What is r in kilometers when the received power is 100 pW. [Parallels Example 1.1]
8. A transmitter transmits an AM signal with a bandwidth of 20 kHz and the signal is received by a receiver at a distance r from the transmitter. When $r = 10$ km the signal power received is 10 nW. When the receiver moves further away from the transmitter the power received drops off as $1/r^2$. What is r in kilometers when the received power is equal to the received noise power of 1 pW? [Parallels Example 1.1]
9. The logarithm to base 2 of a number x is 0.38 (i.e., $\log_2(x) = 0.38$). What is x?
10. The natural logarithm of a number x is 2.5 (i.e., $\ln(x) = 2.5$). What is x?
11. The logarithm to base 2 of a number x is 3 (i.e., $\log_2(x) = 3$). What is $\log_2(\sqrt[2]{x})$?
12. What is $\log_3(10)$?
13. What is $\log_{4.5}(2)$?
14. Without using a calculator evaluate $\log\{[\log_3(3x) - \log_3(x)]\}$.
15. A 50 Ω resistor has a sinusoidal voltage across it with a peak voltage of 0.1 V. The RF voltage is $0.1\cos(\omega t)$, where ω is the radian frequency of the signal and t is time.
    (a) What is the power dissipated in the resistor in watts?
    (b) What is the power dissipated in the resistor in dBm?
16. The power of an RF signal is 10 mW. What is the power of the signal in dBm?
17. The power of an RF signal is 40 dBm. What is the power of the signal in watts?
18. An amplifier has a power gain of 2100.
    (a) What is the power gain in decibels?
    (b) If the input power is -5 dBm, what is the output power in dBm? [Parallels Example 1.2]
19. An amplifier has a power gain of 6. What is the power gain in decibels? [Parallels Example 1.2]
20. A filter has a loss factor of 100. [Parallels Example 1.2]
    (a) What is the loss in decibels?
    (b) What is the gain in decibels?
21. An amplifier has a power gain of 1000. What is the power gain in dB? [Parallels Example 1.2]
22. An amplifier has a gain of 14 dB. The input to the amplifier is a 1 mW signal, what is the output power in dBm?
23. An RF transmitter consists of an amplifier with a gain of 20 dB, a filter with a loss of 3 dB and then that is then followed by a lossless transmit antenna. If the power input to the amplifier is 1 mW, what is the total power radiated by the antenna in dBm? [Parallels Example 1.4]
24. The final stage of an RF transmitter consists of an amplifier with a gain of 30 dB and a filter with a loss of 2 dB that is then followed by a transmit antenna that looses half of the RF power as heat. [Parallels Example 1.4]
    (a) If the power input to the amplifier is 10 mW, what is the total power radiated by the antenna in dBm?
    (b) What is the radiated power in watts?
25. A 5 mW-RF signal is applied to an amplifier that increases the power of the RF signal by a factor of 200. The amplifier is followed by a filter that losses half of the power as heat.
    (a) What is the output power of the filter in watts?
    (b) What is the output power of the filter in dBW?
26. The power of an RF signal at the output of a receive amplifier is 1 μW and the noise power at the output is 1 nW. What is the output signal-to-noise ratio in dB?

27. The power of a received signal is 1 pW and the received noise power is 200 fW. In addition the level of the interfering signal is 100 fW. What is the signal-to-noise ratio in dB? Treat interference as if it is an additional noise signal.age gain of 1 has an input impedance of 100 Ω, a zero output impedance, and drives a 5 Ω load. What is the power gain of the amplifier?
28. A transmitter transmits an FM signal with a bandwidth of 100 kHz and the signal power received by a receiver is 100 nW. In the same bandwidth as that of the signal the receiver receives 100 pW of noise power. In decibels, what is the ratio of the signal power to the noise power, i.e. the signal-to-noise ratio (SNR) received by the receiver?
29. An amplifier with a voltage gain of 20 has an input resistance of 100 Ω and an output resistance of 50 Ω. What is the power gain of the amplifier in decibels? [Parallels Example 1.0]
30. An amplifier with a voltage gain of 1 has an input resistance of 100 Ω and an output resistance of 5 Ω. What is the power gain of the amplifier in decibels? Explain why there is a power gain of more than 1 even though the voltage gain is 1. [Parallels Example 1.0]
31. An amplifier has a power gain of 1900.
    (a) What is the power gain in decibels?
    (b) If the input power is -8 dBm, what is the output power in dBm? [Parallels Example 1.2]
32. An amplifier has a power gain of 20.
    (a) What is the power gain in decibels?
    (b) If the input power is -23 dBm, what is the output power in dBm? [Parallels Example 1.2]
33. An amplifier has a voltage gain of 10 and a current gain of 100.
    (a) What is the power gain as an absolute number?
    (b) What is the power gain in decibels?
    (c) If the input power is -30 dBm, what is the output power in dBm?
    (c) What is the output power in mW?
34. An amplifier with 50 Ω input impedance and 50 Ω load impedance has a voltage gain of 100. What is the (power) gain in decibels?
35. An attenuator reduces the power level of a signal by 75%. What is the (power) gain of the attenuator in decibels?

1.7.1 Exercises By Section

†challenging

§1.2	$1, 2, 3, 4, 5, 6, 7, 8$	$18, 19, 20, 21, 22, 23^\dagger, 24^\dagger, 25^\dagger$
§1.3	$9, 10, 11, 12, 13, 14, 15, 16, 17$	$26, 27, 28, 29, 30, 31, 32, 33, 34, 35$

1.7.2 Answers to Selected Exercises

4	3.25 cm	17	10 W	23	50.12 mW
12	2.096	19	7.782 dB	24(b)	3.162 W
16	10 dBm	22	1.301		

CHAPTER 2

Antennas and the RF Link

2.1 Introduction

An antenna interfaces circuits with free-space with a transmit antenna converting a guided wave signal on a transmission line to an electromagnetic (EM) wave propagating in free space, while a receive antenna is a transducer that converts a free-space EM wave to a guided wave on a transmission line and eventually to a receiver circuit. Together the transmit and receive antennas are part of the RF link. The RF link is the path between the output of the transmitter circuit and the input of the receiver circuit (see Figure 2-1). Usually this path includes the cable from the transmitter to the transmit antenna, the transmit antenna itself, the propagation path, the receive antenna, and the transmission line connecting the receive antenna to the receiver circuit. The received signal is much smaller than the transmitted signal. The overwhelming majority of the loss is from the propagation path as the EM signal spreads out, and usually diffracts, reflects, and is partially blocked by objects such as hills and buildings.

The first half of this chapter is concerned with the properties of antennas. One of the characteristics of antennas is that the energy can be focused in a particular direction, a phenomenon captured by the concept of antenna gain, which can partially compensate for path loss. The second half of this chapter considers modeling the RF link and the geographical arrangement of antennas that manage interference from other radios while providing support for as many users as possible.

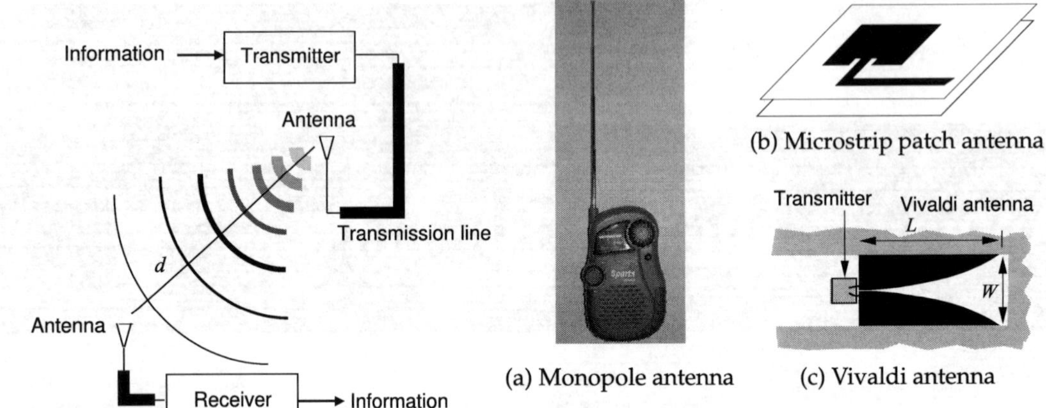

Figure 2-1: RF link.

Figure 2-2: Representative resonant, (a) and (b), and traveling-wave, (c), antennas.

EXAMPLE 2.1 Interference

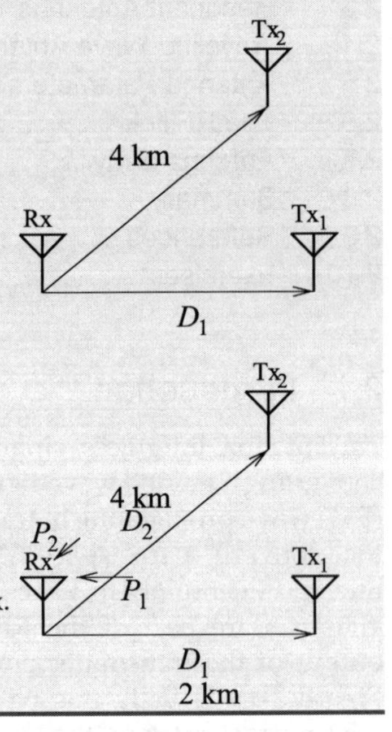

In the figure there are two transmitters, Tx_1 and Tx_2, operating at the same power level, and one receiver, Rx. Tx_1 is an intentional transmitter and its signal is intended to be received at Rx. Tx_1 is separated from Rx by $D_1 = 2$ km. Tx_2 uses the same frequency channel as Tx_1, and as far as RX is concerned it transmits an interfering signal. Assume that the antennas are omnidirectional (i.e., they transmit and receive signals equally in all directions) and that the transmitted power density drops off as $1/d^2$, where d is the distance from the transmitter. Calculate the signal-to-interference ratio (SIR) at Rx.

Solution:

$D_1 = 2$ km and $D_2 = 4$ km.
P_1 is the signal power transmitted by Tx1 and received at Rx.
P_2 is the interference power transmitted by Tx2 and received at Rx.

So $\text{SIR} = \dfrac{P_1}{P_2} = \left(\dfrac{D_2}{D_1}\right)^2 = 4 = 6.02\text{ dB}.$

2.2 RF Antennas

Antennas are of two fundamental types: **resonant** and **traveling-wave antennas**, see Figure 2-2. Resonant antennas establish a standing wave of current with required resonance usually established when the antenna section is either a quarter- or half-wavelength long. These antennas are also known as standing-wave antennas. Resonant antennas are inherently narrowband because of the resonance required to establish a large standing wave of current. Figures 2-2(a and b) show two representative resonant antennas. The physics of the operation of a resonant antenna is understood by considering the time domain. First consider the physical operation of a transmit antenna. When a sinusoidal voltage is applied to the conductor

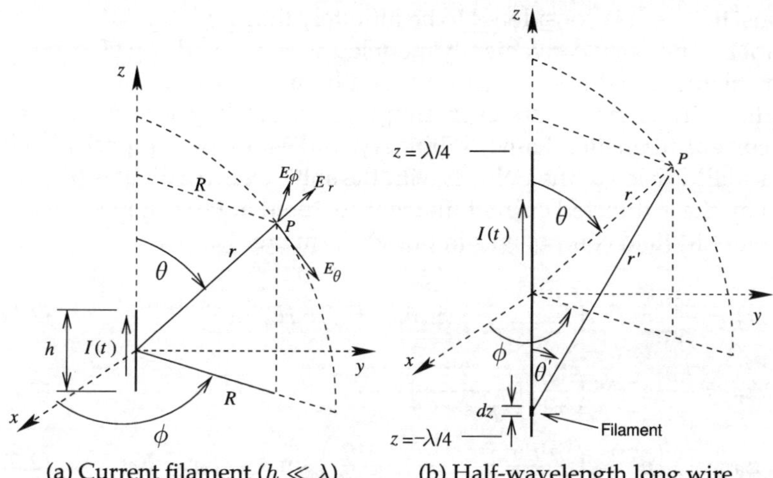

(a) Current filament ($h \ll \lambda$) (b) Half-wavelength long wire

Figure 2-3: Wire antennas. The distance from the center of an antenna to the field point P is r. $R = r \sin\theta$.

of an antenna, charges, i.e. free electrons, accelerate or decelerate under the influence of an applied voltage source which typically arrives at the antenna from a traveling wave voltage on a transmission line. When the charges accelerate (or decelerate) they produce an EM field which radiates away from the antenna. At a point on the antenna there is a current sinusoidally varying in time, and the net acceleration of the charges (in charge per second per second) is also sinusoidal with an amplitude that is directly proportional to the amplitude of the current sinewave. Hence having a large standing wave of current, when the antenna resonates sinusoidally, results in a large charge acceleration and hence large radiated fields.

When an EM field impinges upon a conductor the field causes charges to accelerate and hence induce a voltage that propagates on a transmission line connected to the receive antenna. Traveling-wave antennas, an example of one is shown in Figures 2-2(c), operate as extended lines that gradually flare out so that a traveling wave on the original transmission line transitions into free space. Traveling-wave antennas tend to be two or more wavelengths long at the lowest frequency of operation. While relatively long, they are broadband, many 3 or more octaves wide (e.g. extending from 500 MHz to 4 GHz or more). These antennas are sometimes referred to as **aperture antennas**.

2.3 Resonant Antennas

With a resonant antenna the current on the antenna is directly related to the amplitude of the radiated EM field. Resonance ensures that the standing wave current on the antenna is high.

2.3.1 Radiation from a Current Filament

The fields radiated by a resonant antenna are most conveniently calculated by considering the distribution of current on the antenna. The analysis begins by considering a short filament of current, see Figure 2-3(a). Considering the sinusoidal steady state at radian frequency ω, the current on the filament with phase χ is $I(t) = |I_0| \cos(\omega t + \chi)$, so that $I_0 = |I_0| e^{-\jmath\chi}$ is the phasor of the current on the filament. The length of the filament is h, but it has no other dimensions, that is, it is considered to be infinitely thin.

Resonant antennas are conveniently modeled as being made up of an array of current filaments with spacings and lengths being a tiny fraction of a wavelength. Wire antennas are even simpler and can be considered to be a line of current filaments. Ramo, Whinnery, and Van Duzer [1] calculated the spherical EM fields at the point P with the spherical coordinates (ϕ, θ, r) generated by the z-directed current filament centered at the origin in Figure 2-3. The total EM field components in phasor form are

$$H_\phi = \frac{I_0 h}{4\pi} \mathrm{e}^{-\jmath kr} \left(\frac{\jmath k}{r} + \frac{1}{r^2} \right) \sin\theta, \quad \bar{H}_\phi = H_\phi \hat{\phi} \tag{2.1}$$

$$E_r = \frac{I_0 h}{4\pi} \mathrm{e}^{-\jmath kr} \left(\frac{2\eta}{r^2} + \frac{2}{\jmath\omega\varepsilon_0 r^3} \right) \cos\theta, \quad \bar{E}_r = E_r \hat{\mathbf{r}} \tag{2.2}$$

$$E_\theta = \frac{I_0 h}{4\pi} \mathrm{e}^{-\jmath kr} \left(\frac{\jmath\omega\mu_0}{r} + \frac{1}{\jmath\omega\varepsilon_0 r^3} + \frac{\eta}{r^2} \right) \sin\theta, \quad \bar{E}_\theta = E_\theta \hat{\theta}, \tag{2.3}$$

where η is the free-space characteristic impedance. The variable k is called the **wavenumber** and $k = 2\pi/\lambda = \omega\sqrt{\mu_0\varepsilon_0}$. The $\mathrm{e}^{-\jmath kr}$ terms describe the variation of the phase of the field as the field propagates away from the filament. Equations (2.1)–(2.3) are the complete fields with the $1/r^2$ and $1/r^3$ dependence describing the near-field components. In the far field, i.e. $r \gg \lambda$, the components with $1/r^2$ and $1/r^3$ dependence become negligible and the field components left are the propagating components H_ϕ and E_θ:

$$H_\phi = \frac{I_0 h}{4\pi} \mathrm{e}^{-\jmath kr} \left(\frac{\jmath k}{r} \right) \sin\theta, \quad E_r = 0, \quad \text{and} \quad E_\theta = \frac{I_0 h}{4\pi} \mathrm{e}^{-\jmath kr} \left(\frac{\jmath\omega\mu_0}{r} \right) \sin\theta. \tag{2.4}$$

Now consider the fields in the plane normal to the filament, that is, with $\theta = \pi/2$ radians so that $\sin\theta = 1$. The fields are now

$$H_\phi = \frac{I_0 h}{4\pi} \mathrm{e}^{-\jmath kr} \left(\frac{\jmath k}{r} \right) \quad \text{and} \quad E_\theta = \frac{I_0 h}{4\pi} \mathrm{e}^{-\jmath kr} \left(\frac{\jmath\omega\mu_0}{r} \right) \tag{2.5}$$

and the wave impedance is

$$\eta = \frac{E_\theta}{H_\phi} = \frac{I_0 h}{4\pi} \mathrm{e}^{-\jmath kr} \frac{\jmath\omega\mu_0}{r} \left(\frac{I_0 h}{4\pi} \mathrm{e}^{-\jmath kr} \frac{\jmath k}{r} \right)^{-1} = \frac{\omega\mu_0}{k}. \tag{2.6}$$

Note that the strength of the fields is directly proportional to the magnitude of the current. This proves to be very useful in understanding spurious radiation from microwave structures. Now $k = \omega\sqrt{\mu_0\epsilon_0}$, so

$$\eta = \frac{\omega\mu_0}{\omega\sqrt{\mu_0\epsilon_0}} = \sqrt{\frac{\mu_0}{\epsilon_0}} = 377\ \Omega, \tag{2.7}$$

as expected. Thus an antenna can be viewed as having the inherent function of an impedance transformer converting from the lower characteristic impedance of a transmission line (often 50 Ω) to the 377 Ω characteristic impedance of free space.

Further comments can be made about the propagating fields (Equation (2.4)). The EM field propagates in all directions except not directly in line with the filament. For fixed r, the amplitude of the propagating field increases sinusoidally with respect to θ until it is maximum in the direction normal to the filament.

The power radiated is obtained using the **Poynting vector**, which is the cross-product of the propagating electric and magnetic fields. From this the time-average propagating power density is (with the SI units of W/m^2)

$$P_R = \frac{1}{2}\Re\left(E_\theta H_\phi^*\right) = \frac{\eta k^2 |I_0|^2 h^2}{32\pi^2 r^2}\sin^2\theta, \tag{2.8}$$

and the power density is proportional to $1/r^2$. In Equation (2.8) $\Re(\cdots)$ indicates that the real part is taken.

2.3.2 Finite-Length Wire Antennas

The EM wave propagated from a wire of finite length is obtained by considering the wire as being made up of many filaments and the field is then the superposition of the fields from each filament. As an example, consider the antenna in Figure 2-3(b) where the wire is half a wavelength long. As a good approximation the current on the wire is a standing wave and the current on the wire is in phase so that the current phasor is

$$I(z) = I_0\cos(kz). \tag{2.9}$$

From Equation (2.4) and referring to Figure 2-3 the fields in the far field are

$$H_\phi = \int_{-\lambda/4}^{\lambda/4} \frac{I_0\cos(kz)}{4\pi} \mathrm{e}^{-\jmath kr'}\left(\frac{\jmath k}{r'}\right)\sin\theta'\,dz \tag{2.10}$$

$$E_\theta = \int_{-\lambda/4}^{\lambda/4} \frac{I_0\cos(kz)}{4\pi} \mathrm{e}^{-\jmath kr'}\left(\frac{\jmath\omega\mu_0}{r'}\right)\sin\theta'\,dz, \tag{2.11}$$

where θ' is the angle from the filament to the point P. Now $k = 2\pi/\lambda$ and at the ends of the wire $z = \pm\lambda/4$ where $\cos(kz) = \cos(\pm\pi/2) = 0$. Evaluating the equations is analytically involved and will not be done here. The net result is that the fields are further concentrated in the plane normal to the wire. At large r, of at least several wavelengths distant from the antenna, only the field components decreasing as $1/r$ are significant. At large r the phase differences of the contributions from the filaments is significant and results in shaping of the fields. The geometry to be used in calculating the far field is shown in Figure 2-4(a). The phase contribution of each filament, relative to that at $z = 0$, is $(kz\sin\theta)/\lambda$ and Equations (2.10) and (2.11) become

$$H_\phi = I_0\left(\frac{\jmath k}{4\pi r}\right)\sin(\theta)\,\mathrm{e}^{-\jmath kr}\int_{-\lambda/4}^{\lambda/4}\frac{\sin(kz)}{4\pi}\sin(z\sin(\theta))dz \tag{2.12}$$

$$E_\theta = I_0\left(\frac{\jmath\omega\mu_0}{4\pi r}\right)\sin(\theta)\,\mathrm{e}^{-\jmath kr}\int_{-\lambda/4}^{\lambda/4}\sin(kz)\sin(z\sin(\theta))dz. \tag{2.13}$$

Figure 2-4(b) is a plot of the near-field electric field in the y-z plane calculating E_r and E_θ (recall that $E_\phi = 0$) every 90°. Further from the antenna the E_r component rapidly reduces in size, and E_θ dominates.

A summary of the implications of the above equations are, first, that the strength of the radiated electric and magnetic fields are proportional to the

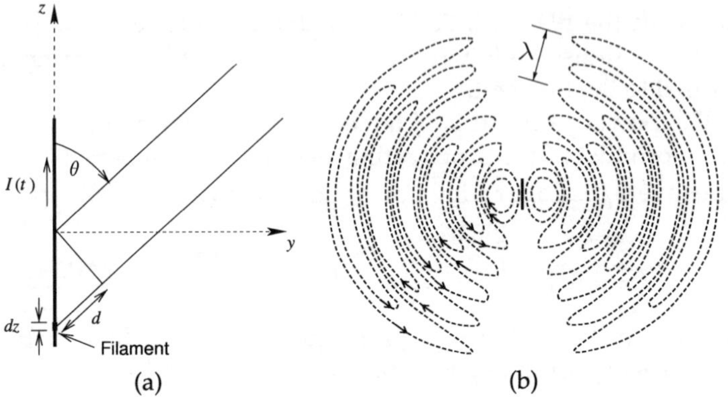

Figure 2-4: Wire antenna: (a) geometry for calculating contributions from current filaments of length dz with cordinate z ($d = -z \sin\theta$); and (b) instantaneous electric field in the y-z plane due to a $\lambda/2$ long current element. There is also a magnetic field.

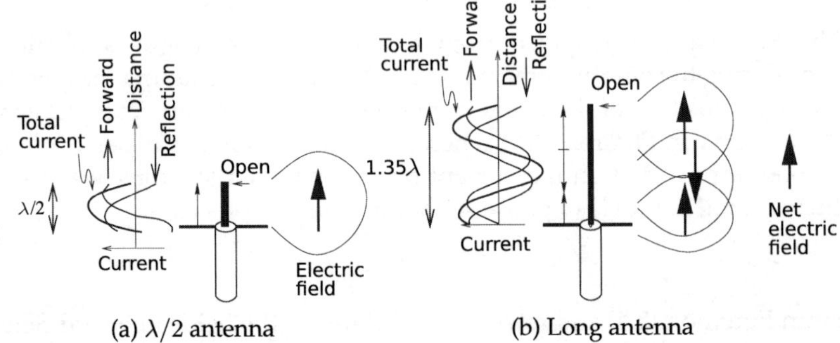

Figure 2-5: Monopole antenna showing total current and forward- and backward-traveling currents: (a) a $\frac{1}{2}\lambda$-long antenna; and (b) a relatively long antenna.

current on the wire antenna. So establishing a standing current wave and hence magnifying the current is important to the efficiency of a wire antenna. A second result is that the power density of freely propagating EM fields in the far field is proportional to $1/r^2$, where r is the distance from the antenna. A third interpretation is that the longer the antenna, the flatter the radiated transmission profile; that is, the radiated energy is more tightly confined to the x-y (i.e. $\Theta = 0$) plane. For the wire antenna the peak radiated field is in the plane normal to the antenna, and thus the wire antenna is generally oriented vertically so that transmission is in the plane of the earth and power is not radiated unnecessarily into the ground or into the sky.

To obtain an efficient resonant antenna, all of the current should be pointed in the same direction at a particular time. One way of achieving this is to establish a standing wave, as shown in Figure 2-5(a). At the open-circuited end, the current reflects so that the total current at the end of the wire is zero. The initial and reflected current waves combine to create a standing wave. Provided that the antenna is sufficiently short, all of the total current—the standing wave—is pointed in the same direction. The optimum length is about a half wavelength. If the wire is longer, the contributions to the field from the oppositely directed current segments cancel (see Figure 2-5(b)).

In Figure 2-5(a) a coaxial cable is attached to the monopole antenna below the ground plane and often a series capacitor between the cable and the antenna provides a low level of coupling leading to a larger standing wave. The capacitor also approximately matches the characteristic impedance of the cable to the input impedance Z_{in} of the antenna. If the length of the monopole is reduced to one-quarter wavelength long it is again resonant, and the input impedance, Z_{in}, is found to be 36 Ω. Then a 50 Ω cable can

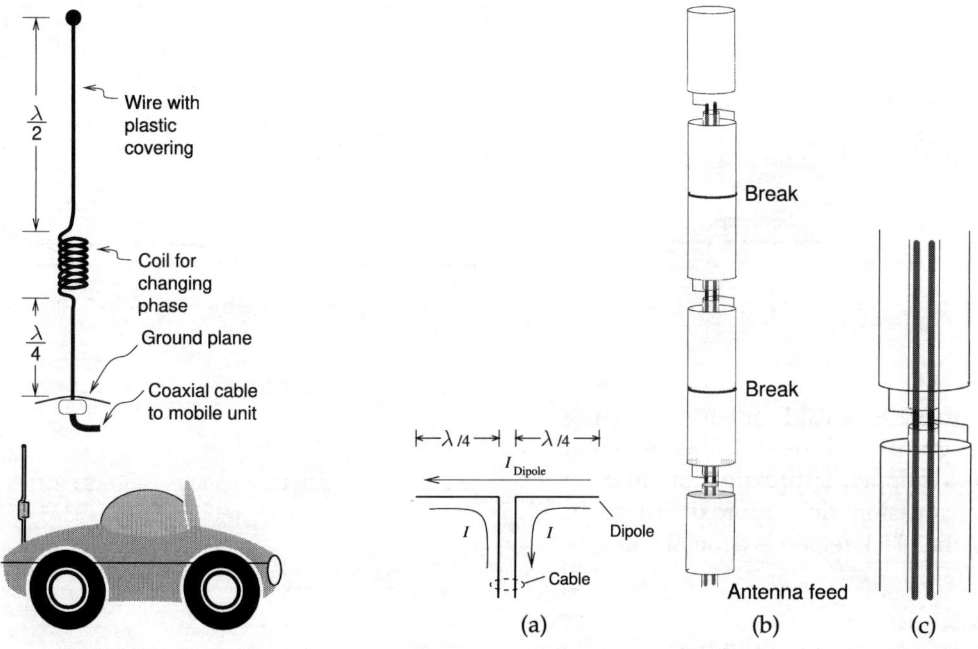

Figure 2-6: Mobile antenna with phasing coil extending the effective length of the antenna.

Figure 2-7: Dipole antenna: (a) current distribution; (b) stacked dipole antenna; and (c) detail of the connection in a stacked dipole antenna.

be directly connected to the antenna and there is only a small mismatch and nearly all the power is transferred to the antenna and then radiated.

Another variation on the monopole is shown in Figure 2-6, where the key component is the phasing coil. The phasing coil (with a wire length of $\lambda/2$) rotates the electrical angle of the current phasor on the line so that the current on the $\lambda/4$ segment is in the same direction as on the $\lambda/2$ segment. The result is that the two straight segments of the loaded monopole radiate a more tightly confined EM field. The phasing coil itself does not radiate (much).

Another ingenious solution to obtaining a longer effective wire antenna with same-directed current (and hence a more tightly confined RF beam) is the stacked dipole antenna (Figure 2-7). The basis of the antenna is a dipole as shown in Figure 2-7(a). The cable has two conductors that have equal amplitude currents, I, but flowing as shown. The wire section is coupled to the cable so that the currents on the two conductors realize a single effective current I_{dipole} on the dipole antenna. The stacked dipole shown in Figure 2-7(b) takes this geometric arrangement further. Now the radiating element is hollow and a coaxial cable is passed through the antenna elements and the half-wavelength sections are fed separately to effectively create a wire antenna that is several wavelengths long with the current pointing in one direction. Most cellular antennas using wire antennas are stacked dipole antennas.

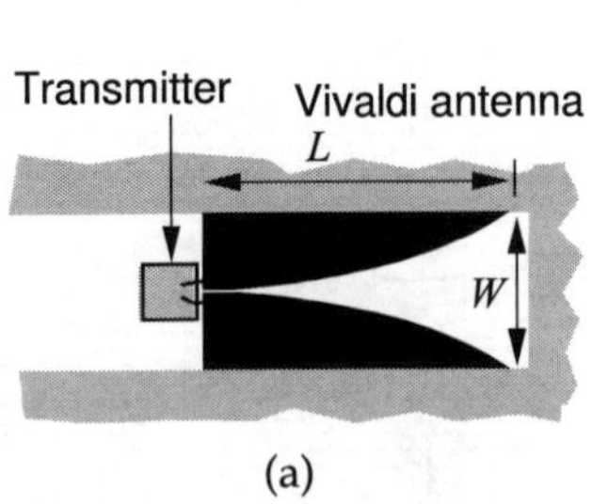

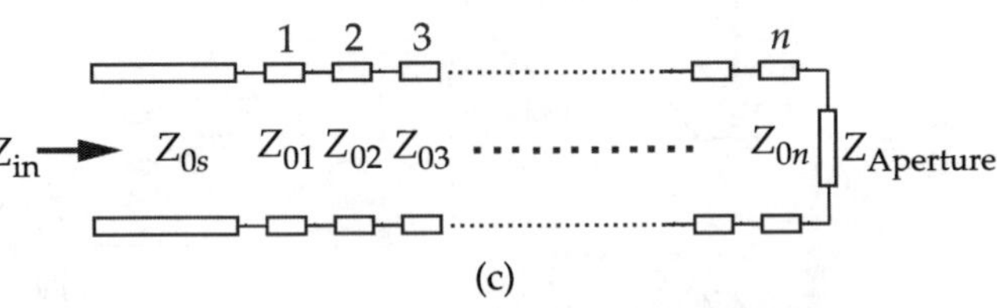

Figure 2-8: Vivaldi antenna showing design procedure: (a) the antenna; (b) a stepped approximation; and (c) transmission line approximation. In (a) the black region is a metal sheet.

Summary

Standing waves of current can be realized by resonant structures other than wires. A microstrip patch antenna, see Figure 2-2(b), is an example, but the underlying principle is that an array of current filaments generates EM components that combine to create a propagating field. Resonant antennas are inherently narrowband because of the reliance on the establishment of a standing wave. A relative bandwidth of 5%–10% is typical.

2.4 Traveling-Wave Antennas

Traveling-wave antennas have the characteristics of broad bandwidth and large size. These antennas begin as a transmission line structure that flares out slowly, providing a low reflection transition from a transmission line to free space. The bandwidth can be very large and is primarily dependent on how gradual the transition is.

One of the more interesting traveling-wave antennas is the **Vivaldi antenna** of Figure 2-8(a). The Vivaldi antenna is an extension of a slotline in which the fields are confined in the space between two metal sheets in the same plane. The slotline spacing increases gradually in an exponential manner, much like that of a Vivaldi violin (from which it gets its name), over a distance of a wavelength or more. A circuit model is shown in Figures 2-8(b and c) where the antenna is modeled as a cascade of many transmission lines of slowly increasing characteristic impedance, Z_0. Since the Z_0 progression is gradual there are low-level reflections at the transmission line interfaces. The forward-traveling wave on the antenna continues to propagate with a negligible reflected field. Eventually the slot opens sufficiently that the effective impedance of the slot is that of free space and the traveling wave continues to propagate in air.

The other traveling-wave antennas work similarly and all are at least a wavelength long, with the central concept being a gradual taper from the characteristic impedance of the originating transmission line to free space. The final aperture is at least one-half wavelength across so that the fields can curl on themselves (i.e. loop back on themselves) and are self-supporting as they leave the antenna.

2.5 Antenna Parameters

This section introduces a number of antenna metrics that are used to characterize antenna performance.

2.5.1 Radiation Density and Radiation Intensity

Antennas do not radiate equally in all directions concentrating radiated power in one direction called the **main** (or **major**) **lobe** of the antenna. This focusing effect is called **directivity**. The power in a particular direction is characterized by the radiation density and the radiation intensity metrics. The radiation density, S_r, is the power per unit area with the SI units of $\mathrm{W/m^2}$, and will be maximum in the main lobe. Referring to Figure 2-9 with an antenna located at the center of the sphere of radius r and radiating a total power P_r, S_r is the incremental radiated power dP_r passing through the incremental shaded region of area, dA:

$$S_r = \frac{dP_r}{dA}. \tag{2.14}$$

S_r reduces with distance falling off as $1/r^2$ in free space. For a practical antenna S_r will vary across the surface of the sphere. The total power radiated is the closed integral over the surface S of the sphere:

$$P_r = \oint_S dP_r = \oint_S S_r dA. \tag{2.15}$$

An alternative measure of power concentration is the radiation intensity U which is in terms of the incremental solid angle $d\Omega$ subtended by dA so that $d\Omega = dA/r^2$ and (with the SI units of W/steradian or W/sr)

$$U = \frac{dP_r}{d\Omega} = \frac{dP_r}{dA}r^2 = r^2 S_r. \tag{2.16}$$

Isotropic Antenna

It is useful to reference the directivity of an antenna with respect to a fictitious isotropic antenna that has no loss and radiates equally in all directions so that S_r is only a function of r. Then integrating over the surface of the sphere

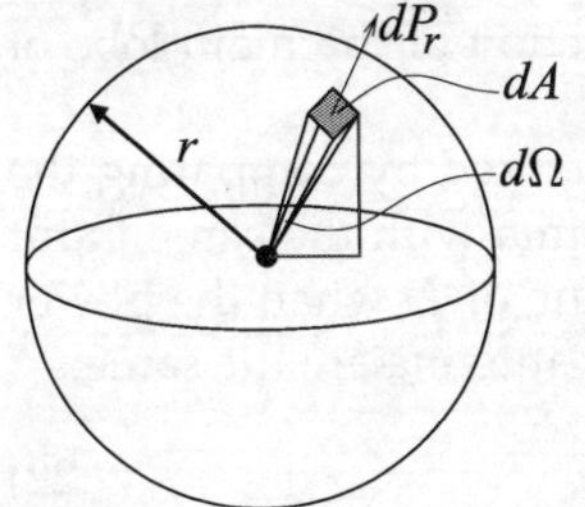

Figure 2-9: Free-space spreading loss. The incremental power, dP_r, intercepted by the shaded region of incremental area dA is proportional to $1/r^2$. The solid angle subtended by the shaded area is the incremental solid angle $d\Omega$. The integral of dA over the surface of the sphere, i.e. the area of the sphere is $4\pi r^2$. The total solid angle subtended by the sphere is the integral of $d\Omega$ over the sphere and is 4π steradians (or 4π sr).

yields the total radiated power

$$P_r|_{\text{Isotropic}} = \oint_S dP_r = \oint_S S_r dA = S_r \oint_S dA = S_r 4\pi r^2 = 4\pi U. \tag{2.17}$$

Since the isotropic antenna has no loss the power input to the antenna P_{IN} is equal to the power radiated $P_r = P_{\text{IN}}$. Thus for an isotropic antenna

$$S_r = \frac{P_r}{4\pi r^2} = \frac{P_{\text{IN}}}{4\pi r^2} \quad (2.18) \quad \text{and} \quad U|_{\text{Isotropic}} = r^2 S_r = \frac{P_{\text{IN}}}{4\pi r^2}. \quad (2.19)$$

Antenna Efficiency

Antenna efficiency, sometimes called the **radiation efficiency**, describes losses in an antenna principally due to resistive (I^2R) losses. Resonant antennas work by creating a large current that is maximized through the generation of a standing wave at resonance. There is a lot of current, and even just a little resistance results in substantial resistive loss. The power that is reflected from the input of the antenna is usually small. The total radiated power (in all directions), P_r, is the power input to the antenna less losses. The antenna efficiency, η_A is therefore defined as

$$\eta_A = P_r / P_{\text{IN}}, \tag{2.20}$$

where P_{IN} is the power input to the antenna and $\eta_A < 1$ and usually expressed as a percentage. Antenna efficiency is very close to one for many antennas, but can be 50% for microstrip patch antennas.

Antenna loss refers to the same mechanism that gives rise to antenna efficiency. Thus an antenna with an antenna efficiency of 50% has an antenna loss of 3 dB. Generally losses are resistive due to I^2R loss and mismatch loss of the antenna that occurs when the input impedance is not matched to the impedance of the cable connected to the antenna. Because of confusion with antenna gain (they are not the opposite of each other) the use of the term 'antenna loss' is discouraged and instead 'antenna efficiency' preferred.

2.5.2 Directivity and Antenna Gain

The directivity of an antenna, D, is the ratio of the radiated power density to that of an isotropic antenna with the same total radiated power P_r:

$$D = \frac{S_r}{S_r|_{\text{Isotropic}}} = \frac{U}{U|_{\text{Isotropic}}} \tag{2.21}$$

where S_r and U refer to the actual antenna and the power densities and intensities are measured at the same distance from the antennas. For an actual antenna D is dependent on the direction from the antenna, see Figure 2-10. The maximum value of D will be in the direction of the main lobe of the antenna and this is called **directivity gain**.

The focusing property of an antenna is characterized by comparing the radiated power density to that of an isotropic antenna with the same input power. The antenna gain, G_A, is the maximum value of D when the power input $P_{\text{IN}} = P_r/\eta_A$ to the antenna and the isotropic antenna are the same:

$$G_A = \eta_A \max(D). \tag{2.22}$$

Antenna	Type	Figure	Gain (dBi)	Notes
Lossless isotropic antenna			0	
$\lambda/2$ dipole	Resonant	2-7(a)	2	$R_{\text{in}} = 73\ \Omega$
3λ diameter parabolic dish	Traveling	—	38	$R_{\text{in}} = \text{match}$
Patch	Resonant	2-2(b)	9	$R_{\text{in}} = \text{match}$
Vivaldi	Traveling	2-2(c)	10	$R_{\text{in}} = \text{match}$
$\lambda/4$ monopole on ground	Resonant	2-5(a)	2	$R_{\text{in}} = 36\ \Omega$
$5/8\lambda$ monopole on ground	Resonant	2-5(a)	3	Matching required

Table 2-1: Several antenna systems. R_{in} = match for resonant antennas indicates that the antenna can be designed to have an input impedance matching that of a feed cable. Traveling-wave antennas are intrinsically matched.

Losses in the antenna are accounted for by the efficiency term η_A.

In Equation (2.22) G_A is a gain factor and is often expressed in terms of decibels (taking 10 times the log of G_A) but dBi (with 'i' standing for 'with-respect-to isotropic') is used to indicate that it is not a power gain in the same sense as amplifier gain. G_A instead is the ratio of power densities for two different antennas. For example, an antenna that focuses power in one direction increasing the peak radiated power density by a factor of 20 relative to that of an isotropic antenna thus has an antenna gain of 13 dBi. With care G_A can be often used in calculations of power as as with amplifier gain.

Since it is almost impossible to calculate internal antenna losses, antenna gain is invariably only measured. The input power to an antenna can be measured and the peak radiated power density, $P_D|_{\text{Maximum}}$, measured in the far field at several wavelengths distant (at $r \gg \lambda$). This is compared to the power density from an ideal isotropic antenna at the same distance with the same input power. Antenna gain is determined from

$$G_A = \frac{\text{Maximum radiated power per unit area}}{\text{Maximum radiated power per unit area for an isotropic antenna}} \tag{2.23}$$

$$\begin{aligned} &= \frac{S_r|_{\text{Maximum}}}{S_r|_{\text{Isotropic}}} = 4\pi r^2 \frac{P_D|_{\text{Maximum}}}{P_{\text{IN}}} \\ &= 4\pi \frac{\text{Maximum radiated power per unit solid angle}}{\text{Total input power to the antenna}} \\ &= 4\pi \frac{(dP_r/d\Omega)|_{\text{Maximum}}}{P_{\text{IN}}} = 4\pi r^2 \frac{(dP_r/dA)|_{\text{Maximum}}}{P_{\text{IN}}}. \end{aligned} \tag{2.24}$$

The antenna gains of common resonant and traveling-wave antennas are given in Table 2-1. In free space the antenna gain determined using Equation (2.22) is independent of distance. Antenna gain is measured on an antenna range using a calibrated receive antenna and care taken to avoid reflections from objects, especially from the ground.

The losses of an antenna are incorporated in the antenna gain which is defined in terms of the power input to the antenna, see Equation (2.24). Thus in calculations of radiated power using antenna gain, there is no need to separately account for resistive losses in the antenna.

In summary, antennas concentrate the radiated power in one direction so that the density of the power radiated in the direction of the peak field is higher than the power density from an isotropic antenna. Power radiated from a base station antenna, such as that shown in Figure 2-11, is concentrated in a region that looks like a toroid or, more closely, a balloon squashed

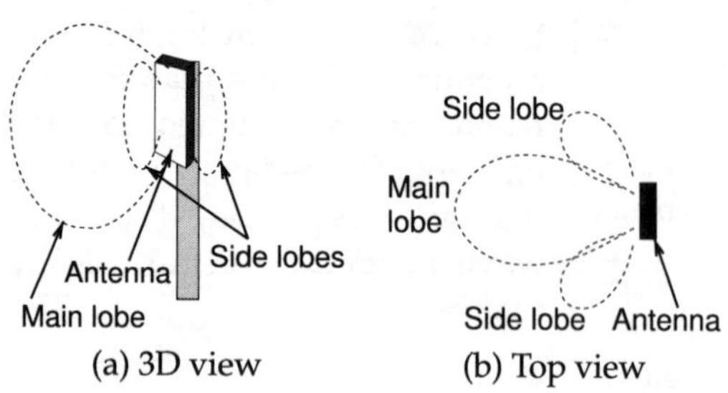

Figure 2-10: Field pattern produced by a microstrip antenna.

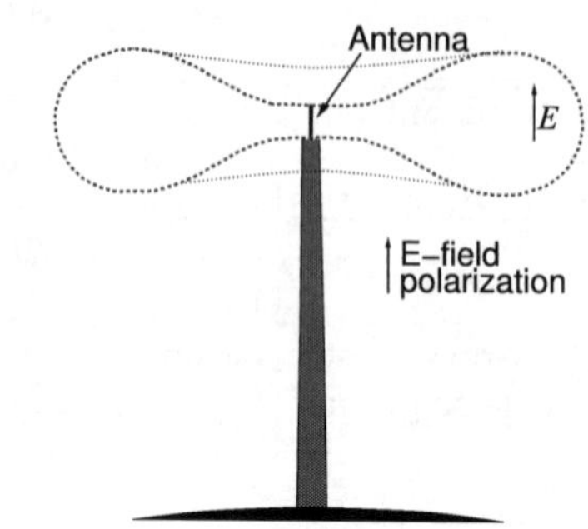

Figure 2-11: A base station transmitter pattern.

at its north and south poles. Then the antenna does not radiate much power into space and will concentrate power in a region skimming the surface of the earth. Antenna gain is a measure of the effectiveness of an antenna to concentrate power in one direction. Thus, in free space where power spreads out by $1/r^2$, the maximum power density (in SI units of $\mathrm{W/m^2}$) at a distance r is

$$P_D = \frac{G_A P_{\mathrm{IN}}}{4\pi d^2}, \tag{2.25}$$

where $4\pi d^2$ is the area of a sphere of radius d and P_{IN} is the input power.

Measurements of antenna gain are used to derive antenna efficiency. It is impossible to measure or simulate the resistive and dielectric losses of an antenna directly. Antenna efficiency is obtained using theoretical calculations of antenna gain assuming no losses in the antenna itself. This is compared to the measured antenna gain yielding the antenna efficiency.

EXAMPLE 2.2 Antenna Gain

A base station antenna has an antenna gain, G_A, of 11 dBi and a 40 W input. The transmitted power density falls off with distance d as $1/d^2$. What is the peak power density at 5 km?

Solution:

A sphere of radius 5 km has an area $A = 4\pi r^2 = 3.142 \cdot 10^8\ \mathrm{m}^2$; $G_A = 11$ dBi = 12.6.
In the direction of peak radiated power, the power density at 5 km is

$$P_D = \frac{P_{\mathrm{in}} G_A}{A} = \frac{40 \cdot 12.6\ \mathrm{W}}{3.142 \cdot 10^8 \mathrm{m}^2} = 1.603\ \mu\mathrm{W/m^2}.$$

EXAMPLE 2.3 Antenna Efficiency

A antenna has an antenna gain of 13 dBi and an antenna efficiency of 50% and all of the loss is due to resistive losses and resistance of metals is proportional to temperature. The RF signal input to the antenna has a power of 40 W.

(a) What is the input power in dBm?

$P_{\text{in}} = 40\text{ W} = 46.02\text{ dBm}$.

(b) What is the total power transmitted in dBm?

$P_{\text{Radiated}} = 50\%$ of $P_{\text{IN}} = 20$ W or 43.01 dBm.
Alternatively, $P_{\text{Radiated}} = 46.02\text{ dBm} - 3\text{ dB} = 43.02\text{ dBm}$.

(c) If the antenna is cooled to near absolute zero so that it is lossless, what would the antenna gain be?

The antenna gain would increase by 3 dB and antenna gain incorporates both directivity and antenna losses. So the gain of the cooled antenna is 16 dBi.

2.5.3 Effective Isotropic Radiated Power

A transmit antenna does not radiate power equally in all directions and for a receiver in the main lobe of the transmit antenna it is as though there is an isotropic transmit antenna with a much higher input power. This concept is incorporated in the effective isotropic radiated power (**EIRP**):

$$\text{EIRP} = P_{\text{IN}} G_A. \tag{2.26}$$

This is the total power that would be radiated by an isotropic antenna producing the same (peak) power density as the actual antenna.

2.5.4 Effective Aperture Size

Effective aperture size is defined so that the power density at a receive antenna when multiplied by its effective aperture size, A_R, yields the power output from the antenna at its connector. An antenna has an effective size that is more than its actual physical size because of its influence on the EM fields around it. The effective aperture size of an antenna is the area of the surface that captures all of the power passing through it and delivers this power to the output terminals of the antenna.

The effective aperture area of a receive antenna, A_R, is related to the receive antenna gain, G_R, as follows [2, 3] (note that A_e is often used if it is not necessary to distinguish antennas):

$$A_R = \frac{G_R \lambda^2}{4\pi}, \tag{2.27}$$

where λ is the wavelength of the radio signal. The effective aperture area of an antenna can have little to do with its physical size; e.g., a wire antenna has almost no physical size but has a significant effective aperture size.

If S_r is the transmitted power density at the receive antenna, the power received is

$$P_R = P_D A_R = P_D \frac{G_R \lambda^2}{4\pi}. \tag{2.28}$$

The power density at a distance d (ignoring multipath effects),is

$$S_r = \frac{P_T G_T}{4\pi d^2}, \tag{2.29}$$

where P_T is the power input to the transmit antenna with antenna gain G_T. The power delivered by the receive antenna is

$$P_R = S_r A_R = \frac{P_T G_T}{4\pi d^2}\frac{G_R \lambda^2}{4\pi} = P_T G_T G_R \left(\frac{\lambda}{4\pi d}\right)^2. \tag{2.30}$$

2.5.5 Summary

This section introduced several metrics for characterizing antennas:

Metric	Equation	Description
S_r	(2.14)	Radiated power density, W/m^2
U	(2.16)	Radiation intensity W/sr
η_A	(2.27)	Antenna efficiency
D	(2.21)	Antenna directivity
G_A	(2.23)	Antenna gain,used with a transmit antenna
A_e	(2.27)	Effective aperture area, used with a receive antenna
EIRP	(2.26)	Equivalent isotropic radiated power

EXAMPLE 2.4 Point-to-Point Communication

In a point-to-point communication system, a parabolic receive antenna has an antenna gain of 60 dBi. If the signal is 60 GHz and the power density at the receive antenna is 1 pW/cm^2, what is the power at the output of the receive antenna connected to the RF electronics?

Solution:

The first step is determining the effective aperture area, A_R, of the antenna. At 60 GHz $\lambda = 5$ mm. Note that $G_R = 60$ dBi $= 10^6$. From Equation (2.27),

$$A_R = \frac{G_R \lambda^2}{4\pi} = \frac{10^6 \cdot 0.005^2}{4\pi} = 1.989 \text{ m}^2. \tag{2.31}$$

Using Equation (2.28) $P_D = 1 \text{ pW/cm}^2 = 10 \text{ nW/m}^2$, the total power delivered to the RF receiver electronics (at the output of the receive antenna) is

$$P_R = P_D A_R = 10 \text{ nW} \cdot \text{m}^{-2} \cdot 1.989 \text{ m}^2 = 19.89 \text{ nW}. \tag{2.32}$$

2.6 The RF Link

The RF link is between a transmit antenna and a receive antenna. Sometimes the RF link includes the antenna, this will be clear from the context, but usually it includes the antennas. The principle source of link loss is the spreading out of the EM field as it propagates. In the absence obstructions the power density reduces as $1/d^2$, where d is distance, and this is the **line-of-sight** (**LOS**) situation. In this section the propagation path is first described along with its impairments including propagation on multiple paths between a transmit antenna and a receive antenna.

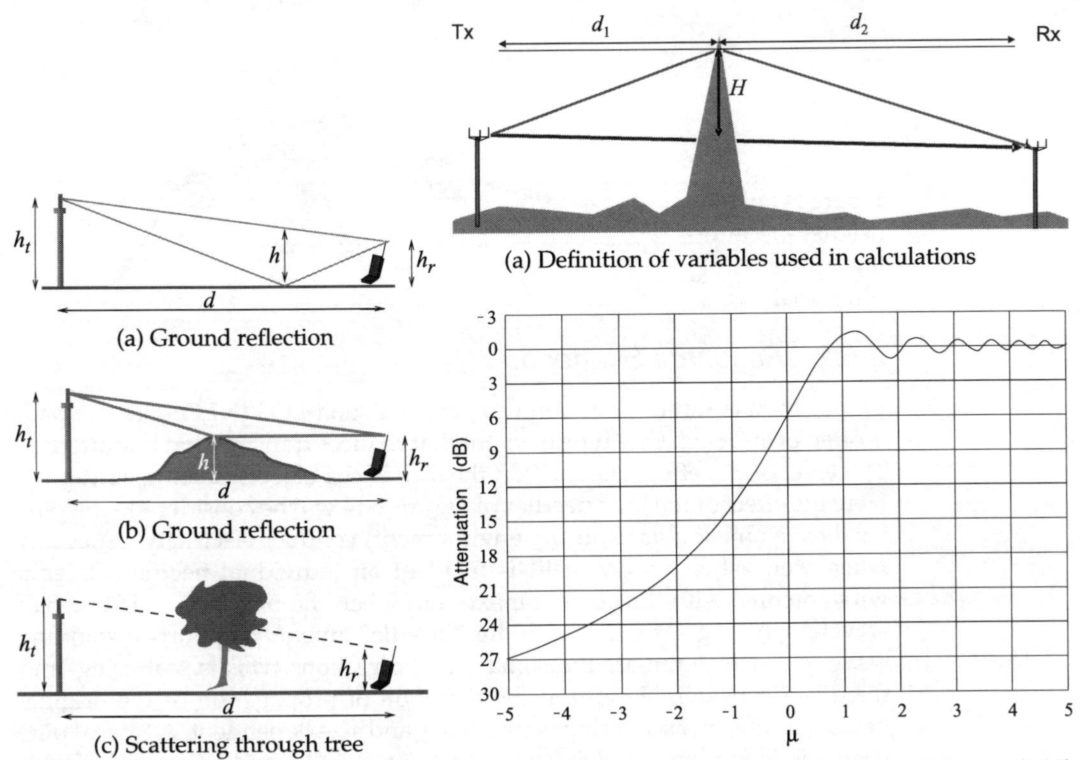

Figure 2-12: Common paths contributing to multipath propagation.

Figure 2-13: Knife-edge diffraction.

2.6.1 Propagation Path

When the radiated signal reflects and diffracts there are multiple propagation paths that result in fading as the paths constructively and destructively combine at the receiver. Of these destructive combining is much worse as it can reduce a signal level below what it would be if propagation was in free space. In urban areas, 10 or 20 paths can have significant powers [4].

Common paths encountered in cellular radio are shown in Figure 2-12. As a rough guide, in the single-digit gigahertz range each diffraction and scattering event reduces the signal received by 20 dB. The knife-edge diffraction scenario is shown in more detail in Figure 2-13. This case is fairly easy to analyze and can be used to estimate the effects of individual obstructions. The diffraction model is derived from the theory of half-infinite screen diffraction [5]. First, calculate the parameter ν from the geometry of the path using

$$\nu = -H\sqrt{\frac{2}{\lambda}\left(\frac{1}{d_1}+\frac{1}{d_2}\right)}. \tag{2.33}$$

Next, consult the plot in Figure 2-13(b) to obtain the diffraction loss (or attenuation). This loss should be added (using decibels) to the otherwise determined path loss to obtain the total path loss.

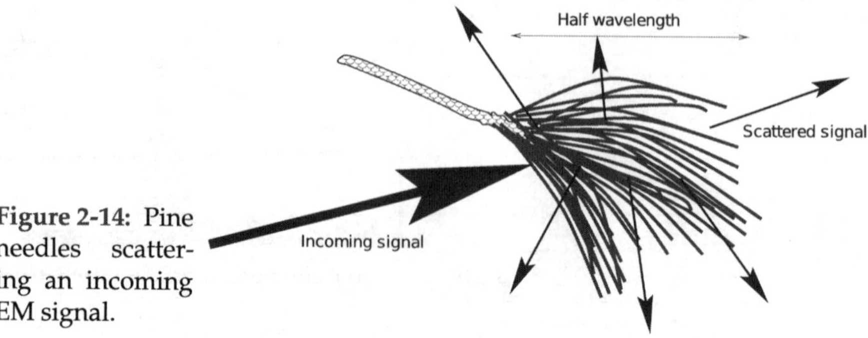

Figure 2-14: Pine needles scattering an incoming EM signal.

2.6.2 Resonant Scattering

Propagation is rarely from point to point, i.e. non-LOS (**NLOS**), as the path is often obstructed. One type of event that reduces transmission is scattering. The level of the effect depends on the size of the objects causing scattering. Here the effect of the pine needles of Figure 2-14 will be considered. The pine needles (as most objects in the environment) conduct electricity, especially when wet. When an EM field is incident an individual needle acts as a wire antenna, with the current maximum when the pine needle is one-half wavelength long. At this length, the "needle" antenna supports a standing wave and will re-radiate the signal in all directions. This is scattering, and there is a considerable loss in the direction of propagation of the original fields. The effect of scattering is frequency and size dependent. A typical pine needle is 15 cm long, which is exactly $\lambda/2$ at 1 GHz, and so a stand of pine trees have an extraordinary impact on cellular communications at 1 GHz. As a rough guide, 20 dB of a signal is lost when passing through a small stand of pine trees.

2.6.3 Fading

Fading refers to the variation of the received signal with time or when the position of transmit or receive antennas is changed. The most important fading types are flat fading, multipath fading, and rain fading.

Flat Fading

Temperature variations of the atmosphere between the transmit and receive antennas give rise to what is called flat fading and sometimes called **thermal fading**. This fading is called flat because it is independent of frequency. One form of flat fading is due to refraction, which occurs when different layers of the atmosphere have different densities and thus dielectric permittivities increasing or decreasing away from the surface of the earth. The temperature profile can increase away from the earth surface or reduce depending on whether the temperature of the earth is higher than that of the air and is commonly associated with the beginning and end of the day. This causes refraction of the propagating wave, , see Figure 2-15(a). **Temperature inversions** can also occur where the temperature profile producing a layer with a relatively higher permittivity. RF energy gets trapped in this layer, reflecting from the top and bottom of the inversion layer, see Figure 2-15(b). This is called **ducting**. In point-to-point communication systems the transmit and receive antennas are mounted high on towers and then

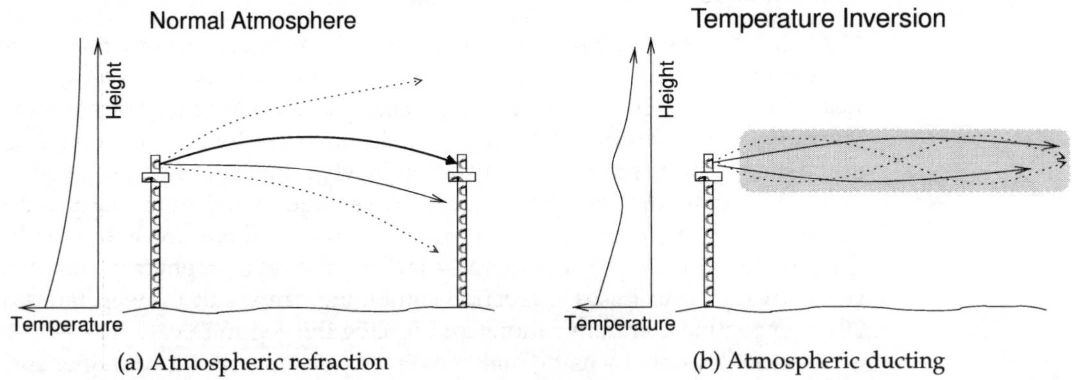

Figure 2-15: Fading resulting from ducting: (a) normal atmospheric refraction (normally the temperature of air drops with increasing height and the lower refractive index at high heights results in a concave refraction); and (b) atmospheric ducting (resulting from temperature inversion inducing an air layer with higher dielectric constant than the surrounding air).

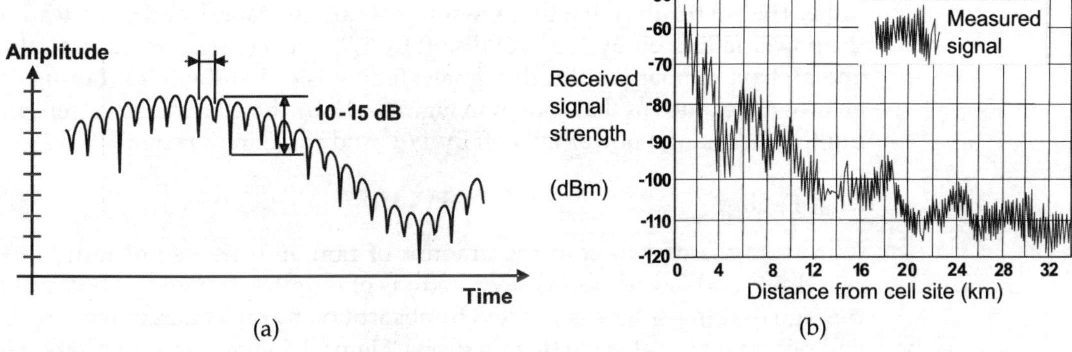

Figure 2-16: Fast and slow fading: (a) in time as a radio and obstructions move; and (b) in distance.

reflection from ground objects is often small. In such cases flat fading is the most commonly observed phenomena and minor fluctuations of several decibels in receive signal level are common throughout the day. However, when temperature variations are extreme, ducting can severely impact communications reducing signal levels by up to 20 dB.

Shadow Fading

Shadow fading occurs when the LOS path is blocked by an obstruction such as a building or hill. This results in relatively slow fades with the amplitude response varying in time and distance as shown in Figure 2-16. This figure shows both fast fades that are 10 to 15 dB deep and slow or shadow fades that are 20 to 30 dB deep.

Multipath Fading

Multipath describes the situation situations where there are many reflections that combine destructively and constructively. Multipath fading is also called **fast fading**, as the characteristics of the channel can change significantly in a few milliseconds. Multipath fading of 20 dB can occur for a small percentage of the time on time scales of many seconds when there are few propagation paths (e.g. in a rural area) to a large percentage of the time many times per second in a dense urban environment where there are many paths. Constructive combining does increase the signal level momentarily, but there is no advantage to this. Destructive combining can result in deep fades of 20 dB impacting communications and forcing the communication system to accommodate either by using higher average powers or using strategies such as multiple antennas or spreading the communication signal over a wide bandwidth since fades tend to be 500 kHz to 1 MHz wide at all frequencies.

When the signal on one of the paths dominates, this is usually the LOS path, fading is called **Rician fading**. With LOS and a single ground reflection, the situation is the classic Rician fading shown in Figure 2-17(a). Here the ground changes the phase of the signal upon reflection by 180°. When the receiver is a long way from the base station, the lengths of the two paths are almost identical and the level of the signals in the two paths are almost the same. The net result is that these two signals almost cancel, and so instead of the power falling off by $1/d^2$, it falls off by $1/d^3$. When there are many paths and all have similar amplitude signals, fading is called **Rayleigh fading**. In an urban area such as that shown in Figure 2-17(b), there are many significant multipaths and the power falls off by $1/d^4$ and sometimes faster.

Rain Fading

Rain fading is due to both the amount of rain and the size of individual rain drops and fading occurs over periods of minutes to hours. Propagation through the atmosphere is affected by absorption by molecules in the air, fog, and rain, and by scattering by rain drops. Figure 1-2 shows the attenuation in decibels per kilometer from 3 GHz to 300 GHz. The attenuation due to rain increases with frequency and this derives largely from scattering.

Summary of Fading

The fades of most concern in a mobile wireless system are the deep fades resulting from destructive interference of multiple reflections. These fades vary rapidly (over a few milliseconds) if a handset is moving at vehicular speeds but occur slowly when the transmitter and receiver are fixed. Fades can be viewed as deep amplitude modulation, and so so modulation is restricted to phase shift keying schemes when a transmitter and receive

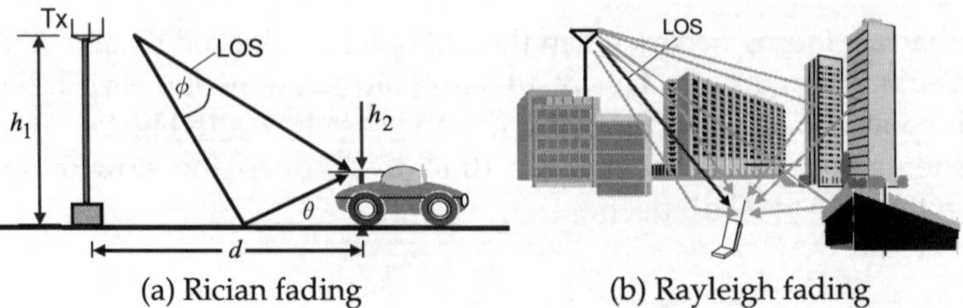

(a) Rician fading (b) Rayleigh fading

Figure 2-17: Multipath propagation: (a) line of sight (LOS) and ground reflection paths only; and (b) in an urban environment.

moving at vehicular speeds relative to each other.

2.6.4 Link Loss and Path Loss

With transmit and receive antennas included in the RF link, the usual case, link loss is defined as the ratio of the power input to the transmit antenna, P_T, to the power delivered by the receive antenna, P_R. Rearranging Equation (2.30) the total line-of-sight (LOS) link loss, $L_{\text{LINK,LOS}}$, between the input of the transmit antenna and the output of the receive antenna separated by distance d is (in decibels):

$$L_{\text{LINK,LOS}}|_{\text{dB}} = 10\log\left(\frac{P_T}{P_R}\right) = 10\log\left(\frac{P_T}{P_D A_R}\right) \tag{2.34}$$

$$= 10\log\left[P_T\left(\frac{4\pi d^2}{P_T G_T}\right)\left(\frac{4\pi}{\lambda^2 G_R}\right)\right] \tag{2.35}$$

$$= 10\log\left[\left(\frac{1}{G_T G_R}\right)\left(\frac{4\pi d}{\lambda}\right)^2\right] \tag{2.36}$$

$$= -10\log G_T - 10\log G_R + 20\log\left(\frac{4\pi d}{\lambda}\right). \tag{2.37}$$

The last term includes d and is called the LOS path loss (in decibels):

$$L_{\text{PATH,LOS}}|_{\text{dB}} = 20\log\left(\frac{4\pi d}{\lambda}\right). \tag{2.38}$$

This is the preferred form of the expression for path loss, as it can be used directly in calculating link loss using the antenna gains of the transmit and receive antennas without the exercise of calculating the **effective aperture size** of the receive antenna.

Multipath effects result in losses that are proportional to d^n [6, 7] so that the general path loss, including multipath effects, is (in decibels)

$$\begin{aligned} L_{\text{PATH}}|_{\text{dB}} &= L_{\text{PATH,LOS}}|_{\text{dB}} + \text{excess loss}|_{\text{dB}} \\ &= 20\log\left(\frac{4\pi d}{\lambda}\right) + 10(n-2)\log\left(\frac{d}{1\text{ m}}\right) \\ &= 20\log\left[\frac{4\pi(1\text{ m})}{\lambda}\right] + 10(2)\log\left(\frac{d}{1\text{ m}}\right) + 10(n-2)\log\left(\frac{d}{1\text{ m}}\right) \\ &= 10n\log[d/(1\text{ m})] + C, \end{aligned} \tag{2.39}$$

where the distance d and wavelength λ are in meters, and C is a constant that captures the effect of wavelength. Here,

$$C = 20\log\left[4\pi(1\text{ m})/\lambda\right]. \tag{2.40}$$

Combining this with Equation (2.37) yields the link loss:

$$L_{\text{LINK}}|_{\text{dB}} = -G_T|_{\text{dB}} - G_R|_{\text{dB}} + 10n\log[d/(1\text{ m})] + C. \tag{2.41}$$

EXAMPLE 2.5 Link Loss

A 5.6 GHz communication system uses a transmit antenna with an antenna gain G_T of 35 dB and a receive antenna with an antenna gain G_R of 6 dB. If the distance between the antennas is 200 m, what is the link loss if the power density reduces as $1/d^3$? The link loss here is between the input to the transmit antenna and the output from the receive antenna.

Solution:

The link loss is provided by Equation (2.41),

$$L_{\text{LINK}}|_{\text{dB}} = -G_T - G_R + 10n\log[d/(1\text{ m})] + C,$$

and C comes from Equation (2.40), where $\lambda = 5.36$ cm. So

$$C = 20\log\left(\frac{4\pi}{\lambda}\right) = 20\log\left(\frac{4\pi}{0.0536}\right) = 47.4\text{ dB}.$$

With $n = 3$ and $d = 200$ m,

$$L_{\text{LINK}}|_{\text{dB}} = -35 - 6 + 10\cdot 3\cdot\log(200) + 47.4\text{ dB} = 75.4\text{ dB}.$$

EXAMPLE 2.6 Radiated Power Density

In free space, radiated power density drops off with distance d as $1/d^2$. However, in a terrestrial environment there are multiple paths between a transmitter and a receiver, with the dominant paths being the direct LOS path and the path involving reflection off the ground. Reflection from the ground partially cancels the signal in the direct path, and in a semi-urban environment results in an attenuation loss of 40 dB per decade of distance (instead of the 20 dB per decade of distance roll-off in free space). Consider a transmitter that has a power density of 1 W/m^2 at a distance of 1 m from the transmitter.

(a) The power density falls off as $1/d^n$, where d is distance and n is an index. What is n?
(b) At what distance from the transmit antenna will the power density reach 1 μW·m^{-2}?

Solution:

(a) Power drops off by 40 dB per decade of distance. 40 dB corresponds to a factor of 10,000 (= 10^4). So, at distance d, the power density $P_D(d) = k/d^n$ (k is a constant).
At a decade of distance, $10d$, $P_D(10d) = k/(10d)^n = P_D(d)/10000$, thus

$$\frac{k}{10^n d^n} = \frac{1}{10,000}\frac{k}{d^n}; \qquad 10^n = 10,000 \Rightarrow n = 4.$$

(b) At $d = 1$ m, $P_D(1\text{ m}) = 1\text{ W/m}^2$. At a distance x,

$$P_D(x) = 1\ \mu\text{W/m}^2 = \frac{k}{x^4}\text{m}^2 = \frac{k}{x^4} \rightarrow x^4 = \frac{1}{10^{-6}} \quad \text{and so} \quad x = 31.6\text{ m}.$$

2.6.5 Propagation Model in the Mobile Environment

RF propagation in the mobile environment cannot be accurately derived. Instead, a fit to measurements is often used. One of the models is the Okumura–Hata model [8], which calculates the path loss as

$$L_{\text{PATH}}|_{(\text{dB})} = 69.55 + 26.16\log f - 13.82\log H + (44.9 - 6.55\log H)\cdot\log d + c, \tag{2.42}$$

where f is the frequency (in MHz), d is the distance between the base station and terminal (in km), H is the effective height of the base station antenna (in m), and c is an environment correction factor ($c = 0$ dB in a dense urban area,

$c = -5$ dB in an urban area, $c = -10$ dB in a suburban area, and $c = -17$ dB in a rural area, for $f = 1$ GHz and H=1.5 m).

There are many propagation models for different frequency ranges and different environments. The considerable effort put into developing reliable models is because being able to predict signal coverage is essential to efficient design of basestation layout.

2.7 Antenna Array

An antenna array comprises multiple radiating elements, i.e. individual antennas, and focuses a transmit beam in a desired direction. In Figure 2-20 the field pattern in the plane of the earth produced by an array of 30 antenna elements arranged horizontally is shown. The fields from each antenna element combined to narrow and strengthen the main beam. Side (or grating) lobes are produced and these will result in some interference but their level here is 40 dB below that of the main beam. This is another way of managing interference in a cellular system but the direction to the mobile unit must be known. Antenna arrays are used in 4G and 5G. The affect of the array is to increase the power density of the main beam and the density relative to that of an isotropic antenna is called the directional gain of the array, D_{Array}, and is the product of the antenna gain, G_A, of an individual antenna element and the array gain, G_{Array}:

$$D_{\text{Array}} = G_{\text{Array}} G_A \tag{2.43}$$

The maximum value of G_{Array} is N for an N element array. So the maximum directional gain of the array in dBi is

$$D_{\text{Array}}|_{\text{dBi}} = G_A|_{\text{dBi}} + 10 \log N. \tag{2.44}$$

2.7.1 Multiple Input, Multiple Output

Multiple input, multiple output (**MIMO**, pronounced my-moe) technology uses multiple antennas to transmit and receive signals. The MIMO concept was developed in the 1990s [9, 10] and implemented in 4G and 5G, and a variety of WLAN systems. MIMO relies on signals traveling on multiple paths between an array of transmit antennas and an array of receive antennas In MIMO these paths are used to carry more information with each path propagates an image of one transmitted signal (from one antenna) that differs in both amplitude and phase from the images following other paths. Effectively there are multiple connections between each transmit antenna and each receive antenna, see Figure 2-18. Here a high-speed data-stream is split into several slower data streams, shown in Figure 2-18 as the a, b, and c bitstreams. The distinct bitstreams are separately modulated and sent from their own transmit antenna, with the constellation diagrams of the transmitted modulated signals labeled A, B, and C. The signals from each of the transmit antenna reaches all of the receive antennas by following different uncorrelated paths.

The output of each receive antenna is a linear combination of the multiple transmitted data streams, with the sampled RF phasor diagrams labeled M, N, and O. (It is not really appropriate to call these constellation diagrams.) That is, each receive antenna has a different linear combination of the multiple images. In effect, the output from each receive antenna can be

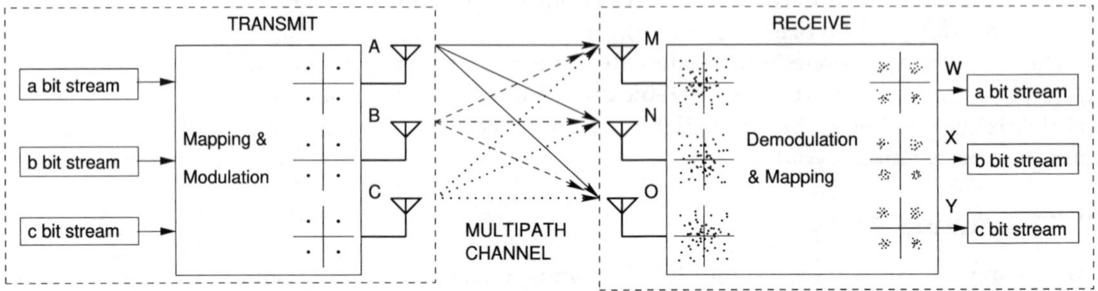

Figure 2-18: A MIMO system showing multiple paths between each transmit antenna and each receive antenna.

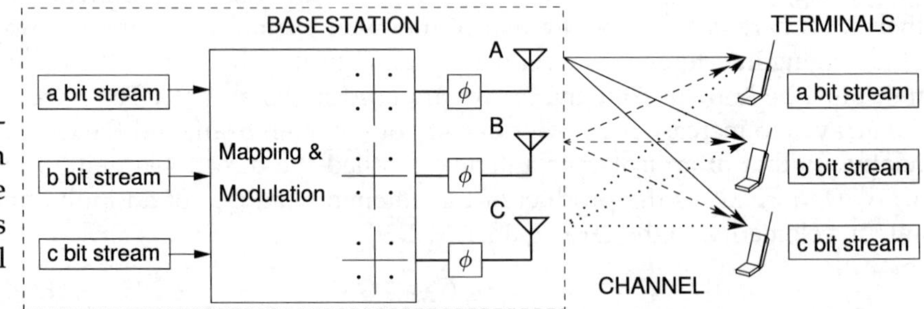

Figure 2-19: A massive MIMO system with beams from the basestation antennas to each terminal unit.

thought of as the solution of linear equations, with each transmit antenna-receive antenna link corresponding to an equation. Continuing the analogy, the signal from each transmit antenna represents a variable. So a set of simultaneous equations can be solved to obtain the original bitstreams. This is accomplished by demodulation and mapping using knowledge of the channel characteristics to yield the original transmitted signals modified by interference. The result is that the constellation diagrams W, X, and Y are obtained. The composite channel can be characterized using known test signals.

The capacity of a MIMO system with high SIR scales approximately linearly with the minimum of M and N, $\min(M, N)$, where M is the number of transmit antennas and N is the number of receive antennas (provided that there is a rich set of paths) [11, 12]. So a system with $M = N = 4$ will have four times the capacity of a system with just one transmit antenna or one receive antenna.

2.7.2 Massive MIMO

MIMO in 4G uses multiple antennas at the basestation and at the mobile terminals to increase overall data rates provided that there are multiple paths between the transmitter and receiver. In 5G there can be a very large number of transmit antennas even though there are few receive antennas per terminal unit as multiple terminal units mean that there can effectively be a very large number of receive antennas, see Figure 2-19.

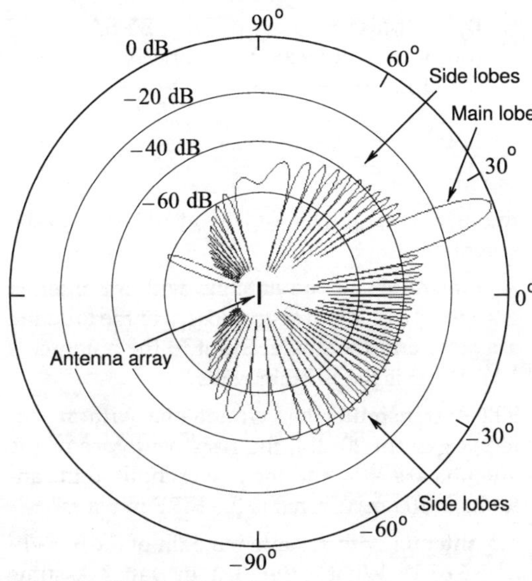

Figure 2-20: Electric field pattern from a 30 element array of antennas spaced 0.65λ appart. The sidelobe levels are about 40 dB below the power level of the main lobe. The same signal is presented to the antenna elements except that phases of the signal at each antennas is adjusted to produce a main beam directed at 20 degrees. The signals to each antenna are thus correlated. After [13].

2.8 Summary

This chapter discussed the impact of imperfections in the RF link from the output of the transmitter to the input of the receiver. Prior to cellular communications becoming so important, only the LOS communication path was considered and the impact of fading, multiple reflections, and delay spread was regarded as detrimental. While many aspects of RF propagation are random, concepts and statistical models have been developed that enable design choices to be made that permit digital communication systems to operate in what would have been regarded as a hostile environment.

2.9 References

[1] S. Ramo, J. Whinnery, and T. Van Duzer, *Fields and Waves in Communication Electronics*. John Wiley & Sons, 1965.

[2] C. A. Balanis, "Antenna theory: A review," *Proc. IEEE*, vol. 80, no. 1, pp. 7–23, 1992.

[3] C. Balanis, *Antenna Theory: Analysis and Design*. John Wiley & Sons, 2005.

[4] J. Doble, *Introduction to Radio Propagation for Fixed and Mobile Communications*. Norwood, MA, USA: Artech House, Inc., 1996.

[5] M. Born and E. Wolf, "Two-dimensional diffraction of a plane wave by a half-plane,," in *Principles of Optics: Electromagnetic Theory of Propagation, Interference, and Diffraction of Light*, 7th ed. Cambridge University Press, 1999, section 11.5.

[6] J. Andersen, T. Rappaport, and S. Yoshida, "Propagation measurements and models for wireless communications channels," *IEEE Communications Magazine*, vol. 33, no. 1, pp. 42–49, Jan. 1995.

[7] T. Sarkar, Z. Ji, K. Kim, A. Medouri, and M. Salazar-Palma, "A survey of various propagation models for mobile communication," *IEEE Antennas and Propagation Magazine*, vol. 45, no. 3, pp. 51–82, Jun. 2003.

[8] J. Seybold, *Introduction to RF Propagation*. John Wiley & Sons, 2005.

[9] G. Foschini, "Layered space-time architecture for wireless communication in a environment when using multi-element antennas," *Bell Labs Technical J.*, vol. 1, no. 2, pp. 41–59, Autumn 1996.

[10] G. Raleigh and J. Cioffi, "Spatio-temporal coding for wireless communications," in *Global Telecommunications Conf., 1996)*, vol. 3, Nov. 1996, pp. 1809–1814.

[11] A. Goldsmith, S. Jafar, N. Jindal, and S. Vishwanath, "Capacity limits of mimo channels," *IEEE J. on Selected Areas in Communications*, vol. 21, no. 5, pp. 684–702, Jun. 2003.

[12] D. Gesbert, M. Shafi, D. shan Shiu, P. Smith, and A. Naguib, "From theory to practice: an overview of mimo space-time coded wireless systems," *IEEE J. on Selected Areas in Communications*, vol. 21, no. 3, pp. 281–302, Apr.

2003.

[13] Phased array radiation pattern, Phased_array_radiation_pattern.gif, By Maxter315 [CC BY-SA 4.0 (https://creativecommons.org/licenses/by-sa/4.0)], from Wikimedia Commons.

2.10 Exercises

1. An antenna only radiates 45% of the power input to it. The rest is lost as heat. What input power (in dBm) is required to radiate 30 dBm?
2. The output stage of an RF front end consists of an amplifier followed by a filter and then an antenna. The amplifier has a gain of 27 dB, the filter has a loss of 1.9 dB, and of the power input to the antenna, 35% is lost as heat due to resistive losses. If the power input to the amplifier is 30 dBm, calculate the following:
   (a) What is the power input to the amplifier in watts?
   (b) Express the loss of the antenna in dB.
   (c) What is the total gain of the RF front end (amplifier + filter)?
   (d) What is the total power radiated by the antenna in dBm?
   (e) What is the total power radiated by the antenna in mW?
3. The output stage of an RF front end consists of an amplifier followed by a filter and then an antenna. The amplifier has a gain of 27 dB, the filter has a loss of 1.9 dB, and of the power input to the antenna, 45% is lost as heat due to resistive losses. If the power input to the amplifier is 30 dBm, calculate the following:
   (a) What is the power input to the amplifier in watts?
   (b) Express the loss of the antenna in decibels.
   (c) What is the total gain of the RF front end (amplifier + filter)?
   (d) What is the total power radiated by the antenna in dBm?
   (e) What is the total power radiated by the antenna in milliwatts?
4. Thirty five percent of the power input to an antenna is lost as heat, what is the loss of the antenna in dB.
5. Only 65% of the power input to an antenna is radiated with the rest lost to dissipation in the antenna, what is the gain of the antenna in dB? (This is not the antenna gain.)
6. The efficiency of an antenna is 66%. If the power input to the antenna is 10 W what is the power radiated by the antenna in dBm?
7. An antenna with an input of 1 W operates in free space and has an antenna gain of 12 dBi. What is the maximum power density at 100 m from the antenna?
8. A transmitter has an antenna with an antenna gain of 10 dBi, the resistive losses of the antenna are 50%, and the power input to the antenna is 1 W. What is the EIRP in watts?

   90.] A transmitter has an antenna with an antenna gain of 20 dBi, the resistive losses of the antenna are 50%, and the power input to the antenna is 100 mW. What is the EIRP in watts?
10. An antenna with an antenna gain of 8 dBi radiates 6.67 W. What is the EIRP in watts? Assume that the antenna is 100% efficient.
11. An antenna has an antenna gain of 10 dBi and a 40 W input signal. What is the EIRP in watts?
12. An antenna with 5 W of input power has an antenna gain of 20 dBi and an antenna efficiency of 25% and all of the loss is due to resistive losses in the antenna. [Parallels Example 2.3]
    (a) How much power in dBm is lost as heat in the antenna?
    (b) How much power in dBm is radiated by the antenna?
    (c) What is the EIRP in dBW?
13. An antenna with an efficiency of 50% has an antenna gain of 12 dBi and radiates 100 W. What is the EIRP in watts?
14. An antenna with an efficiency of 75% and an antenna gain of 10 dBi. If the power input to the antenna is 100 W,
    (a) what is the total power in dBm radiated by the antenna?
    (b) what is the EIRP in dBm?
15. The output stage of an RF front end consists of an amplifier followed by a filter and then an antenna. The amplifier has a gain of 27 dB, the filter has a loss of 1.9 dB, and of the power input to the antenna, 45% is lost as heat due to resistive losses. If the power input to the amplifier is 30 dBm, calculate the following:
    (a) What is the power input to the amplifier in watts?
    (b) Express the loss of the antenna in decibels.
    (c) What is the total gain of the RF front end (amplifier + filter)?
    (d) What is the total power radiated by the antenna in dBm?

(e) What is the total power radiated by the antenna in milliwatts?

16. A communication system operating at 2.5 GHz includes a transmit antenna with an antenna gain of 12 dBi and a receive antenna with an effective aperture area of 20 cm^2. The distance between the two antennas is 100 m.

    (a) What is the antenna gain of the receive antenna?

    (b) If the input to the transmit antenna is 1 W, what is the power density at the receive antenna if the power falls off as $1/d^2$, where d is the distance from the transmit antenna?

    (c) Thus what is the power delivered at the output of the receive antenna?

17. Consider a point-to-point communication system. Parabolic antennas are mounted high on a mast so that ground effects do not exist, thus power falls off as $1/d^2$. The gain of the transmit antenna is 20 dBi and the gain of the receive antenna is 15 dBi. The distance between the antennas is 10 km. The effective area of the receive antenna is 3 cm^2. If the power input to the transmit antenna is 600 mW, what is the power delivered at the output of the receive antenna?

18. Consider a 28 GHz point-to-point communication system. Parabolic antennas are mounted high on a mast so that ground effects do not exist, thus power falls off as $1/d^2$. The gain of the transmit antenna is 20 dBi and the gain of the receive antenna is 15 dBi. The distance between the antennas is 10 km. If the power output from the receive antenna is 10 pW, what is the power input to the transmit antenna?

19. An antenna has an effective aperture area of 20 cm^2. What is the antenna gain of the antenna at 2.5 GHz?

20. An antenna operating at 28 GHz has an antenna gain of 50 dBi. What is the effective aperture area of the antenna?

21. A 15 GHz receive antenna has an antenna gain of 20 dBi. If the power density at the receive antenna is 1 nW/cm^2, what is the power at the output of the antenna? [Parallels Example 2.6]

22. Two identical antennas are used in a point-to-point communication system, each having a gain of 50 dBi. The system has an operating frequency of 28 GHz and the antennas are at the top of masts 100 m tall. The RF link between the antennas consists only of the direct line-of-sight path.

    (a) What is the effective aperture area of each antenna?

    (b) How does the power density of the propagating signal rolloff with distance.

    (c) If the separation of the transmit and receive antennas is 10 km, what is the path loss in decibels?

    (d) If the separation of the transmit and receive antennas is 10 km, what is the link loss in decibels?

23. A transmitter and receiver operating at 2 GHz are at the same level, but the direct path between them is blocked by a building and the signal must diffract over the building for a communication link to be established. This is a classic knife-edge diffraction situation. The transmit and receive antennas are each separated from the building by 4 km and the building is 20 m higher than the antennas (which are at the same height). Consider that the building is very thin. It has been found that the path loss can be determined by considering loss due to free-space propagation and loss due to diffraction over the knife edge.

    (a) What is the additional attenuation (in decibels) due to diffraction?

    (b) If the operating frequency is 100 MHz, what is the attenuation (in decibels) due to diffraction?

    (c) If the operating frequency is 10 GHz, what is the attenuation (in decibels) due to diffraction?

24. A hill is 1 km from a transmit antenna and 2 km from a receive antenna. The receive and transmit antennas are at the same height and the hill is 20 m above the height of the antennas. What is the additional loss caused by diffraction over the top of the hill? Treat the hill as causing knife-edge diffraction and the operating frequency is 1 GHz.

25. Two identical antennas are used in a point-to-point communication system, each having a gain of 30 dBi. The system has an operating frequency of 14 GHz and the antennas are at the top of masts 100 m tall. The RF link between the antennas consists only of the direct LOS path.

    (a) What is the effective aperture area of each antenna?

    (b) How does the power density of the propagating signal rolloff with distance?

    (c) If the separation of the transmit and receive antennas is 10 km, what is the path loss? Ignore atmospheric loss.

26. The three main cellular communication bands are centered around 450 MHz, 900 MHz, and 2 GHz. Compare these three bands in terms of

multipath effects, diffraction around buildings, object (such as a wall) penetration, scattering from trees and parts of trees, and ability to follow the curvature of hills. Complete the table below with the relative attributes: high, medium, and low.

Characteristic	450 MHz	900 MHz	2 GHz
Multipath			
Scattering			
Penetration			
Following curvature			
Range			
Antenna size			
Atmospheric loss			

27. Describe the difference in multipath effects in a central city area compared to multipath effects in a desert. Your description should be approximately 4 lines long and not use a diagram

28. Wireless LAN systems can operate at 2.4 GHz, 5.6 GHz, 40 GHz and 60 GHz. Contrast with explanation the performance of these schemes inside a building in terms of range.

29. At 60 GHz the atmosphere strongly attenuates a signal. Discuss the origin of this and indicate an advantage and a disadvantage.

30. Short answer questions. Each part requires a short paragraph of about five lines and a figure, where appropriate, to illustrate your understanding.
    (a) Cellular communications systems use two frequency bands to communicate between the basestation and the mobile unit. The bands are generally separated by 50 MHz or so. Which band (higher or lower) is used for the downlink from the basestation to the mobile unit and what are the reasons behind this choice?
    (b) Describe at least two types of interference in a cellular system from the perspective of a mobile handset.

31. The three main cellular communication bands are centered around 450 MHz, 900 MHz, and 2 GHz. Compare these three bands in terms of multipath effects, diffraction around buildings, object (such as a wall) penetration, scattering from trees and parts of trees, and the ability to follow the curvature of hills. Use a table and indicate the relative attributes: high, medium, and low.

32. Describe Rayleigh fading in approximately 4 lines and without using a diagram.

33. In several sentences and using a diagram describe Rayleigh fading and the impact it has on radio communications.

34. A transmitter and receiver operate at 100 MHz, are at the same level, and are separated by 4 km. The signal must diffract over a building half way between the antennas that is 20 m higher than the direct path between the antennas. What is the attenuation (in decibels) due to diffraction?

35. A transmitter and receiver operate at 10 GHz, are at the same level, and are 4 km apart. The signal must diffract over a building that is half way between the antennas and is 20 m higher than the line between the antennas. What is the attenuation (in dB) due to diffraction?

2.10.1 Exercises By Section

†challenging

§2.3 $1^{\dagger}$

§2.5 $2^{\dagger}, 3^{\dagger}, 4, 5, 6^{\dagger}, 7^{\dagger}, 8^{\dagger}, 19, 10,$ $11, 12, 13, 14, 15, 16, 17, 18, 19, 20, 21$

§2.6 $22^{\dagger}, 23^{\dagger}, 24^{\dagger}, 25^{\dagger}, 27^{\dagger}, 28, 29, 30^{\dagger}, 31, 32, 33, 34, 35$

2.10.2 Answers to Selected Exercises

2(e) 53.23 dBm

16 0.251 μW

17 14.3 pW

CHAPTER 3

Transmission Lines

3.1 Introduction

A transmission line stores electric and magnetic energy. As such a line has a circuit form that combines inductors, Ls (for the magnetic energy), capacitors, Cs (for the electric energy), and resistors, Rs (modeling losses), whose values depend on the line geometry and material properties.

The transmission lines considered in this chapter are restricted to just two parallel conductors, as shown in Figure 3-1, with the distance between the two wires (i.e., in the transverse direction) being substantially smaller than the wavelengths of the signals on the line. The correct physical interpretation is that the conductors of a transmission line confine and guide an EM field. The EM field contains the energy of the signal and not the current on the line. However, with electrically small transverse dimensions, a two conductor line

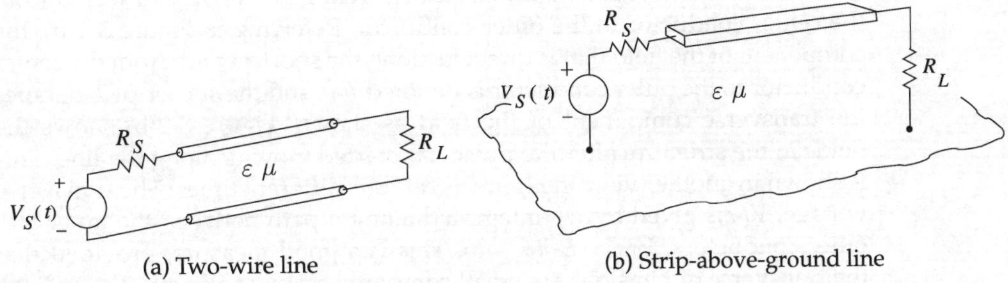

Figure 3-1: Two conductor transmission lines.

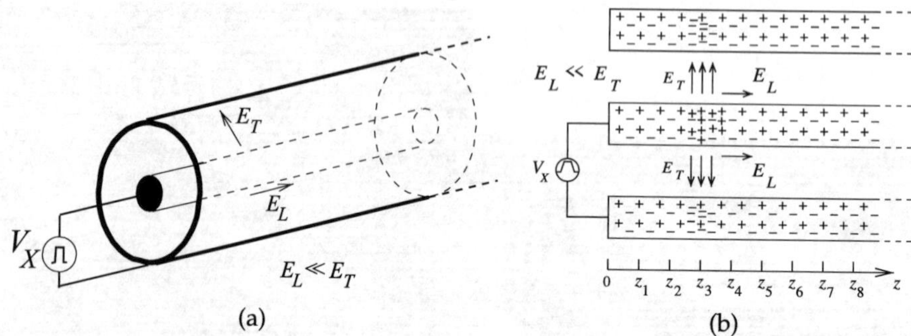

Figure 3-2: A coaxial transmission line: (a) three-dimensional view; (b) the line with pulsed voltage source showing the electric fields at an instant in time as a voltage pulse travels down the line.

may be satisfactorily analyzed on the basis of voltages and currents.

Frequency-domain analysis is the best way to understand transmission lines. This transmission line theory with modern developments is presented in Section 3.2 and useful formulas and concepts are developed in Section 3.3 for lossless transmission lines. Section 3.4 presents several configurations of lossless lines that are particularly useful in microwave circuit design and used in many places in this book series.

3.1.1 When Must a Line be Considered a Transmission Line

The key determinant of whether a transmission line can be considered as an invisible connection between two points is whether the signal anywhere along the interconnect has the same value at a particular instant. If the value of the signal (say, voltage) varies along the line (at an instant), then it may be necessary to consider transmission line effects. A typical criterion used is that if the length of the interconnect is less than 1/20th of the wavelength of the highest-frequency component of a signal, then transmission line effects can be safely ignored and the circuit can be modeled as a single RLC circuit [1].

3.1.2 Movement of a Signal on a Transmission Line

When a positive voltage pulse is applied to the center conductor of a coaxial line, as shown in Figure 3-2(a), an electric field results that is directed from the center conductor to the outer conductor. Referring to Figure 3-2(b), the component of the field that is directed along the shortest path from the center conductor to the outer conductor is denoted E_T, and the subscript T denotes the transverse component of the field as shown. Figure 3-2(b) shows the fields in the structure after the pulse has started moving along the line. This is shown in another view in Figure 3-3 at four different times. The transverse voltage, V_T, is given by E_T integrated along a path between the inner and outer conductors: $V_T \approx E_T(a - b)$. This is a good measure, provided that the transverse dimensions are small compared to a wavelength. The voltage pulse exciting the line has a trapezoidal shape and the voltage between the inner and outer conductor has the same form as E_T. As indicated by Maxwell's equations, a change in time of the electric field results in a spatial

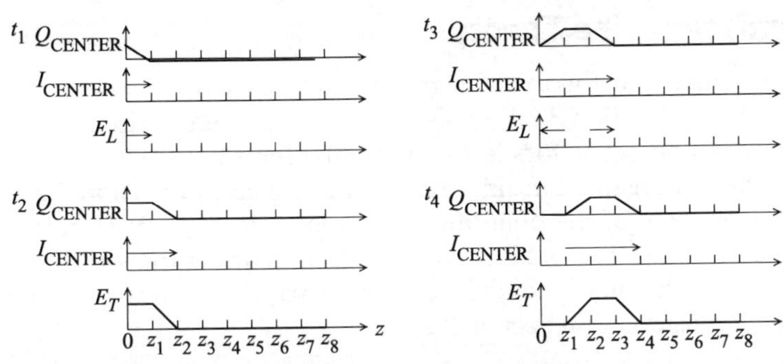

Figure 3-3: Fields, currents, and charges on the coaxial transmission line of Figure 3-2 at times $t_4 > t_3 > t_2 > t_1$. Q_{CENTER} is the net free charge on the center conductor. I_{CENTER} is the current on the center conductor.

change in the magnetic field and hence current. As a result there is a variation of the current in time and this results in a spatial change of the electric field. The chasing from a time variation to a spatial variation and then back to a time variation causes the pulse to move down the line.

The pulse moves down the line at the **group velocity**, which for a lossless coaxial line is the same as the **phase velocity**, v_p.[1] This is determined by the physical properties of the region between the conductors. The permittivity, ε, describes energy storage associated with the electric field, E, and the energy storage associated with the magnetic field, H, is described by the permeability, μ. It has been determined that

$$v_p = 1/\sqrt{\mu\varepsilon}. \tag{3.1}$$

In free space $v_p = c = 1/\sqrt{\mu_0\varepsilon_0} = 3 \times 10^8$ m/s. The free-space wavelengths, $\lambda_0 = c/f$, at several frequencies, f, are

f	100 MHz	1 GHz	10 GHz
λ_0	3 m	30 cm	3.0 cm

Here λ_0 is used to indicate the wavelength in free space and λ_g, the guide wavelength, is used to denote the wavelength on a transmission line.

EXAMPLE 3.1 Transmission Line Wavelength

A length of coaxial line is filled with a dielectric having a relative permittivity, ε_r, of 20 and is designed to be 1/4 wavelength long at a frequency, f, of 1.850 GHz. ($\mu_r = 1$)

(a) What is the free-space wavelength?
(b) What is the wavelength of the signal in the dielectric-filled coaxial line?
(c) How long is the line?

Solution:

(a) $\lambda_0 = c/f = 3 \times 10^8/1.85 \times 10^9 = 0.162$ m $= 16.2$ cm.
(b) Note that for a dielectric-filled line with $\mu_r = 1$, $\lambda = v_p/f = c/(\sqrt{\varepsilon_r}f) = \lambda_0/\sqrt{\varepsilon_r}$, so $\lambda = \lambda_0/\sqrt{\varepsilon_r} = 16.2\text{ cm}/\sqrt{20} = 3.62$ cm.
(c) $\lambda_g/4 = 3.62\text{ cm}/4 = 9.05$ mm.

[1] The phase velocity is the apparent velocity of a point of constant phase on a sinewave and is almost frequency independent for a low-loss coaxial line of small transverse dimensions (less than $\lambda/10$).

3.2 Transmission Line Theory

This section develops the theory of signal propagation on transmission lines. The first section, Section 3.2.1, makes the argument that a circuit with resistors, inductors, and capacitors is a good model for a transmission line. The development of transmission line theory is presented in Section 3.2.2. The dimensions of some of the quantities that appear in transmission line theory are discussed in Section 3.2.3. Section 3.2.4 summarizes the important parameters of a lossless line and then a particularly important line, the microstrip line, is considered in Section 3.2.5.

3.2.1 Transmission Line RLGC Model

Regardless of the actual structure, a segment of uniform transmission line (i.e., a line with constant cross section along its length) as shown in Figure 3-4(a) can be modeled by the circuit shown in Figure 3-4(b) with

Resistance along the line	$= R$	
Inductance along the line	$= L$	all specified
Conductance shunting the line	$= G$	per unit length.
Capacitance shunting the line	$= C$	

R, L, G, and C are referred to as resistance, inductance, conductance, and capacitance per unit length. (Sometimes p.u.l. is used as shorthand for **per unit length**.) In the metric system, ohms per meter (Ω/m), henries per meter (H/m), siemens per meter (S/m) and farads per meter (F/m), respectively, are used. The values of R, L, G, and C are affected by the geometry of the transmission line and by the electrical properties of the dielectrics and conductors. C describes the ability to store electrical energy and is mostly due to the properties of the dielectric. G describes loss in the dielectric which derives from conduction in the dielectric and from dielectric relaxation. Most microwave substrates have negligible conductivity so dielectric loss dominates. Dielectric relaxation loss results from the movement of charge centers which result in distortion of the dielectric lattice (if a crystal) or molecular structure. The periodic variation of the E field transfers energy from the EM field to mechanical vibrations. R is due to ohmic loss in the metal more than anything else. L describes the ability to store magnetic energy and is mostly a function of geometry, as most materials used with transmission lines have $\mu_r = 1$ (so no more magnetic energy is stored than in a vacuum).

For most lines the effects due to L and C dominate because of the relatively low series resistance and shunt conductance. The propagation characteristics of the line are described by its loss-free, or lossless, equivalent line, although in practice some information about R or G is necessary to determine power losses. The lossless concept is just a useful and good approximation.

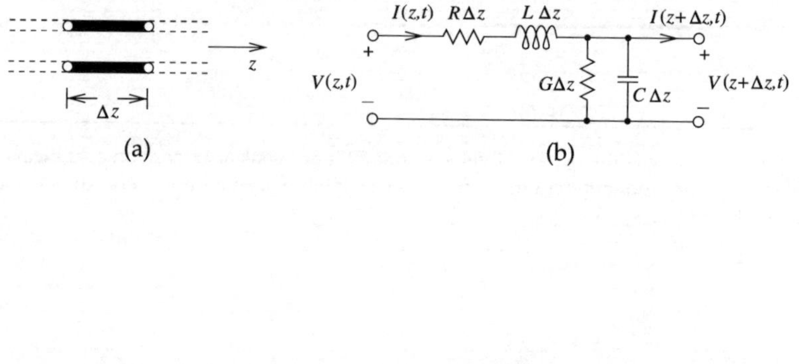

Figure 3-4: Transmission line segment: (a) of length Δz; and (b) lumped-element model.

3.2.2 Derivation of Transmission Line Properties

In this section the differential equations governing the propagation of signals on a transmission line are derived. These are coupled first-order differential equations and are akin to Maxwell's Equations in one dimension. Solution of the differential equations describes how signals propagate, and leads to the extraction of a few parameters that describe transmission line properties.

Applying **Kirchoff's laws** applied to the model in Figure 3-4(b) and taking the limit as $\Delta z \to 0$ the **transmission line equations**

$$\frac{\partial v(z,t)}{\partial z} = -Ri(z,t) - L\frac{\partial i(z,t)}{\partial t} \tag{3.2}$$

$$\frac{\partial i(z,t)}{\partial z} = -Gv(z,t) - C\frac{\partial v(z,t)}{\partial t}. \tag{3.3}$$

In sinusoidal steady-state using cosine-based phasors these become

$V(z)$ is a phasor and $v(z,t) = \Re\{V(z)e^{\jmath\omega t}\}$. $\Re\{w\}$ denotes the real part of w, a complex number.

$$\frac{dV(z)}{dz} = -(R + \jmath\omega L)I(z) \quad (3.4) \quad \text{and} \quad \frac{dI(z)}{dz} = -(G + \jmath\omega C)V(z). \quad (3.5)$$

Eliminating $I(z)$ in the above yields the wave equation for $V(z)$:

$$\frac{d^2V(z)}{dz^2} - \gamma^2 V(z) = 0. \quad (3.6) \quad \text{Similarly} \quad \frac{d^2I(z)}{dz^2} - \gamma^2 I(z) = 0, \quad (3.7)$$

where the propagation constant is

$$\gamma = \alpha + \jmath\beta = \sqrt{(R + \jmath\omega L)(G + \jmath\omega C)}, \tag{3.8}$$

with SI units of m^{-1} and where α is the attenuation coefficient and has units of nepers per meter (Np/m), and β is the phase-change coefficient, or phase constant, and has units of radians per meter (rad/m or radians/m). Nepers and radians are dimensionless units, but serve as prompts for what is being referred to.

Equations (3.6) and (3.7) are second-order differential equations that have solutions of the form

$$V(z) = V_0^+ e^{-\gamma z} + V_0^- e^{\gamma z} \quad (3.9) \quad \text{and} \quad I(z) = I_0^+ e^{-\gamma z} + I_0^- e^{\gamma z}. \quad (3.10)$$

The physical interpretation of these solutions is that $V^+(z) = V_0^+ e^{-\gamma z}$ and $I^+(z) = I_0^+ e^{-\gamma z}$ are forward-traveling waves (moving in the $+z$ direction) and $V^-(z) = V_0^- e^{\gamma z}$ and $I^-(z) = I_0^- e^{\gamma z}$ are backward-traveling waves (moving in the $-z$ direction). $V(z)$, V_0^+, V_0^-, $I(z)$, I_0^+ and I_0^- are all phasors. Substituting Equation (3.9) in Equation (3.4) results in

$$I(z) = \frac{\gamma}{R + \jmath\omega L}\left[V_0^+ e^{-\gamma z} - V_0^- e^{\gamma z}\right]. \tag{3.11}$$

Then from Equations (3.11) and (3.10)

$$I_0^+ = \frac{\gamma}{R + \jmath\omega L} V_0^+ \quad \text{and} \quad I_0^- = \frac{\gamma}{R + \jmath\omega L}(-V_o^-). \tag{3.12}$$

The characteristic impedance is defined as

$$Z_0 = \frac{V_0^+}{I_0^+} = \frac{-V_0^-}{I_0^-} = \frac{R + \jmath\omega L}{\gamma} = \sqrt{\frac{R + j\omega L}{G + j\omega C}}, \tag{3.13}$$

with the SI unit of ohms (Ω). Equations (3.9) and (3.10) can be rewritten as

$$V(z) = V_0^+ \mathrm{e}^{-\gamma z} + V_0^- \mathrm{e}^{\gamma z} \quad (3.14) \quad \text{and} \quad I(z) = \frac{V_0^+}{Z_0}\mathrm{e}^{-\gamma z} - \frac{V_0^-}{Z_0}\mathrm{e}^{\gamma z}. \quad (3.15)$$

Converting back to the time domain:

$$v(z,t) = \left|V_0^+\right| \cos(\omega t - \beta z + \varphi^+)\mathrm{e}^{-\alpha z} + \left|V_0^-\right| \cos(\omega t + \beta z + \varphi^-)\mathrm{e}^{\alpha z}, \quad (3.16)$$

where φ^+ and φ^- are phases of the forward- and backward-traveling waves, respectively. The phasors of the traveling voltage waves are

$$V_0^+(z) = \left|V_0^+\right| \mathrm{e}^{\jmath\varphi^+}\mathrm{e}^{-\jmath\beta z} \quad \text{and} \quad V_0^-(z) = \left|V_0^-\right| \mathrm{e}^{\jmath\varphi^-}\mathrm{e}^{\jmath\beta z}. \quad (3.17)$$

The following quantities are defined:

$$\begin{aligned}
\text{Characteristic impedance:} &\quad Z_0 = \sqrt{(R+\jmath\omega L)/(G+\jmath\omega C)} && (3.18)\\
\text{Propagation constant:} &\quad \gamma = \sqrt{(R+\jmath\omega L)(G+\jmath\omega C)} && (3.19)\\
\text{Attenuation constant:} &\quad \alpha = \Re\{\gamma\} && (3.20)\\
\text{Phase constant:} &\quad \beta = \Im\{\gamma\} && (3.21)\\
\text{Wavenumber:} &\quad k = -\jmath\gamma && (3.22)\\
\text{Phase velocity:} &\quad v_p = \omega/\beta && (3.23)\\
\text{Wavelength:} &\quad \lambda = \frac{2\pi}{|\gamma|} = \frac{2\pi}{|k|}, && (3.24)
\end{aligned}$$

where $\omega = 2\pi f$ is the radian frequency and f is the frequency with the SI units of hertz (Hz). The wavenumber k as defined here is used in electromagnetics and where wave propagation is concerned. Considering one of the traveling waves, the phase velocity refers to the apparent velocity of which a point of constant phase on the sinewave appears to move.

The important result here is that a voltage wave (and a current wave) can be defined on a transmission line. One more parameter needs to be introduced: the group velocity,

$$v_g = \frac{\partial\omega}{\partial\beta}. \quad (3.25)$$

The group velocity is the velocity of a modulated waveform's envelope and describes how fast information propagates. It is the velocity at which the energy (i.e. information) in the waveform moves. Thus group velocity can never be more than the speed of light in a vacuum, c. Phase velocity, however, can be more than c. If the speed at which information moves varies with frequency, then a signal such as a pulse will spread out. Such a line is said to have dispersion. For a lossless, dispersionless line, the group and phase velocity are the same. If the phase velocity is frequency independent, then β is linearly proportional to ω.

Electrical length is used in designs prior to establishing the physical length of a line. The electrical length is expressed either as a fraction of a wavelength or in degrees (or radians), where a wavelength corresponds to 360° (or 2π radians). If ℓ is its physical length, the electrical length of the line in radians is $\beta\ell$.

EXAMPLE 3.2 Physical and Electrical Length

A transmission line is 10 cm long and at the operating frequency the phase constant β is 30 rad/m. What is the electrical length of the line?

Solution:

The physical length of the line is ℓ = 10 cm = 0.1 m. Then the electrical length of the line is $\ell_e = \beta\ell = (30\ \text{rad/m}) \times 0.1\ \text{m} = 3$ radians. The electrical length can also be expressed in terms of wavelength noting that 360° corresponds to 2π radians, which also corresponds to λ. Thus $\ell_e = (3\ \text{radians}) = 3 \times 360/(2\pi) = 171.9°$ or $\ell_e = 3/(2\pi)\ \lambda = 0.477\ \lambda$.

EXAMPLE 3.3 *RLGC* Parameters

A transmission line has the *RLGC* parameters $R = 100\ \Omega/\text{m}$, $L = 80\ \text{nH/m}$, $G = 1.6\ \text{S/m}$, and $C = 200\ \text{pF/m}$. Consider a traveling wave at 2 GHz on the line.

(a) What is the attenuation constant?
(b) What is the phase constant?
(c) What is the phase velocity?
(d) What is the characteristic impedance of the line?
(e) What is the group velocity?

Solution:

(a) α: $\gamma = \alpha + \jmath\beta = \sqrt{(R + \jmath\omega L)(G + \jmath\omega C)}$; $\omega = 12.57 \cdot 10^9$ rad/s

$$\gamma = \sqrt{(100 + \jmath\omega \cdot 80 \cdot 10^{-9})(1.6 + \jmath\omega 200 \times 10^{-12})} = (17.94 + \jmath 51.85)\ \text{m}^{-1}$$
$$\alpha = \Re\{\gamma\} = 17.94\ \text{Np/m}$$

(b) Phase constant: $\beta = \Im\{\gamma\} = 51.85$ rad/m

(c) Phase velocity:
$$v_p = \frac{\omega}{\beta} = \frac{2\pi f}{\beta} = \frac{12.57 \times 10^9 \text{rad} \cdot \text{s}^{-1}}{51.85\ \text{rad} \cdot \text{m}^{-1}} = 2.42 \times 10^8\ \text{m/s}$$

(d) $Z_0 = (R + \jmath\omega L)/\gamma = (100 + \jmath\omega \cdot 80 \cdot 10^{-9})/(17.94 + \jmath 51.85) = (17.9 + j4.3)\ \Omega$
Note also that $Z_0 = \sqrt{(R + \jmath\omega L)/(G + j\omega C)}$, which yields the same answer.

(e) Group velocity:
$$v_g = \left.\frac{\partial\omega}{\partial\beta}\right|_{f=2\ \text{GHz}}$$

Numerical derivatives will be used, thus $v_g = \Delta\omega/\Delta\beta$. Now β is already known at 2 GHz. At 1.9 GHz, $\gamma = 17.884 + \jmath 49.397\ \text{m}^{-1}$, and so $\beta = 49.397$ rad/m.

$$v_g = \frac{2\pi(2\ \text{GHz} - 1.9\ \text{GHz})}{\beta(2\ \text{GHz}) - \beta(1.9\ \text{GHz})} = \frac{2\pi(2 - 1.9)10^9\ \text{Hz}}{(51.85 - 49.397)\ \text{m}^{-1}} = 2.563 \times 10^8\ \text{m/s}.$$

(Note that Hz = s^{-1}. Note that $v_g \neq v_p$, and so the transmission line has dispersion.)

3.2.3 Dimensions of γ, α, and β

The SI unit of γ are inverse meters (m^{-1}) and the attenuation constant, α, and the phase constant, β, have, strictly speaking, the same units. However, the convention is to introduce the dimensionless quantities Neper and radian to convey additional information. Thus the attenuation constant α has the units of Nepers per meter (Np/m) and the phase constant β has the units radians per meter (rad/m). The unit Neper comes from the name of the e (= 2.7182818284590452354...) symbol (written in upright font and

The name for e derives from John **Napier**, who developed the theory of logarithms [2]. e is sometimes called Euler's constant.

not italics since it is a constant), which is called the Neper. The Neper is used in calculating transmission line signal levels, as in Equations (3.9) and (3.10). The attenuation and phase constants are often separated and then the attenuation constant describes the decrease in signal amplitude as the signal travels down a transmission line. So when $\alpha\ell = 1$ Np, where ℓ is the length of the line, the signal has decreased to $1/\mathrm{e}$ of its original value, and the power drops to $1/\mathrm{e}^2$ of its original value. The decrease in signal level represents loss and the units of decibels per meter (dB/m) are used with 1 Np = $20 \log \mathrm{e}$ = 8.6858896381 dB. So expressing α as 1 Np/m is the same as saying that the attenuation loss is 8.6859 dB/m. To convert from dB to Np multiply by 0.1151. Thus $\alpha = x$ dB/m = $x \times 0.1151$ Np/m.

In engineering $\log x \equiv \log_{10} x$ and $\ln x \equiv \log_{\mathrm{e}} x$.

EXAMPLE 3.4 Transmission Line Characteristics

A line has an attenuation of 10 dB/m and a phase constant of 50 radians/m at 2 GHz.

(a) What is the complex propagation constant of the transmission line?

(b) If the capacitance of the line is 100 pF/m and the conductive loss is zero (i.e., $G = 0$), what is the characteristic impedance of the line?

Solution:

(a) $\alpha|_{\mathrm{Np}} = 0.1151 \times \alpha|_{\mathrm{dB}} = 0.1151 \times$ (10 dB/m) = 1.151 Np/m, $\beta = 50$ rad/m
Propagation constant, $\gamma = \alpha + \jmath\beta = (1.151 + \jmath 50)\ \mathrm{m}^{-1}$

(b) $\gamma = \sqrt{(R + \jmath\omega L)(G + \jmath\omega C)}$, and $Z_0 = \sqrt{(R + \jmath\omega L)/(G + \jmath\omega C)}$,
therefore $Z_0 = \gamma/(G + \jmath\omega C)$; $\omega = 2\pi \cdot 2 \times 10^9\ s^{-1}$; $G = 0$; $C = 100 \times 10^{-12}$ F,
so $Z_0 = 39.8 - \jmath 0.916\ \Omega$.

3.2.4 Lossless Transmission Line

If the conductor and dielectric are ideal (i.e., lossless), then $R = 0 = G$ and the equations for the transmission line characteristics simplify. The transmission line parameters from Equations (3.13) and (3.19)–(3.24) are then

$$Z_0 = \sqrt{\frac{L}{C}} \qquad (3.26) \qquad\qquad \alpha = 0 \qquad (3.27)$$

$$\beta = \omega\sqrt{LC} \qquad (3.28) \qquad\qquad v_p = 1/\sqrt{LC} \qquad (3.29)$$

$$\lambda_g = \frac{2\pi}{\omega\sqrt{LC}} = \frac{v_p}{f}. \qquad (3.30)$$

3.2.5 Microstrip Line

A microstrip line is shown in Figure 3-5(a). This is a commonly used transmission line, as it can be cheaply fabricated using printed circuit board techniques. This line consists of a metal-backed substrate of relative permittivity ε_r on top of which is a metal strip. Above that is air. The

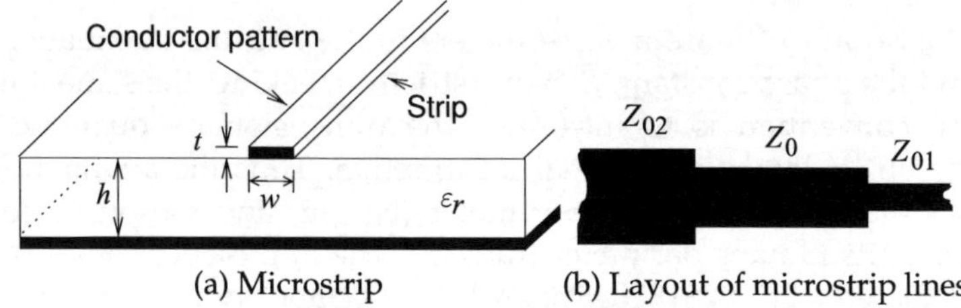

Figure 3-5: Microstrip transmission line. The layout (or top) view is commonly used with circuit designs using microstrip. This is the pattern of the strip where (b) shows three lines of different width.

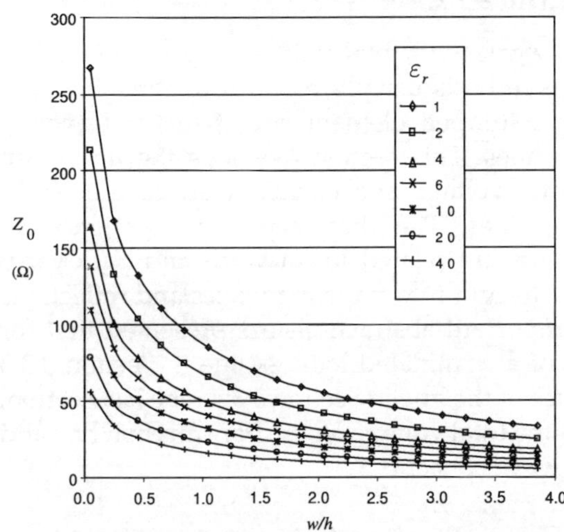

Figure 3-6: Dependence of Z_0 of a microstrip line at 1 GHz for various ε_r and aspect (w/h) ratios. Calculated using EM simulation.

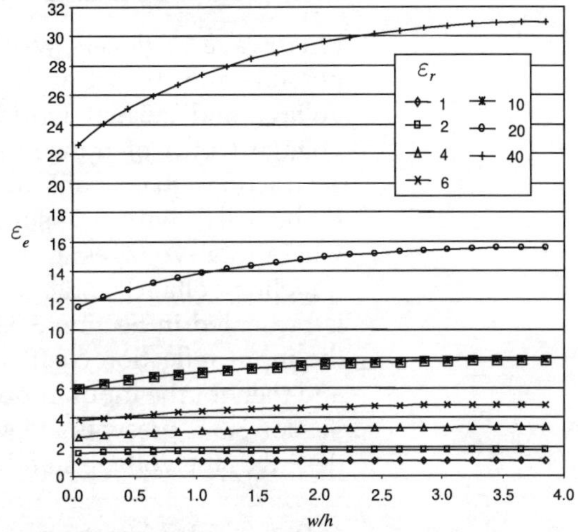

Figure 3-7: Dependence of effective relative permittivity ε_e of a microstrip line at 1 GHz for various permittivities and aspect ratios (w/h).

width of the strip determines the characteristic impedance of the line. The characteristic impedance of microstrip lines having various strip widths is shown in Figure 3-6 for several substrate permittivities. So the wider the strip and the higher the substrate permittivity, the lower the characteristic impedance of the line. The EM fields are partly in air and partly in the dielectric and an effective permittivity must be used when calculating the electrical length of the line. The results of field simulations of the effective permittivity of lines of various widths and with various substrate permittivities are shown in Figure 3-7, where it can be seen that the effective relative permittivity, ε_e, increases for wide strips. This is because more of the EM field is in the substrate. Microstrip transmission line structures are often drawn showing just the layout of the strip, as shown in Figure 3-5(b), where the three lines have different characteristic impedances. The next chapter presents detailed analyses of microstrip and other planar transmission lines.

3.2.6 Summary

The important takeaway from this section is that a signal moves on a transmission line as forward- and backward-traveling waves. The energy transferred is in the traveling waves. The total voltage and current at a point on the line is the sum of the traveling voltage and current waves, respectively, but the total voltage/current view is not sufficient to describe how a transmission line works. Transmission line theory is developed in terms of traveling voltages and current waves and these are akin to a one-dimensional form of Maxwell's equations.

3.3 The Lossless Terminated Line

Microwave engineers want to work with total voltage and current when possible and the art of design synthesis usually requires relating the total voltage and current world of a lumped element circuit to the traveling voltage world of transmission lines. This section develops the important abstractions that enable the total voltage and current view of the world to be used with transmission lines. The first step is in Section 3.3.1 where total voltages and currents are related to forward- and backward-traveling voltages and currents. Insight into traveling waves and reflections is presented in Section 3.3.2. Important abstractions are presented first for the input reflection coefficient of a terminated lossless line in Section 3.3.3 and then for the input impedance of the line in Section 3.3.4. The last section, Section 3.3.5, presents a view of the total voltage on the transmission line and describes the voltage standing wave concept.

3.3.1 Total Voltage and Current on the Line

Consider the terminated line shown in Figure 3-8(a). Assume an incident or forward-traveling wave, with traveling voltage $V_0^+ \, \mathrm{e}^{-\jmath\beta z}$ and current $I_0^+ \, \mathrm{e}^{-\jmath\beta z}$ propagating toward the load Z_L at $z = 0$. The characteristic impedance of the transmission line is the ratio of the voltage and current traveling waves so that

$$\frac{V_0^+(z)}{I_0^+(z)} = \frac{V_0^+ \, \mathrm{e}^{-\jmath\beta z}}{I_0^+ \, \mathrm{e}^{-\jmath\beta z}} = \frac{V_0^+(0)}{I_0^+(0)} = \frac{V_0^+}{I_0^+} = Z_0. \tag{3.31}$$

The reflected wave has a similar relationship (but note the sign change):

$$\frac{V_0^- \, \mathrm{e}^{\jmath\beta z}}{-I_0^- \, \mathrm{e}^{\jmath\beta z}} = \frac{V_0^-}{-I_0^-} = Z_0. \tag{3.32}$$

The load Z_L imposes an additional constraint on the relationship of the total voltage and current at $z = 0$:

$$\frac{V_L}{I_L} = \frac{V(z=0)}{I(z=0)} = Z_L. \tag{3.33}$$

When $Z_L \neq Z_0$ there must be a reflected wave with appropriate amplitude to satisfy the above equations. Now the total voltage

$$V(z) = V_0^+ \, \mathrm{e}^{-\jmath\beta z} + V_0^- \, \mathrm{e}^{\jmath\beta z}, \tag{3.34}$$

and the total current, $I(z)$, is related to the traveling current waves by

$$I(z) = \frac{V_0^+}{Z_0} \mathrm{e}^{-\jmath\beta z} - \frac{V_0^-}{Z_0} \mathrm{e}^{\jmath\beta z} = I_0^+ \mathrm{e}^{-\jmath\beta z} + I_0^- \mathrm{e}^{\jmath\beta z}. \tag{3.35}$$

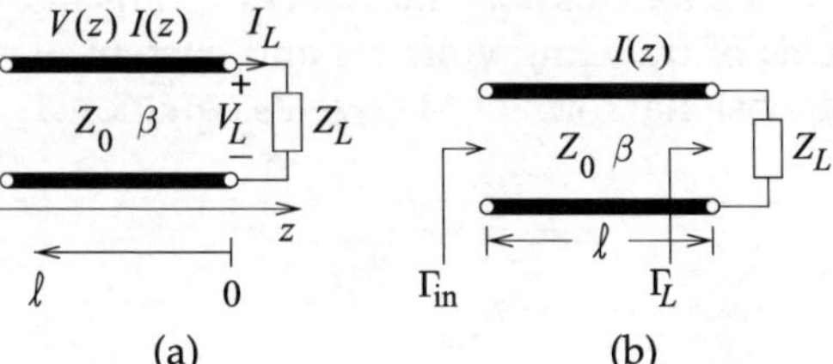

Figure 3-8: A terminated transmission line.

Thus at the termination of the line ($z = 0$),

$$\frac{V(0)}{I(0)} = Z_L = Z_0 \frac{V_0^+ + V_0^-}{V_0^+ - V_0^-}.$$

This can be rearranged as the ratio of the reflected voltage to the incident voltage:

$$\frac{V_0^-}{V_0^+} = \frac{Z_L - Z_0}{Z_L + Z_0}.$$

This ratio is defined as the voltage **reflection coefficient** at the load,

$$\Gamma_L = \Gamma_L^V = \frac{V_0^-(0)}{V_0^+(0)} = \frac{V_0^-}{V_0^+} = \frac{Z_L - Z_0}{Z_L + Z_0}. \tag{3.36}$$

That is, at the load

$$V_0^- = \Gamma_L V_0^+. \tag{3.37}$$

The relationship in Equation (3.36) can be rewritten so that the input load impedance can be obtained from the reflection coefficient:

$$Z_L = Z_0 \frac{1 + \Gamma^V}{1 - \Gamma^V}. \tag{3.38}$$

Similarly, the current reflection coefficient can be written as

$$\Gamma^I = \frac{I_0^-}{I_0^+} = \frac{-Z_L + Z_0}{Z_L + Z_0} = -\Gamma^V. \tag{3.39}$$

The voltage reflection coefficient is used most of the time, so the reflection coefficient, Γ, on its own refers to the voltage reflection coefficient, $\Gamma^V = \Gamma$.

EXAMPLE 3.5 Forward- and Backward-Traveling Waves at an Open Circuit

A lossless transmission line is terminated in an open circuit. What is the relationship between the forward- and backward-traveling voltage waves at the end of the line?

Solution:

At the end of the line the total current is zero, so that $I^+ + I^- = 0$ and so

$$I^- = -I^+. \tag{3.40}$$

The forward- and backward traveling voltages and currents are related to the characteristic impedance by

$$Z_0 = V^+/I^+ = -V^-/I^-, \tag{3.41}$$

Note the change in sign, as a result of the direction of propagation changing but the positive reference for current is in the same direction. Substituting for I^- at the termination,

$$V^+ = -V^- I^+/I^- = -V^- I^+/(-I^+) = V^-. \tag{3.42}$$

Thus the total voltage at the end of the line, V_{TOTAL}, is $V^+ + V^- = 2V^+$. Note that the total voltage at the end of the line is twice the incident (forward-traveling) voltage.

EXAMPLE 3.6 Current Reflection Coefficient

A load consists of a shunt connection of a capacitor of 10 pF and a resistor of 60 Ω. The load terminates a lossless 50 Ω transmission line. The operating frequency is 5 GHz.

(a) What is the impedance of the load?
(b) What is the normalized impedance of the load (normalized to Z_0 of the line)?
(c) What is the reflection coefficient of the load?
(d) What is the current reflection coefficient of the load?

Solution:

(a) $C = 10 \cdot 10^{-12}$ F; $R = 60\ \Omega$; $f = 5 \cdot 10^{9}$ Hz; $\omega = 2\pi f$; $Z_0 = 50\ \Omega$

$$Z_L = R||C = (1/R + \jmath\omega C)^{-1} = 0.168 - \jmath 3.174\ \Omega.$$

(b) $z_L = Z_L/Z_0 = 3.368 \cdot 10^{-3} - \jmath 0.063$.
(c) This is the voltage reflection coefficient. $\Gamma_L = (z_L - 1)/(z_L + 1) = -0.985 - \jmath 0.126 = 0.993\angle 187.3°$.
(d) $\Gamma_L^I = -\Gamma_L = 0.985 + \jmath 0.126 = 0.993\angle(187.3 - 180)° = 0.993\angle 7.3°$.

3.3.2 Forward- and Backward-Traveling Pulses

Reflections at the end of a line produce a backward-traveling signal. Forward- and backward-traveling pulses are shown in Figure 3-9(a) for the situation where the resistance at the end of the line is lower than the

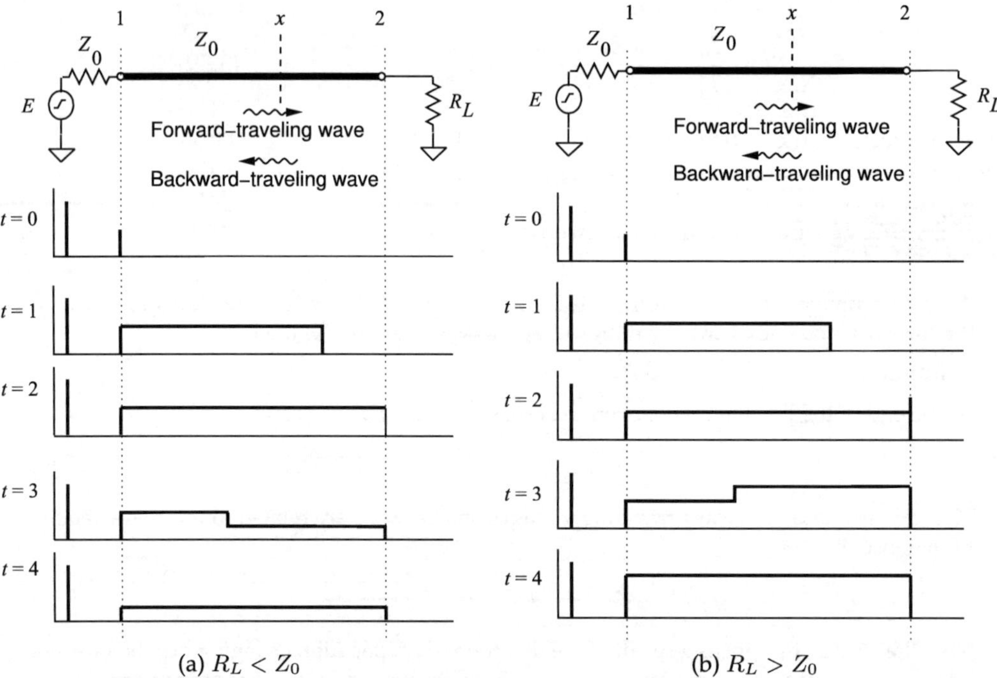

(a) $R_L < Z_0$ (b) $R_L > Z_0$

Figure 3-9: Reflection of a voltage pulse from a load: (a) when the resistance of the load, R_L is lower than the characteristic impedance of the line, Z_0; and (b) when R_L is greater than Z_0.

characteristic impedance of the line ($Z_L < Z_0$). The voltage source is a step voltage that is zero for time $t < 0$. At time $t = 0$, the step is applied to the line and it begins traveling down the line, as shown at time $t = 1$. This voltage step moving from left to right is called the forward-traveling voltage wave.

At time $t = 2$, the leading edge of the step reaches the load, and as the load has lower resistance than the characteristic impedance of the line, the total voltage across the load drops below the level of the forward-traveling voltage step. The reflected wave is called the backward-traveling wave and it must be negative, as it adds to the forward-traveling wave to yield the total voltage. Thus the voltage reflection coefficient, Γ, is negative and the total voltage on the line, which is all that can be directly observed, drops. A reflected, smaller, and opposite step signal travels in the backward direction and adds to the forward-traveling step to produce the waveform shown at $t = 3$. The impedance of the source matches the transmission line impedance so that the reflection at the source is zero. The signal on the line at time $t = 4$, the time for round-trip propagation on the line, therefore remains at the lower value. The easiest way to remember the polarity of the reflected pulse is to consider the situation with a short-circuit at the load. Then the total voltage on the line at the load must be zero. The only way this can occur when a signal is incident is if the reflected signal is equal in magnitude but opposite in sign, in this case $\Gamma = -1$. So whenever $|Z_L| < |Z_0|$, the reflected pulse will tend to subtract from the incident pulse.

The opposite situation occurs when the resistance at the load is higher than the characteristic impedance of the line (Figure 3-9(b)). In this case the reflected pulse has the same polarity as the incident signal. Again, to remember this, think of the open-circuited case. The voltage across the load doubles, as the reflected pulse has the same sign as well as magnitude as that of the incident signal, in this case $\Gamma = +1$. This is required so that the total current is zero.

A more illustrative situation is shown in Figure 3-10, where a more complicated signal is incident on a load that has a resistance higher than that of the characteristic impedance of the line. The peaking of the voltage that results at the load is typically the design objective in many long digital interconnects, as less overall signal energy needs to be transmitted down the line, or equivalently a lower current drive capability of the source is required to achieve first incidence switching. This is at the price of having reflected signals on the interconnects, but these are dissipated through a combination of line loss and absorption of the reflected signal at the driver.

3.3.3 Input Reflection Coefficient of a Lossless Line

The reflection coefficient looking into a line varies with position along the line as the forward- and backward-traveling waves change in relative phase. Referring to Figure 3-11, at a distance ℓ from the load (i.e., $z = -\ell$), the input reflection looking into a terminated lossless line is

$$\Gamma_{\text{in}}|_{z=-\ell} = \frac{V^-(z=-\ell)}{V^+(z=-\ell)} = \frac{V^-(z=0)\mathrm{e}^{-\jmath\beta\ell}}{V^+(z=0)\mathrm{e}^{+\jmath\beta\ell}} = \frac{V^-(z=0)\ \mathrm{e}^{-\jmath\beta\ell}}{V^+(z=0)\ \mathrm{e}^{+\jmath\beta\ell}} = \Gamma_L\mathrm{e}^{-\jmath 2\beta\ell} \tag{3.43}$$

Note that Γ_{in} has the same magnitude as Γ_L but rotates in the clockwise direction (becomes increasingly negative) at twice the rate of increase of the

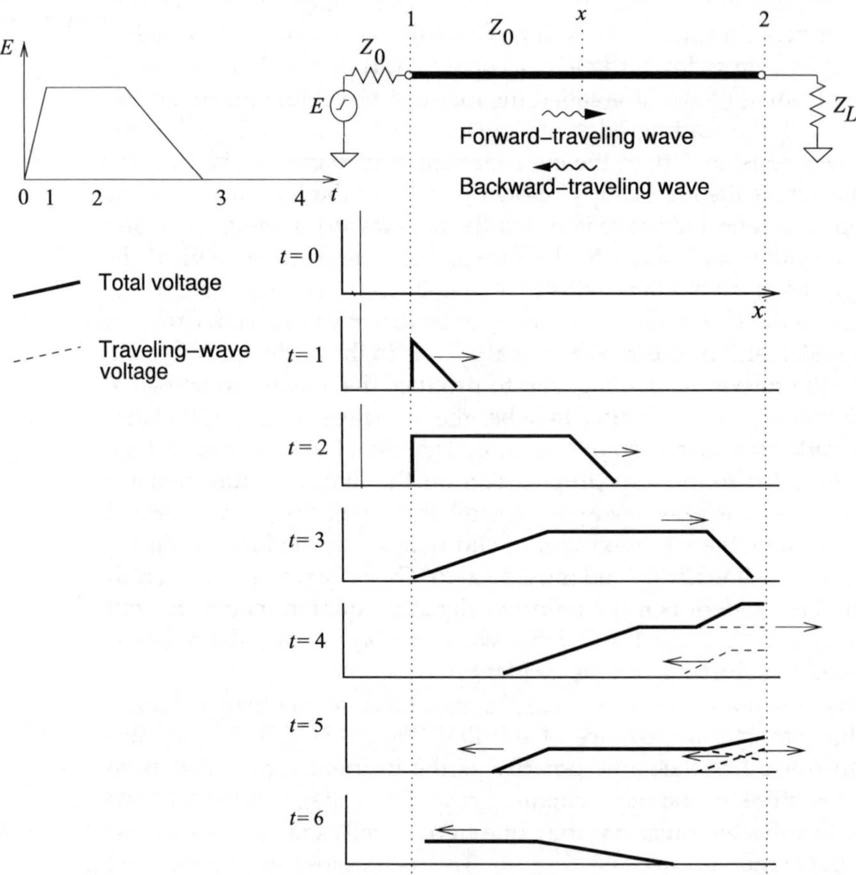

Figure 3-10: Reflection of a pulse from a termination with $Z_L > Z_0$. Z_L and Z_0 are real.

Figure 3-11: Terminated transmission line: (a) a transmission line terminated in a load impedance, Z_L, with an input impedance of Z_{in}; and (b) a transmission line with source impedance Z_G and load Z_L.

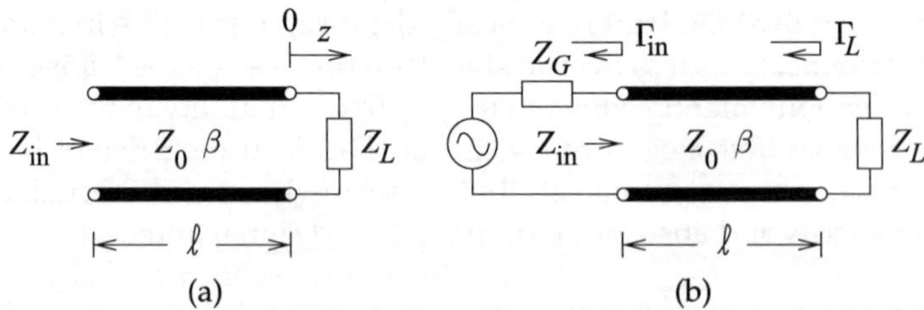

electrical length $\beta\ell$.

It is important to graphical concepts introduced later that there be a full appreciation for the angle of Γ_{in} becoming increasingly negative at twice the rate at which the electrical length of the line increases. Figure 3-12 is a way of visualizing this. The transmission line here is $\lambda/4$ long with an electrical length of 90° and is terminated in a load with reflection coefficient $\Gamma_L = +1$. At position $z = 0$ the forward-traveling voltage wave is $v^+(t, 0) = |V^+|\cos(\omega t)$, and this then propagates down the line in the $+z$ direction. The forward-traveling voltage at point $z = \lambda/8$ at $t = 0$ will be the same as the voltage at $z = 0$ at a time one-eighth of a period in the past. The voltage at $z =$

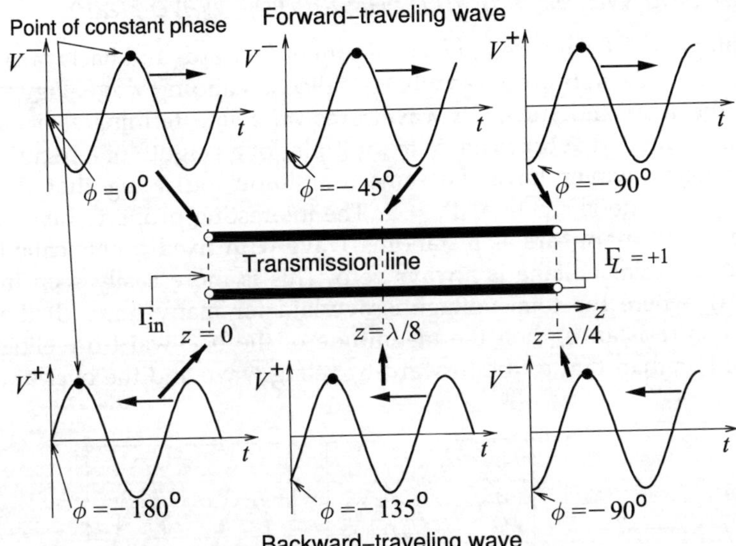

Figure 3-12: The forward-traveling wave $v^+(t,z) = |V^+|\cos(\omega t - \beta z) = |V^+|\cos(\omega t + \phi(z))$ and the backward-traveling wave $v^-(t,z) = |V^+|\cos(\omega t + \beta z) = |V^+|\cos[\omega t + \phi(z)]$. The phase, ϕ, of the forward-traveling wave becomes increasingly negative along the line as z increases, and when reflected the phase ϕ of the backward-traveling wave becomes increasingly negative as the wave moves away from the load (i.e. as z decreases).

$\lambda/8$ is $v^+(t,\lambda/8) = |V^+|\cos(\omega t - 2\pi/8)$, i.e. there is a phase rotation of $-45°$. Then at $z = \lambda/4$, $v^+(t,\lambda/4) = |V^+|\cos(\omega t - 2\pi/4)$, i.e. at time $t = 0$ there is a phase rotation of $-90°$ relative to $v^+(0,0)$, and this is the negative of the electrical length of the line. The voltage wave reflects at the load and becomes a backward-traveling wave. Here $\Gamma_L = +1$ and so, at the load, the phase of the backward- and forward-traveling waves are the same. The backward-traveling wave continues to travel in the $-z$ direction and its phase at $t = 0$ becomes increasingly negative as z gets closer to the input of the line. The phase of the backward-traveling wave at $z = 0$ is rotated $-90°$ with respect to the backward-traveling wave at the load, and has rotated $-180°$ relative to the forward-traveling wave at $z = 0$. For a lossless line, in general, the angle of Γ_{in} = [phase of $V^-(z=0)$ relative to the phase of $V^+(z=0)$] + (the phase of Γ_L) = -2(electrical length of the line) + (the phase of Γ_L).

3.3.4 Input Impedance of a Lossless Line

The impedance looking into a lossless line varies with position, as the forward- and backward-traveling waves combine to yield position-dependent total voltage and current. At a distance ℓ from the load (i.e., $z = -\ell$), the input impedance seen looking toward the load is

$$Z_{in}|_{z=-\ell} = \frac{V(z=-\ell)}{I(z=-\ell)} = Z_0\frac{1+|\Gamma|\,e^{\jmath(\Theta-2\beta\ell)}}{1-|\Gamma|\,e^{\jmath(\Theta-2\beta\ell)}} = Z_0\frac{1+\Gamma_L e^{\jmath(-2\beta\ell)}}{1-\Gamma_L e^{\jmath(-2\beta\ell)}}. \tag{3.44}$$

Another form is obtained by substituting Equation (3.36) in Equation (3.44):

$$\begin{aligned} Z_{in} &= Z_0\frac{(Z_L+Z_0)e^{\jmath\beta\ell}+(Z_L-Z_0)e^{-\jmath\beta\ell}}{(Z_L+Z_0)e^{\jmath\beta\ell}-(Z_L-Z_0)e^{-\jmath\beta\ell}} = Z_0\frac{Z_L\cos(\beta\ell)+\jmath Z_0\cos(\beta\ell)}{Z_0\cos(\beta\ell)+\jmath Z_L\cos(\beta\ell)} \\ &= Z_0\frac{Z_L+\jmath Z_0\tan\beta\ell}{Z_0+\jmath Z_L\tan\beta\ell}. \end{aligned} \tag{3.45}$$

This is the **lossless telegrapher's equation**. The electrical length, $\beta\ell$, is in radians when used in calculations.

3.3.5 Standing Waves and Voltage Standing Wave Ratio

The total voltage on a terminated line is the sum of forward- and backward-traveling waves. This sum produces what is called a standing wave. Figure 3-13 shows the total and traveling waveforms on a line terminated in a reactance and evaluated at times equal to multiples of an eighth of a period. Here the traveling waves have the same amplitude indicating that the termination of the line is reactive, $|\Gamma| = 1$. The interesting property here is that the total voltage appears as a standing wave with fixed points called nodes where the total voltage is always zero. This is more easily seen in Figure 3-14(a), where the total voltage is overlaid for many times. If the termination has resistance, then the magnitude of the backward-traveling wave will be less than that of the forward-traveling wave and the overlaid

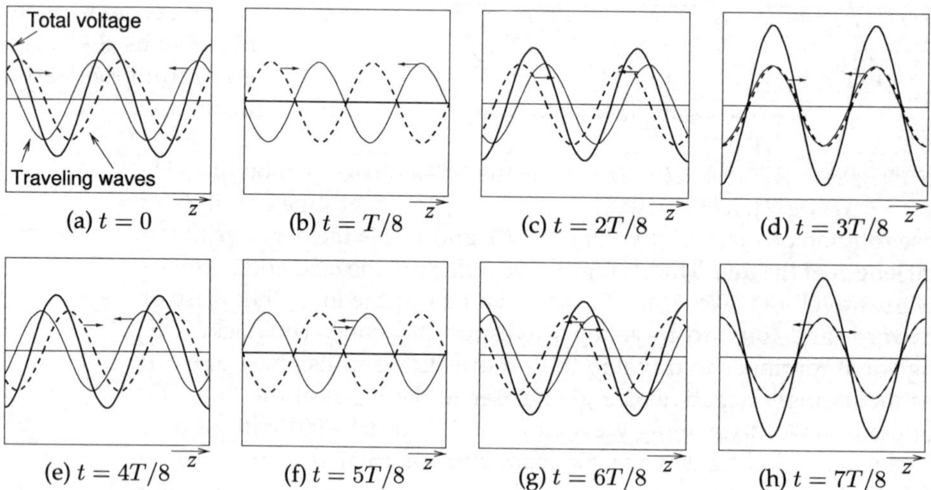

Figure 3-13: Evolution of a standing wave as the sum of forward- and backward-traveling waves (to the right and left, respectively) of equal amplitude evaluated at times t equal to eighths of the period T. At $t = T/8$ and $t = 5T/8$ the total voltage everywhere on the line is zero.

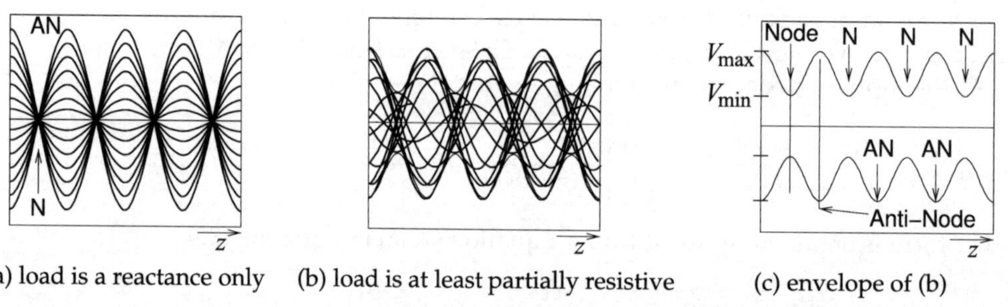

Figure 3-14: Standing waves as an overlay of waveforms at many times: (a) when the forward- and backward-traveling waves have the same amplitude; (b) when the waves have different amplitudes; and (c) the envelope of the standing wave. N is a node (a minimum) and AN is an antinode (a maximum). Nodes, N, are separated by $\lambda/2$. Antinodes, AN, are separated by $\lambda/2$.

total voltage is as shown in Figure 3-14(b). This is still a standing wave, but the minima are now not zero. The envelope of the standing wave is shown in Figure 3-14(c), where there is a maximum amplitude V_{max} and a minimum amplitude V_{min}.

Now this situation will be examined mathematically to relate the standing wave to the reflection coefficient. If $\Gamma = 0$, then the magnitude of the total voltage on the line, $|V(z)|$, is equal to $|V_0^+|$ anywhere on the line. For this reason, such a line is said to be "flat." If there is reflection the magnitude of the total voltage on the line is not constant (see Figure 3-14(b)). Thus from Equation (3.43) :

$$|V(z)| = |V_0^+||1 + \Gamma e^{2\jmath\beta z}| = |V_0^+||1 + \Gamma e^{-2\jmath\beta\ell}|, \tag{3.46}$$

where $z = -\ell$ is the positive distance measured from the load at $z = 0$ toward the generator. Or, setting $\Gamma = |\Gamma|e^{\jmath\Theta}$,

$$|V(z)| = |V_0^+|\left|1 + |\Gamma|e^{\jmath(\Theta - 2\beta\ell)}\right|, \tag{3.47}$$

where Θ is the phase of the reflection coefficient ($\Gamma = |\Gamma|e^{\jmath\Theta}$) at the load. This result shows that the voltage magnitude oscillates with position z along the line. The maximum value occurs when $e^{\jmath(\Theta-2\beta\ell)} = 1$ and is given by

$$V_{\text{max}} = |V_0^+|(1 + |\Gamma|). \tag{3.48}$$

Similarly the minimum value of the total voltage magnitude occurs when the phase term is $e^{\jmath(\Theta-2\beta l)} = -1$, and is given by

$$V_{\text{min}} = |V_0^+|(1 - |\Gamma|). \tag{3.49}$$

A mismatch can be defined by the **voltage standing wave ratio (VSWR)**:

$$\text{VSWR} = \frac{V_{\text{max}}}{V_{\text{min}}} = \frac{(1 + |\Gamma|)}{(1 - |\Gamma|)}. \tag{3.50}$$

Also

$$|\Gamma| = \frac{\text{VSWR} - 1}{\text{VSWR} + 1}. \tag{3.51}$$

Notice that in general Γ is complex, but VSWR is necessarily always real and $1 \le \text{VSWR} \le \infty$. For the matched condition, $\Gamma = 0$ and $\text{VSWR} = 1$, and the closer VSWR is to 1, the closer the load is to being matched to the line and the more power is delivered to the load. The magnitude of the reflection coefficient on a line with a short-circuit or open-circuit load is 1, and in both cases the VSWR is infinite.

To determine the position of the standing wave maximum, ℓ_{max}, consider Equation (3.47) and note that at the maximum

$$\Theta - 2\beta\ell_{\text{max}} = 2n\pi, \quad n = 0, 1, 2, \ldots. \tag{3.52}$$

Here Θ is the angle of the reflection coefficient at the load:

$$\Theta - 2n\pi = 2\frac{2\pi}{\lambda_g}\ell_{\text{max}}. \tag{3.53}$$

Thus the position of the voltage maxima, l_{max}, normalized to wavelength is

$$\frac{\ell_{\text{max}}}{\lambda_g} = \frac{1}{2}\left(\frac{\Theta}{2\pi} - n\right), \quad n = 0, -1, -2, \ldots. \tag{3.54}$$

Similarly the position of the voltage minima is (using Equation (3.47)):

$$\Theta - 2\beta\ell_{\text{min}} = (2n+1)\pi. \tag{3.55}$$

After rearranging the terms,

$$\frac{\ell_{\text{min}}}{\lambda_g} = \frac{1}{2}\left(\frac{\Theta}{2\pi} - n + \frac{1}{2}\right), \quad n = 0, -1, -2, \ldots. \tag{3.56}$$

Summarizing from Equations (3.54) and (3.56):

1. The distance between two successive maxima is $\lambda_g/2$.
2. The distance between two successive minima is $\lambda_g/2$.
3. The distance between a maximum and an adjacent minimum is $\lambda_g/4$.
4. From the measured VSWR the magnitude of the reflection coefficient $|\Gamma|$ can be found. From the measured ℓ_{max} the angle Θ of Γ can be found. Then from Γ the load impedance can be found.

In a similar manner to that above, the magnitude of the total current on the line is

$$|I(\ell)| = \frac{|V_0^+|}{Z_0}\left|1 - |\Gamma|e^{\jmath(\Theta-2\beta\ell)}\right|. \tag{3.57}$$

Hence the standing wave current is maximum where the standing-wave voltage amplitude is minimum, and minimum where the standing-wave voltage amplitude is maximum.

Now Z_{in} in Equation (3.45) is a periodic function of ℓ with period $\lambda/2$ and it varies between Z_{max} and Z_{min}, where

$$Z_{\text{max}} = \frac{V_{\text{max}}}{I_{\text{min}}} = Z_0 \times \text{VSWR} \quad \text{and} \quad Z_{\text{min}} = \frac{V_{\text{min}}}{I_{\text{max}}} = \frac{Z_0}{\text{VSWR}}. \tag{3.58}$$

EXAMPLE 3.7 Standing Wave Ratio

In Example 3.6 a load consisted of a capacitor of 10 pF in shunt with a resistor of 60 Ω. The load terminated a lossless 50 Ω transmission line. The operating frequency is 5 GHz.

(a) What is the SWR?

(b) What is the current standing wave ratio (ISWR)? (When SWR is used on its own it is assumed to refer to VSWR.)

Solution:

(a) From Example 3.6 on page 52, $\Gamma_L = 0.993\angle 187.3°$ and so

$$\text{VSWR} = \frac{1+|\Gamma_L|}{1-|\Gamma_L|} = \frac{1+0.993}{1-0.993} = 285.$$

(b) ISWR = VSWR = 285.

EXAMPLE 3.8 Standing Waves

A load has an impedance $Z_L = 45 + \jmath 75\ \Omega$ and the system reference impedance, Z_0, is 100 Ω.

(a) What is the reflection coefficient?
(b) What is the current reflection coefficient?
(c) What is the SWR?
(d) What is the ISWR?
(e) The power available from a source with a 100 Ω Thevenin equivalent impedance is 1 mW. The source is connected directly to the load, Z_L. Use the reflection coefficient to calculate the power delivered to Z_L.
(f) What is the total power absorbed by the Thevenin equivalent source impedance?
(g) Discuss the effect on power flow of inserting a lossless 100 Ω transmission line between the source and the load.

Solution:

(a) The voltage reflection coefficient is

$$\begin{aligned}\Gamma_L &= (Z_L - Z_0)/(Z_L + Z_0) = (45 + \jmath 75 - 100)/(45 + \jmath 75 + 100)\\ &= (93.0\angle(2.204\ \text{rads}))/(163.2\angle(0.4773\ \text{rads}))\\ &= 0.570\angle(1.726\ \text{rads}) = 0.570\angle 98.9^\circ = -0.0881 + \jmath 0.563 = \Gamma^V.\end{aligned} \tag{3.59}$$

(b) The current reflection coefficient is

$$\Gamma^I = -\Gamma^V = 0.0881 - \jmath 0.563 = 0.570\angle(98.9^\circ - 180^\circ) = 0.570\angle 81.1^\circ. \tag{3.60}$$

(c) The SWR is the VSWR, so

$$\text{SWR} = \text{VSWR} = \frac{V_{\text{max}}}{V_{\text{min}}} = \frac{1 + |\Gamma^V|}{1 - |\Gamma^V|} = \frac{1 + 0.570}{1 - 0.570} = 3.65. \tag{3.61}$$

(d) The current SWR is ISWR = VSWR.

(e) To determine the reflection coefficient of the load, begin by developing the Thevenin equivalent circuit of the load. The power available from the source is P_A = 1 mW, so the Thevenin equivalent circuit is

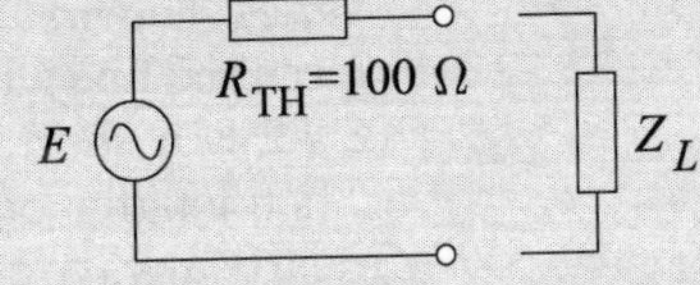

The power reflected by the load is

$$P_R = P_A\left|\Gamma_L^2\right| = 1\ \text{mW} \cdot (0.570)^2 = 0.325\ \text{mW}$$

and the power delivered to the load is

$$P_D = P_A\left(1 - \left|\Gamma_L^2\right|\right) = 0.675\ \text{mW}.$$

(f) It is tempting to think that the power dissipated in R_{TH} is just P_R. However, this is not correct. Instead, the current in R_{TH} must be determined and then the power dissipated in R_{TH} found. Let the current through R_{THcdot} be I, and this is composed of forward- and backward-traveling components:

$$I = I^+ + I^- = (1 + \Gamma_I)I^+,$$

where I^+ is the forward-traveling current wave. Thus

$$P_A = \tfrac{1}{2}|I^+|^2 R_{\text{TH}} = \tfrac{1}{2}|I^+|^2 \times 100 = 1\ \text{mW} = 10^{-3}\ \text{W},$$

so $I^+ = 4.47$ mA, and

$$I = (1 + \Gamma_I)I^+ = (1 + 0.0881 - \jmath 0.563) \times 4.47 \times 10^{-3}\ \text{A}, \qquad |I| = 5.48\ \text{mA}.$$

The power dissipated in R_{TH} is

$$P_{\text{TH}} = \tfrac{1}{2}|I|^2 R_{\text{TH}} = \tfrac{1}{2}(5.48 \times 10^{-3})^2 R_{\text{TH}} = 1.50\ \text{mW}. \tag{3.62}$$

The circuit is that shown in part (e) and so the current in R_{TH} is the same as the current in Z_L. Thus the power delivered to the load Z_L is due to the real part of Z_L:

$$P_D = \frac{1}{2}\left|I^2\right|\Re(Z_L) = \frac{1}{2}(5.48 \times 10^{-3})^2 \times 45 = 0.676 \text{ mW} \tag{3.63}$$

(g) Inserting a transmission line with the same characteristic impedance as the Thevenin equivalent impedance will have no effect on power flow.

3.3.6 Summary

This section related the physics of traveling voltage and current waves on lossless transmission lines to the total voltage and current view. First the input reflection coefficient of a terminated lossless line was developed and from this the input impedance, which is the ratio of total voltage and total current, derived. At any point along a line the amplitude of total voltage varies sinusoidally, tracing out a standing wave pattern along the line and yielding the VSWR metric which is the ratio of the maximum amplitude of the total voltage to the minimum amplitude of that voltage. This is an important metric that is often used to provide an indication of how good a match, i.e. how small the reflection is, with a VSWR $=$ 1 indicating no reflection and a VSWR $= \infty$ indicating total reflection, i.e. a reflection coefficient magnitude of 1.

3.4 Special Lossless Line Configurations

The lossless transmission line configurations considered in this section are used as circuit elements in RF designs and are used elsewhere in this book series. The first element considered in Section 3.4.1 is a short length of short-circuited line which looks like an inductor. The element considered in Section 3.4.2 is a short length of open-circuited line which looks like a capacitor. Then lengths of short-circuited and open-circuited lines, called stubs, used nearly always as shunt elements to introduce an admittance in a circuit, are described in Sections 3.4.3 and 3.4.4. Another type of element, described in Section 3.4.5, is a short length of line with either high or low characteristic impedance realizing a small series inductor or capacitor respectively. The final element described in Section 3.4.6 is a quarter-wave transformer, a quarter-wavelength long line with a particular characteristic impedance which is used in two ways. It can be used to provide maximum power transfer from a source to a load resistance, and it can invert an impedance, e.g. making a capacitor terminating the line look like an inductor.

3.4.1 Short Length of Short-Circuited Line

A transmission line terminated in a short circuit ($Z_L = 0$) has the input impedance (using Equation (3.45))

$$Z_{\text{in}} = \jmath Z_0 \tan(\beta\ell). \tag{3.64}$$

So a short length of short-circuited line, $\ell < \lambda_g/4$, looks like an inductor with inductance L_s,

$$Z_0 \tan(\beta\ell) = \omega L_s, \quad \text{and so} \quad L_s = \frac{Z_0}{\omega}\tan\left(\frac{2\pi\ell}{\lambda_g}\right). \tag{3.65}$$

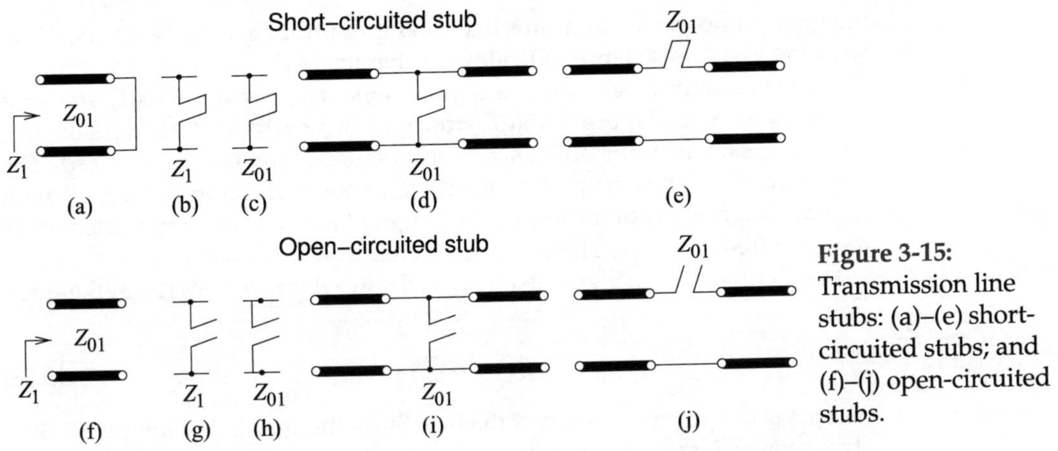

Figure 3-15: Transmission line stubs: (a)–(e) short-circuited stubs; and (f)–(j) open-circuited stubs.

From Equation (3.65) it can be seen that for a given ℓ, L_s is proportional to Z_0. Hence, for a larger values of L_s, sections of transmission line of high characteristic impedance is needed. So microstrip lines with narrow strips can be used to realize inductors in planar microstrip circuits.

3.4.2 Short Length of Open-Circuited Line

An open-circuited line has $Z_L = \infty$ and so (using Equation (3.45))

$$Z_{\text{in}} = -\jmath \frac{Z_0}{\tan \beta\ell} \, . \tag{3.66}$$

For lengths ℓ such that $\ell < \lambda/4$, an open-circuited segment of line realizes a capacitor C_0 for which

$$\frac{1}{\omega C_0} = \frac{Z_0}{\tan \beta\ell} \quad \text{and so} \quad C_0 = \frac{1}{Z_0}\frac{\tan(\beta\ell)}{\omega}. \tag{3.67}$$

From the above relationship, it can be seen that C_0 is inversely proportional to Z_0. Hence, for a larger value of C_0, a section of transmission line with low characteristic impedance is needed.

3.4.3 Short-Circuited Stub

A stub is a section of open-circuited or short-circuited transmission line and is used as a series or shunt element in a microwave circuit. There are several representations. A shorted stub is shown in Figure 3-15(a) as a transmission line with characteristic impedance Z_{01} that is short circuited. The input impedance of the line is Z_1. If the line is lossless, the usual assumption, then Z_{01} will be real and Z_1 will be imaginary. Stubs are commonly used in microwave circuits and generally all stubs in a network have the same length, such as $\lambda/4$ long or $\lambda/8$ long. Which it is is specified in the design. Realistically they do not need to have the same length but there are some special properties for certain lengths, as will become clearer. A cleaner way to indicate a shorted stub is shown in Figure 3-15(b), where the impedance of the stub is as indicated. The absence of a 0 subscript (which would indicate characteristic impedance) means that this is the reactive input impedance of the stub. If a 0 subscript is used, as in Figure 3-15(c), the characteristic impedance of the stub is indicated. If a numerical value is given then an imaginary impedance indicates that the input impedance is being specified, whereas a real impedance indicates the characteristic impedance of the stub. The shorted stub is shown as a shunt element in Figure 3-15(d) and as a series element in Figure 3-15(e). However in nearly all transmission line technologies, including microstrip, only shunt stubs can be realized. The open-circuited stubs with annotations are shown in Figures 3-15(f–j) with similar assignments of meaning. The length of a stub is often indicated by its resonant frequency, f_r. This is the frequency at which the stub is $\lambda/4$ long.

The shorted stub in Figure 3-15(a) has the input impedance (from Equation (3.45))

$$Z_1 = \jmath Z_{01} \tan \beta \ell, \tag{3.68}$$

where ℓ is the physical length of the line. Since the stub is $\lambda/4$ long at f_r, then at frequency f, the input impedance of the stub is

$$Z_1 = \jmath Z_{01} \tan\left(\frac{\pi}{2}\frac{f}{f_r}\right). \tag{3.69}$$

A special situation, and the most commonly used in design, is when the operating frequency is around one-half of the resonant frequency (i.e., $f \approx \frac{1}{2} f_r$). Then the stub is one-eighth of a wavelength long and the argument of the tangent function in Equation (3.69) is approximately $\pi/4$ and Z_1 becomes

$$Z_1 \approx \jmath Z_{01} \tan\left(\frac{\pi}{4}\right) = \jmath Z_{01}, \tag{3.70}$$

thus realizing an inductance at $f = f_r/2$ with a reactance equal to the characteristic impedance of the line.

EXAMPLE 3.9 Short-Circuited Stub

Develop the electrical design of the shunt stub shown with a load of impedance $Z_L = 75 + \jmath 15\ \Omega$ so that the total impedance of the load and stub is real.

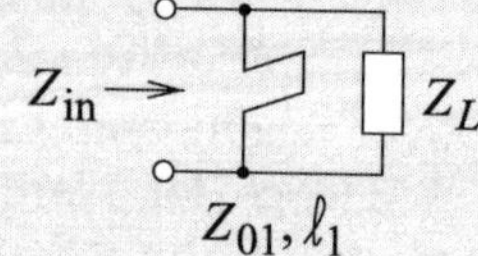

Solution:

The short-circuited stub has characteristic impedance Z_{01} and length ℓ_1. Choose $Z_{01} = 75\ \Omega$ (generally this must be between 15 Ω and 100 Ω for most transmission line technologies). The stub needs to be designed so that the susceptances of the stub and load sum to zero. The admittance of the load $Y_L = 1/Z_L = 0.01282 - \jmath 0.002564$ S. The required admittance of the stub is $Y_{\text{STUB}} = \jmath 0.002564$ S so, using Equation (3.68),

$$Z_{\text{STUB}} = 1/Y_{\text{STUB}} = \jmath Z_{01} \tan \beta \ell_1 = -\jmath 390\ \Omega.$$

Therefore, the electrical length of the stub is

$$\beta \ell_1 = \arctan(-\jmath 390/\jmath 75) = -1.381 + n\pi \text{ radians}, n = 0, 1, 2, \ldots. \tag{3.71}$$

The first positive angle is taken so the stub has the shortest length. So

$$\beta \ell_1 = 1.761 \text{ radians} = 100.9°. \tag{3.72}$$

The complete electrical design of the stub is that it is a shunt short-circuited stub with a characteristic impedance of 75 Ω and with an electrical length of 100.9°. The combined impedance of the stub and load is $Z_X = 1/(\Re\{Y_L\}) = 1/0.01282 = 78.00\ \Omega$.

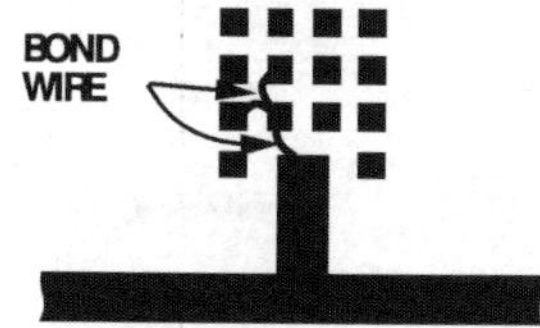

Figure 3-16: Open-circuited stub with variable length realized using wire bonding from the fixed stub to one of the bond pads. The bond pads are on the same layer as the strip metal layer and bonding to them extends the length of the open-circuited stub.

3.4.4 Open-Circuited Stub

An open-circuited transmission line is commonly used as a circuit element called an open stub shown in Figure 3-15(f–j). From Equation (3.45) and noting that $Z_L = \infty$, the open stub input impedance is

$$Z_1 = -\jmath Z_{01}\frac{1}{\tan\beta\ell}. \tag{3.73}$$

With the stub one-quarter wavelength long at the frequency f_r, the input impedance at f_r is a short circuit and the stub is said to be resonant at f_r. Then at a frequency f, the input impedance of the stub is

$$Z_1 = -\jmath Z_{01}\tan^{-1}\left(\frac{\pi}{2}\frac{f}{f_r}\right). \tag{3.74}$$

When $f = \frac{1}{2}f_r$ the stub is one-eighth wavelength long and

$$Z_1 = -\jmath Z_{01}\frac{1}{\tan\left(\frac{\pi}{4}\right)} = -\jmath Z_{01}. \tag{3.75}$$

So a $\lambda/4$ long open-circuited stub realizes a capacitance with a reactance equal to the characteristic impedance of the line.

If the length of a stub can be changed then the stub can be used as a tuning element. A common microstrip tuning technique is shown in Figure 3-16, where bonding to different pads enables a variable length stub to be realized.

EXAMPLE 3.10 Open-Circuited Stub

Develop the electrical design of the open-circuit stub shown with a load of impedance $Z_L = 75 + \jmath 15\ \Omega$ so that the total impedance of the load and stub is real.

$Z_{\text{in}} \rightarrow$ Z_L Z_{01}, ℓ_1

Solution:

The open-circuited stub has characteristic impedance Z_{01} and length ℓ_1. A good choice is to choose Z_{01} around the impedance level of the load as long as it can be realized; so choose $Z_{01} = 75\ \Omega$. The stub needs to be designed so that the susceptances of the stub and load sum to zero. The admittance of the load $Y_L = 1/Z_L = 0.01282 - \jmath 0.002564$ S. The required admittance of the stub is $Y_{\text{STUB}} = \jmath 0.002564$ S, so, using Equation (3.73),

$$Z_{\text{STUB}} = 1/Y_{\text{STUB}} = -\jmath Z_{01}/\tan\beta\ell_1 = -\jmath 390\ \Omega.$$

Therefore, the electrical length of the stub is

$$\beta\ell_1 = \arctan(\jmath 75/\jmath 390) = 0.1900 + n\pi \text{ radians}, n = 0, 1, 2, \ldots. \tag{3.76}$$

The first positive angle is taken so the stub has the shortest length. Then

$$\beta\ell_1 = 0.1900 \text{ radians} = 10.89^\circ. \tag{3.77}$$

The complete electrical design of the stub is that it is a shunt open-circuited stub with a characteristic impedance of 75 Ω and an electrical length of 10.89°. The combined impedance of the stub and load is $Z_X = 1/(\Re\{Y_L\}) = 1/0.01282 = 78.00\ \Omega$.

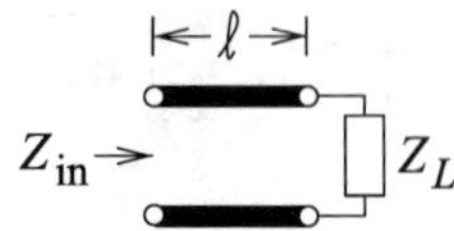

Figure 3-17: An electrically short line.

3.4.5 Electrically Short Lossless Line

Consider the input impedance, Z_{in}, of an electrically short line (i.e., $\beta\ell$ is small) (see Figure 3-17). Using Equation (3.45),

$$Z_{\text{in}} \approx \frac{Z_L + \jmath Z_0(\beta\ell)}{1+\jmath(Z_L/Z_0)(\beta\ell)} \approx [Z_L + \jmath Z_0(\beta\ell)]\left[1 - \jmath\frac{Z_L}{Z_0}(\beta\ell)\right]. \tag{3.78}$$

Since $Z_0\beta = \sqrt{L/C}(\omega\sqrt{LC}) = \omega L$ and $\beta/Z_0 = (\omega\sqrt{LC})/\sqrt{L/C} = \omega C$ (where L and C are the inductance and capacitance per unit length of the line), Equation (3.78) can be written as

$$Z_{\text{in}} \approx Z_L\left[1 + (\beta\ell)^2\right] + \jmath\left[\omega(L\ell) - Z_L^2\omega(C\ell)\right]. \tag{3.79}$$

Since $\beta\ell$ is small, $(\beta\ell)^2$ is very small, and so the $(\beta\ell)^2$ term can be ignored. Then the input impedance of an electrically short line terminated in impedance Z_L is

$$Z_{\text{in}} \approx Z_L + \jmath\left[\omega(L\ell) - Z_L^2\omega(C\ell)\right]. \tag{3.80}$$

Some special cases of this result will be considered in the following examples.

EXAMPLE 3.11 Capacitive Transmission Line Segment

This example demonstrates that a predominantly capacitive behavior can be obtained from a short segment of transmission line, the Z_{01} line here, of low characteristic impedance. Consider the transmission line system shown below with lines having the characteristic impedances, Z_{01} and Z_{02}, $Z_{02} \gg Z_{01}$.

The value of Z_{in} is (treating Z_{02} as the load)

$$Z_{\text{in}} = Z_{01}\frac{Z_{02} + \jmath Z_{01}\tan\beta\ell}{Z_{01} + \jmath Z_{02}\tan\beta\ell}. \tag{3.81}$$

Now $(1+\jmath x)^{-1} \approx 1 - \jmath x - x^2$. Thus for a short line (and so dropping the $\tan^2(\beta\ell)$ term)

$$Z_{\text{in}} \approx Z_{02} - \jmath\frac{Z_{02}^2}{Z_{01}}\tan(\beta\ell) + \jmath Z_{01}\tan(\beta\ell) = Z_{02} + \jmath Z_{01}\tan(\beta\ell)\left[1 - \frac{Z_{02}^2}{Z_{01}^2}\right]. \tag{3.82}$$

For $Z_{02} \gg Z_{01}$ and for a short line, $\tan(\beta\ell) \approx \beta\ell$, and this becomes

$$Z_{\text{in}} \approx Z_{02} - \jmath\frac{Z_{02}^2}{Z_{01}}\tan(\beta\ell) \approx Z_{02} - \jmath\frac{Z_{02}^2}{Z_{01}}\beta\ell, \tag{3.83}$$

which is capacitive. Now consider the circuit to the right where an effective capacitance C_{eff} is in shunt with a load Z_{02}. This has the input impedance

$$Z_x = \left(\jmath\omega C_{\text{eff}} + \frac{1}{Z_{02}}\right)^{-1} = \frac{Z_{02}}{1+\jmath\omega C_{\text{eff}}Z_{02}} = Z_{02}[1 - \jmath\omega C_{\text{eff}}Z_{02} - (\jmath\omega C_{\text{eff}}Z_{02})^2 + \ldots. \tag{3.84}$$

For $\omega C_{\text{eff}}Z_{02} \ll 1$ (i.e. an electrically short line)

$$Z_x \approx Z_{02} - \jmath\omega C_{\text{eff}} Z_{02}^2 \tag{3.85}$$

Equating Equations (3.83) and (3.85), the effective value of the shunt capacitor realized by the short length of low-impedance line, the Z_{01} line, is

$$C_{\text{eff}} = \frac{1}{\omega Z_{02}^2}\frac{Z_{02}^2\beta\ell}{Z_{01}} = \frac{\beta}{\omega}\frac{\ell}{Z_{01}}. \tag{3.86}$$

Thus a shunt capacitor can be realized approximately by a low-impedance line embedded between two high-impedance lines. The microstrip layout of this is shown in the figure on the right. Recall that a wide microstrip line has a low characteristic impedance.

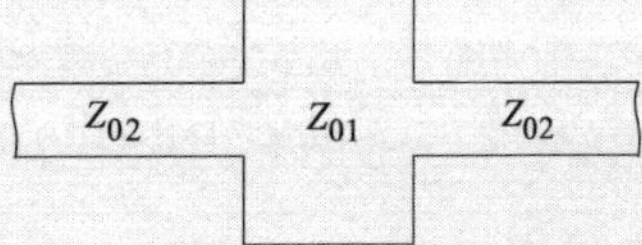

EXAMPLE 3.12 Inductive Transmission Line Segment

This example demonstrates that a (predominantly) inductive behavior can be obtained from a segment of transmission line. Consider the transmission line system shown below with lines having two different characteristic impedances, Z_{01} and Z_{02}, $Z_{02} \ll Z_{01}$.

$Z_{\text{in}} \longrightarrow$ Z_{01} Z_{02} $\leftarrow \ell \rightarrow$

The value of Z_{in} is (using Equation (3.45))

$$Z_{\text{in}} = Z_{01}\frac{Z_{02} + \jmath Z_{01}\tan\beta\ell}{Z_{01} + \jmath Z_{02}\tan\beta\ell}, \tag{3.87}$$

which for a short line can be expressed as

$$Z_{\text{in}} \approx Z_{02}\left[1 + \tan(\beta l)\right] + \jmath Z_{01}\tan(\beta\ell). \tag{3.88}$$

Note that $\jmath Z_{01}\tan(\beta\ell)$ is the dominant part for $\ell < \lambda/8$ and $Z_{02} \ll Z_{01}$.

Thus a microstrip realization of a series inductor is a high-impedance line embedded between two low-impedance lines. A top view of such a configuration in microstrip is shown in the figure. A narrow microstrip line has high characteristic impedance.

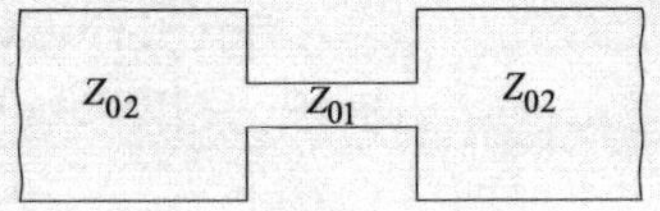

The previous two examples showed how a shunt capacitance and series inductance can be realized using short sections of line, the Z_{01} line here, with a high characteristic impedance. This enables realization of some lumped element circuits in microstrip form. A lumped element lowpass filter is shown in Figure 3-18(a) and this can be realized using wide and narrow microstrip lines, as shown in Figure 3-18(b).

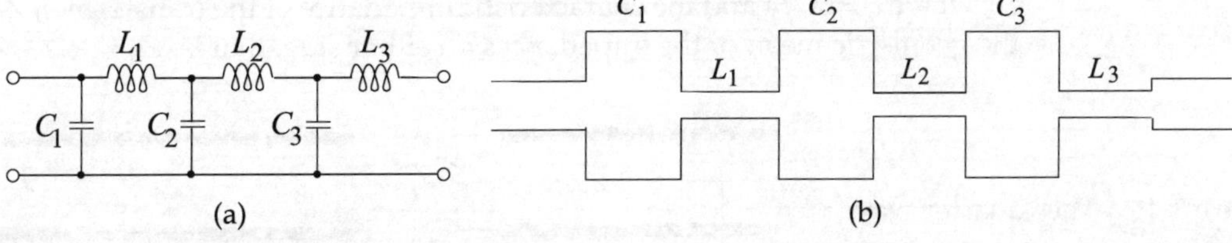

Figure 3-18: A lowpass filter: (a) in the form of an LC ladder network; and (b) realized using microstrip lines.

3.4.6 Quarter-Wave Transformer

Figure 3-19(a) shows a resistive load R_L and a section of transmission line with length $\ell = \lambda_g/4$ (hence the name quarter-wave transformer). The input impedance of the line is

$$Z_{\text{in}} = Z_1 \frac{R_L + \jmath Z_1 \tan(\beta\ell)}{Z_1 + \jmath R_L \tan(\beta\ell)} = Z_1 \frac{R_L + \jmath Z_1 \infty}{Z_1 + \jmath R_L \infty} = \frac{Z_1^2}{R_L}\,. \tag{3.89}$$

The input impedance is matched to the transmission line Z_0 if

$$Z_{\text{in}} = Z_0^* = Z_0, \tag{3.90}$$

since here the characteristic impedance is real. Thus

$$Z_1 = \sqrt{Z_0 R_L} \tag{3.91}$$

and so the one-quarter wavelength long line acts as an ideal impedance transformer.

Another example of the quarter-wave transformer is shown in Figure 3-19(b). The input impedance looking into the quarter-wave transformer (from the left) is given by

$$Z_{\text{in}} = Z_0 \frac{Z_{01} + \jmath Z_0 \tan(\beta\ell)}{Z_0 + \jmath Z_{01} \tan(\beta\ell)} = Z_0 \frac{Z_{01} + \jmath Z_0 \infty}{Z_0 + \jmath Z_{01} \infty} = \frac{Z_0^2}{Z_{01}}. \tag{3.92}$$

Hence a section of transmission line of length $\ell = \lambda_g/4 + n\lambda_g/2$, where $n = 0, 1, 2, \ldots$, can be used to match lines having different impedances, Z_{01} and Z_{02}, by constructing the line so that its characteristic impedance is

$$Z_0 = \sqrt{Z_{01} Z_{02}}. \tag{3.93}$$

Note that for a design center frequency f_0, the matching section provides a perfect match only at the center frequency and at frequencies where $\ell = \lambda_g/4 + n\lambda_g/2$.

A quarter-wave transformer has an interesting property that is widely used. Examine the final result in Equation (3.92), which is repeated here:

$$Z_{\text{in}} = \frac{Z_0^2}{Z_{01}}\,. \tag{3.94}$$

Equation (3.94) indicates that a one-quarter wavelength long line is an impedance inverter presenting, at Port 1, the inverse of the impedance presented at port 2, Z_{01}. This result also applies to complex impedances replacing Z_{01}. This impedance inversion is scaled by the square of the characteristic impedance of the line. This inversion holds in the reverse direction as well.

The layout of a microstrip quarter-wave transformer is shown in Figure 3-20, where $\ell = \lambda_g/4$ and the characteristic impedance of the transformer, Z_0, is the geometric mean of the impedances on either side, that is, $Z_0 = \sqrt{Z_{01} Z_{02}}$.

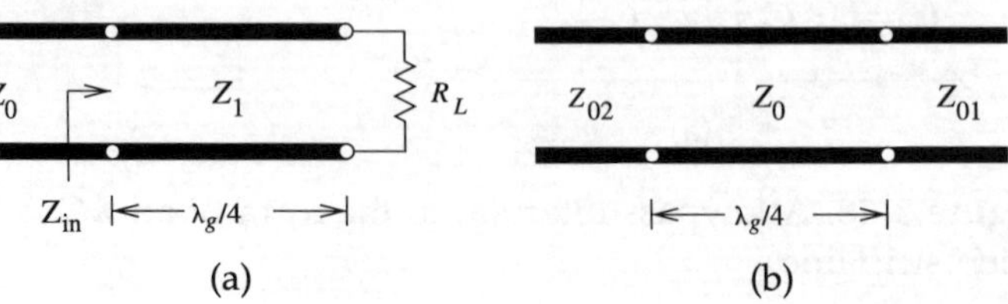

Figure 3-19: The quarter-wave transformer line: (a) transforming a load; and (b) interfacing two lines.

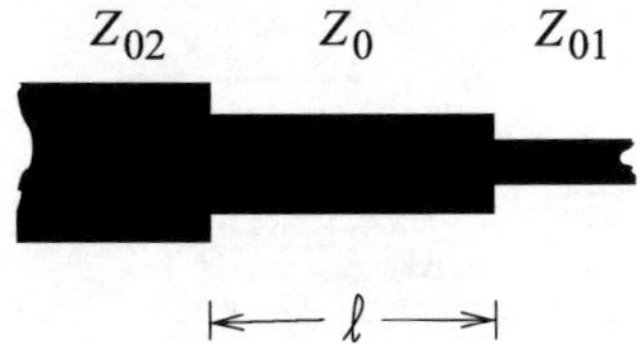

Figure 3-20: Layout of a microstrip quarter-wave transformer.

3.4.7 Summary

The lossless transmission line configurations considered in this section are those most commonly used in microwave circuit design. It is important to note that the stub line is almost always used in shunt configuration to provide an admittance in a circuit. Most transmission line technologies, including coaxial lines and microstrip, only permit shunt stubs. The quarter-wave transformer is a particularly interesting element enabling maximum power transfer from a source to a load that may be different. An interesting feature that is widely exploited is that the quarter-wave transformer inverts an impedance. For example, turning a small resistance into a large resistance, or even turning a small capacitor into a large inductance. These transformations are valid over a moderate bandwidth.

3.4.8 Summary

An earlier section developed the input reflection coefficient and input impedance of a lossless line. Several circuit elements based on a a transmission line were introduced: stubs, short-sections of line have either high or low characteristic impedance, and the quarter-wave transformer.

3.5 Circuit Models of Transmission Lines

Circuit models of transmission lines are required if they are to be used in a circuit simulator. RF and microwave engineering uses two types of simulators. Spice-like simulators use lumped-element transmission line models in which an $RLGC$ model of a short segment of line is replicated for the length of the line. If the ground plane is treated as a universal ground, then the model of a segment of length Δz is as shown in Figure 3-21(a). In this segment $r = R\Delta z$, $l = L\Delta z$, $g = G\Delta z$, and $c = C\Delta z$, where R, L, G, and C are the per unit length parameters of the line. Cascading the segments to get the length of the line yields the complete lumped-element model of the line, as shown in Figure 3-21(b).

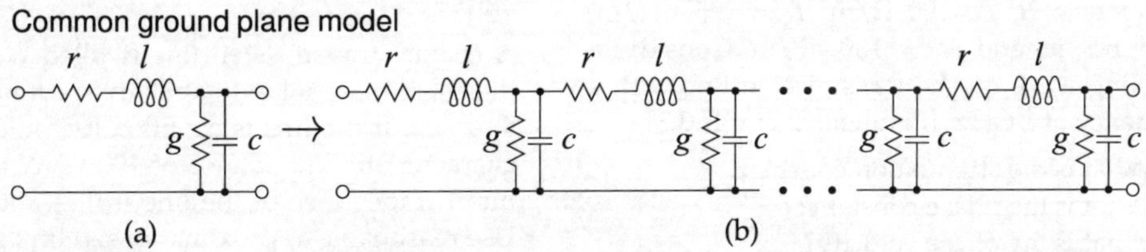

Figure 3-21: Lumped-element transmission line models: (a) model of a short segment (e.g. $\lambda/20$ long); and (b) complete model of a transmission line.

3.6 Summary

In this chapter a classical treatment of transmission lines was presented. Transmission lines are distributed elements and form the basis of microwave circuits. A distinguishing feature is that they support forward- and backward-traveling waves and they can be used to implement circuit functions.

The most important of the formulas presented in this chapter are listed here. Reflection coefficients are referenced to an impedance Z_0, the load impedance is Z_L, and a line has a characteristic impedance Z_0, physical length ℓ, and propagation constant γ (or electrical length in radians of $\beta\ell$ where ℓ is the physical length of the line.

Reflection coefficient of a load impedance Z_L:

$$\Gamma = \Gamma^V = \frac{Z_L - Z_{\text{REF}}}{Z_L + Z_{\text{REF}}} \quad (3.36)$$

Load impedance in terms of reflection coefficient Γ:

$$Z_L = Z_{\text{REF}}\frac{1+\Gamma}{1-\Gamma} \quad (3.38)$$

Input reflection coefficient of a lossless line of length ℓ

$$\Gamma_{\text{in}} = \Gamma_L e^{-\jmath 2\beta\ell} \quad (3.43)$$

Reflection coefficient in terms of VSWR

$$|\Gamma| = \frac{\text{VSWR} - 1}{\text{VSWR} + 1} \quad (3.51)$$

Input impedance of a lossless line

$$Z_{\text{in}} = Z_0\frac{Z_L + \jmath Z_0 \tan\beta\ell}{Z_0 + \jmath Z_L \tan\beta\ell} \quad (3.45)$$

VSWR in terms of reflection coefficient

$$\text{VSWR} = \frac{(1+|\Gamma|)}{(1-|\Gamma|)} \quad (3.50)$$

3.7 References

[1] T. Edwards and M. Steer, *Foundations for Microstrip Circuit Design*. John Wiley & Sons, 2016.

[2] C. Boyer and U. Merzbach, "Invention of logarithms," in *A History of Mathematics*, 2nd ed. John Wiley & Sons, 1991.

3.8 Exercises

1. A coaxial line is short-circuited at one end and is filled with a dielectric with a relative permittivity of 64. [Parallels Example 3.1]
   (a) What is the free-space wavelength at 18 GHz?
   (b) What is the wavelength in the dielectric-filled coaxial line at 18 GHz?
   (c) The first resonance of the coaxial resonator is at 18 GHz. What is the physical length of the resonator?

2. A transmission line has the following $RLGC$ parameters: $R = 100\ \Omega/\text{m}$, $L = 85\ \text{nH/m}$, $G = 1\ \text{S/m}$, and $C = 150\ \text{pF/m}$. Consider a traveling wave on the transmission line with a frequency of 1 GHz. [Parallels Example 3.3]
   (a) What is the attenuation constant?
   (b) What is the phase constant?
   (c) What is the phase velocity?
   (d) What is the characteristic impedance of the line?
   (e) What is the group velocity?

3. A transmission line has the per-unit length parameters $L = 85\ \text{nH/m}$, $G = 1\ \text{S/m}$, and $C = 150\ \text{pF/m}$. Use a frequency of 1 GHz. [Parallels Example 3.3]
   (a) What is the phase velocity if $R = 0\ \Omega/\text{m}$?
   (b) What is the group velocity if $R = 0\ \Omega/\text{m}$?
   (c) If $R = 10\ \text{k}\Omega/\text{m}$ what is the phase velocity?
   (d) If $R = 10\ \text{k}\Omega/\text{m}$ what is the group velocity?

4. A line is 10 cm long and at the operating frequency the phase constant β is 40 rad/m. What is the electrical length of the line? [Parallels Example 3.2]

5. A coaxial transmission line is filled with lossy dielectric material with a relative permittivity of $5 - \jmath 0.2$. If the line is air-filled it would have a characteristic impedance of 100 Ω. What is the input impedance of the line if it is 1 km long? Use reasonable approximations. [Hint: Does the termination matter?]

6. A transmission line has the per unit length parameters $R = 2\ \Omega/\text{cm}$, L=100 nH/m, $G =$

1 mS/m, $C = 200$ pF/m.

(a) What is the propagation constant of the line at 5 GHz?
(b) What is the characteristic impedance of the line at 5 GHz?
(c) Plot the magnitude of the characteristic impedance versus frequency from 100 MHz to 10 GHz.

7. A line is 20 cm long and at 1 GHz the phase constant β is 20 rad/m. What is the electrical length of the line in degrees?

8. What is the electrical length of a line that is a quarter of a wavelength long,
(a) in degrees?
(b) in radians?

9. A lossless transmission line has an inductance of 8 nH/cm and a capacitance of 40 pF/cm.
(a) What is the characteristic impedance of the line?
(b) What is the phase velocity on the line at 1 GHz?

10. A transmission line has an attenuation of 2 dB/m and a phase constant of 25 radians/m at 2 GHz. [Parallels Example 3.4]
(a) What is the complex propagation constant of the transmission line?
(b) If the capacitance of the line is 50 pF·m^{-1} and $G = 0$, what is the characteristic impedance of the line?

11. A very low-loss microstrip transmission line has the following per unit length parameters: $R = 2\ \Omega$/m, $L = 80$ nH/m, $C = 200$ pF/m, and $G = 1\ \mu$S/m.
(a) What is the characteristic impedance of the line if loss is ignored?
(b) What is the attenuation constant due to conductor loss?
(c) What is the attenuation constant due to dielectric loss?

12. A lossless transmission line carrying a 1 GHz signal has the following per unit length parameters: $L = 80$ nH/m, $C = 200$ pF/m.
(a) What is the attenuation constant?
(b) What is the phase constant?
(c) What is the phase velocity?
(d) What is the characteristic impedance of the line?

13. A transmission line has a characteristic impedance Z_0 and is terminated in a load with a reflection coefficient of $0.8\angle 45°$. A forward-traveling voltage wave on the line has a power of 1 dBm.
1. How much power is reflected by the load?
2. What is the power delivered to the load?

14. A transmission line has an attenuation of 0.2 dB/cm and a phase constant of 50 radians/m at 1 GHz.
(a) What is the complex propagation constant of the transmission line?
(b) If the capacitance of the line is 100 pF/m and $G = 0$, what is the complex characteristic impedance of the line?
(c) If the line is driven by a source modeled as an ideal voltage and a series impedance, what is the impedance of the source for maximum transfer of power to the transmission line?
(d) If 1 W is delivered (i.e. in the forward-traveling wave) to the transmission line by the generator, what is the power in the forward-traveling wave on the line at 2 m from the generator?

15. A lossless transmission line is driven by a 1 GHz generator having a Thevenin equivalent impedance of 50 Ω. The transmission line is lossless, has a characteristic impedance of 75 Ω, and is infinitely long. The maximum power that can be delivered to a load attached to the generator is 2 W.
(a) What is the total (phasor) voltage at the input to the transmission line?
(b) What is the magnitude of the forward-traveling voltage wave at the generator side of the line?
(c) What is the magnitude of the forward-traveling current wave at the generator side of the line?

16. A transmission line is terminated in a short circuit. What is the ratio of the forward- and backward-traveling voltage waves at the termination? [Parallels Example 3.5]

17. A 50 Ω transmission line is terminated in a 40 Ω load. What is the ratio of the forward- to the backward-traveling voltage waves at the termination? [Parallels Example 3.5]

18. A 50 Ω transmission line is terminated in an open circuit. What is the ratio of the forward- to the backward-traveling voltage waves at the termination? [Parallels Example 3.5]

19. A line has a characteristic impedance Z_0 and is terminated in a load with a reflection coefficient of 0.8. A forward-traveling voltage wave on the line has a power of 1 W.
(a) How much power is reflected by the load?
(b) What is the power delivered to the load?

20. A load consists of a shunt connection of a capacitor of 10 pF and a resistor of 25 Ω. The load terminates a lossless 50 Ω transmission line. The operating frequency is 1 GHz. [Parallels Example 3.6]
    (a) What is the impedance of the load?
    (b) What is the normalized impedance of the load (normalized to the characteristic impedance of the line)?
    (c) What is the reflection coefficient of the load?
    (d) What is the current reflection coefficient of the load?
    (e) What is the standing wave ratio (SWR)?
    (f) What is the current standing wave ratio (ISWR)?
21. An amplifier is connected to a load by a transmission line matched to the amplifier. If the SWR on the line is 1.5, what percentage of the available amplifier power is absorbed by the load?
22. A load has a reflection coefficient of 0.5 when referred to 50 Ω. The load is at the end of a line with a 50 Ω characteristic impedance.
    (a) If the line has an electrical length of 45°, what is the reflection coefficient calculated at the input of the line?
    (b) What is the VSWR on the 50 Ω line?
23. A 100 Ω resistor in parallel with a 5 pF capacitor terminates a 100 Ω transmission line. Calculate the SWR on the line at 2 GHz.
24. A lossless 50 Ω transmission line has a 50 Ω generator at one end and is terminated in 100 Ω. What is the VSWR on the line?
25. A lossless 75 Ω line is driven by a 75 Ω generator. The line is terminated in a load that with a reflection coefficient (referred to 50 Ω) of $0.5 + \jmath 0.5$. What is the VSWR on the line?
26. A load with a 20 pF capacitor in parallel with a 50 Ω resistor terminates a 25 Ω line. The operating frequency is 5 GHz. [Parallels Example 3.7]
    (a) What is the VSWR?
    (b) What is ISWR?
27. A load $Z_L = 55 - \jmath 55\ \Omega$ and the system reference impedance, Z_0, is 50 Ω. [Parallels Example 3.8]
    (a) What is the load reflection coefficient Γ_L?
    (b) What is the current reflection coefficient?
    (c) What is the VSWR on the line?
    (d) What is the ISWR on the line?
    (e) Now consider a source connected directly to the load. The source has a Thevenin equivalent impedance $Z_G = 60\ \Omega$ and an available power of 1 W. Use Γ_L to find the power delivered to Z_L.
    (f) What is the total power absorbed by Z_G?
28. A load of 100 Ω is to be matched to a transmission line with a characteristic impedance of 50 Ω. Use a quarter-wave transformer. What is the characteristic impedance of the quarter-wave transformer?
29. Determine the characteristic impedance of a quarter-wave transformer used to match a load of 50 Ω to a generator with a Thevenin equivalent impedance of 75 Ω.
30. A transmission line is to be inserted between a 5 Ω line and a 50 Ω load so that there is maximum power transfer to the 50 Ω load at 20 GHz.
    (a) How long is the inserted line in terms of wavelengths at 20 GHz?
    (b) What is the characteristic impedance of the line at 20 GHz?

3.8.1 Exercises By Section

†challenging, ‡very challenging

§3.1 1, $12^\dagger$, 13

§3.2 $2^\dagger$, 3, 4, 5, 6, 7, 8, 9, 10, 11,

§3.3 $14^\dagger$, 15, 16, 17, 18, $19^\dagger$, $20^\dagger$, $21^\dagger$, 22, 23, 24, 25, 26, 27

§3.4 28, 29, 30

3.8.2 Answers to Selected Exercises

10 $0.23 + \jmath 25\ \mathrm{m}^{-1}$

29 61.2 Ω

CHAPTER 4

Planar Transmission Lines

4.1 Introduction

The majority of transmission lines used in high-speed digital, RF, and microwave circuits are planar, as these can be defined using masks, photoresist, and etching of metal sheets. Such lines are called planar transmission lines. A common planar line is the microstrip line shown in Figure 4-1 and in cross section in Figure 4-2. This cross section is typical of what would be found on a semiconductor or **printed circuit board (PCB)**. Current flows in both the top and bottom conductor, but in opposite directions. The physics is such that if there is a signal current on the top

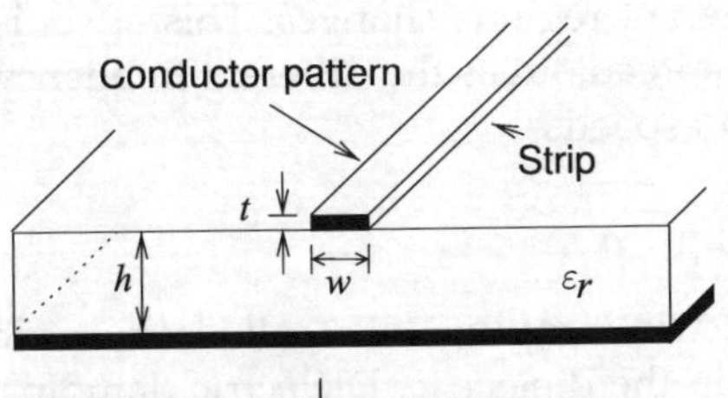

Figure 4-1: Microstrip transmission line.

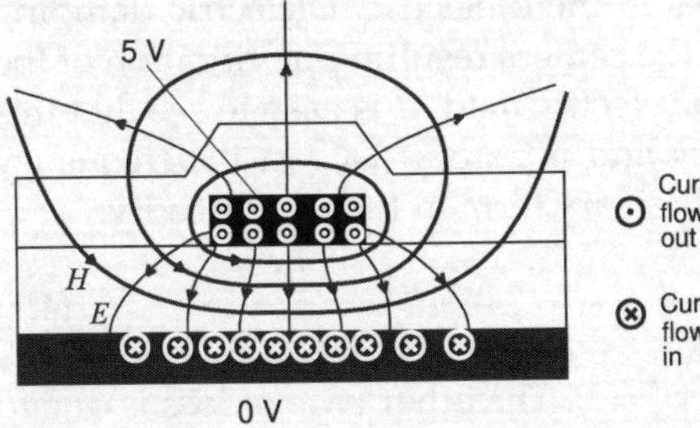

Figure 4-2: Cross-sectional view of a microstrip line showing electric and magnetic field lines and current flow. The electric and magnetic fields are in two mediums—the dielectric and air. If the line is homogeneous (the same dielectric everywhere) the electric and magnetic fields are only in the transverse plane, a field configuration known as the transverse electromagnetic mode (TEM).

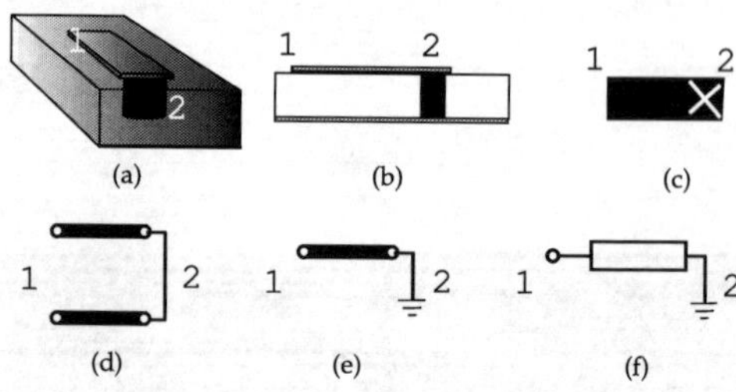

Figure 4-3: Representations of a shorted microstrip line with a short (or via) at port 2: (a) three-dimensional (3D) view indicating via; (b) side view; (c) top view with via indicated by X; (d) schematic representation of transmission line; (e) alternative schematic representation; and (f) representation as a circuit element.

conductor, there must be a return signal current, which will be as close to the signal current as possible to minimize stored energy. The provision of a signal return path is important in maintaining the integrity of (i.e., a predictable signal waveform on) an interconnect.

In the microstrip line, electric field lines start on one of the conductors and finish on the other and are located almost entirely in the plane transverse to the long length of the line. The magnetic field is also mostly confined to the transverse plane, and so this line is referred to as a **transverse electromagnetic (TEM)** line. More accurately it is called a **quasi-TEM** line, as the longitudinal fields, particularly in the air region, are not negligible.

Various schematic representations of a microstrip line are used. Consider the representations in Figure 4-3 of a length of microstrip line shorted by a via at the end denoted by "2" (specifically the "2" refers to Port 2).

4.2 Substrates

Planar line design involves choosing both the transmission line structure to use and the substrate. In this section the electrical and magnetic properties of substrate materials will be discussed.

4.2.1 Dielectric Effect

When the fields are in more than one medium (a **nonhomogeneous transmission line**), as for the microstrip line, the effective relative permittivity, $\varepsilon_{r,e}$ (or usually just $\varepsilon_e = \varepsilon_{r,e}$), is used. The characteristics of the line are then more or less the same as for the same structure with a uniform dielectric of permittivity, $\varepsilon_{\text{eff}} = \varepsilon_e \varepsilon_0$. The ε_{eff} changes with frequency as the proportion of energy stored in the different regions changes. This effect is called **dispersion** and causes a pulse to spread out as the different frequency components of a signal travel at different speeds.

4.2.2 Dielectric Loss Tangent, $\tan\delta$

Loss in a dielectric comes from (a) **dielectric damping** (also called **dielectric relaxation**), and (b) conduction losses in the dielectric. Dielectric damping originates from the movement of charge centers resulting in vibration of the lattice and thus energy is lost from the electric field. It is easy to see that this loss increases linearly with frequency and is zero at DC. In the frequency domain loss is incorporating in an imaginary term in the **permittivity**:

$$\varepsilon = \varepsilon_r \varepsilon_0 = \varepsilon' - \jmath \varepsilon'' = \varepsilon_0 \left(\varepsilon_r' - \jmath \varepsilon_r'' \right). \tag{4.1}$$

If there is no dielectric damping loss, $\varepsilon'' = 0$. The other type of loss is due to the movement of charge carriers in the dielectric. The ability to move charges

Material	$10^4 \tan\delta$ (at 10 GHz)	ε_r
Air (dry)	≈ 0	1
Alumina, 99.5%	1–2	10.1
Sapphire	0.4–0.7	9.4, 11.6
Glass, typical	20	5
Polyimide	50	3.2
Quartz (fused)	1	3.8
FR4 circuit board	100	4.3–4.5
RT-duroid 5880	5–15	2.16–2.24
RT-duroid 6010	10–60	10.2–10.7
AT-1000	20	10.0–13.0
Si (high resistivity)	10–100	11.9
GaAs	6	12.85
InP	10	12.4
SiO_2 (on-chip)	—	4.0–4.2
LTCC (typical, green tape(TM) 951)	15	7.8

Table 4-1: Properties of common substrate materials. The dielectric loss tangent is scaled. For example, for glass, $\tan\delta$ is typically 0.002.

is described by the conductivity, σ, and this loss is independent of frequency. So the energy lost in the dielectric is proportional to $\omega\varepsilon'' + \sigma$ and the energy stored in the electric field is proportional to $\omega\varepsilon'$. Thus a loss tangent, $\tan\delta$, is introduced:

$$\tan\delta = \frac{\omega\varepsilon'' + \sigma}{\omega\varepsilon'}. \tag{4.2}$$

Also the relative permittivity can be redefined as

$$\varepsilon_r = \varepsilon_r' - \jmath\left(\varepsilon_r'' + \frac{\sigma}{\omega\varepsilon_0}\right). \tag{4.3}$$

With the exception of silicon, the loss tangent is very small for dielectrics that are useful at RF and microwave frequencies and so most of the time

$$|\varepsilon_r| \approx \varepsilon_r'. \tag{4.4}$$

Thus

$$\varepsilon_r = \varepsilon_r' - \jmath\left(\varepsilon_r'' + \sigma/(\omega\varepsilon_0)\right) \approx \varepsilon_r'\left(1 - \jmath\tan\delta\right). \tag{4.5}$$

4.2.3 Magnetic Material Effect

Except for very special circumstances substrates used for microstrip lines are non magnetic so $\mu = \mu_0$ and the relative permeability, μ_r, is defined so that

$$\mu = \mu_r\mu_0. \tag{4.6}$$

4.2.4 Substrates for Planar Transmission Lines

The properties of common substrate materials are given in Table 4-1. Crystal substrates have very good dimensional tolerances and uniformity of electrical properties. Many other substrates have high surface roughness and electrical properties that can vary. For example, FR4 is the most common type of PCB substrate and is a weave of fiberglass embedded in resin. So the material is not uniform and there is an unpredictable localized variation in the proportion of resin and glass. High-performance microwave circuit boards have ceramic particles embedded in the resin.

4.3 Planar Transmission Line Structures

Two planar transmission line structures are shown in Figure 4-4. The reason these are so popular is that they can be mass produced. For the microstrip line in Figure 4-4(a) the fabrication process begins with a dielectric sheet with solid metal layers on the top and bottom. One of these is covered with a photosensitive material, called a photoresist, exposed to a prepared pattern that defines the interconnect line network, then the photoresist is developed and the unexposed (or exposed, depending on whether the photoresist is positive or negative) metal on one side is etched away. The stripline in Figure 4-4(b) is fabricated similarly to microstrip but followed by one more step in which a dielectric sheet with a ground plane only is bonded on top.

The most important planar transmission line structures are shown in Figure 4-5. With the homogeneous lines virtually all of the fields are in the plane transverse to the direction of propagation (i.e., the longitudinal direction). Transmission lines where the longitudinal fields are almost insignificant are referred to as supporting a TEM mode, and they are called TEM lines.

The most important inhomogeneous lines are shown in Figure 4-5(a–c). The main difference between the two sets of configurations (homogeneous and inhomogeneous) is the frequency-dependent variation of the EM field distributions with inhomogeneous lines. With inhomogeneous lines, the EM fields are not confined entirely to the transverse plane even if the conductors are perfect. However, they are largely confined to the transverse plane and so these lines are called **quasi-TEM lines**. The actual choice of structure depends on several factors, including operating frequency and the type of substrate and metallization system available. This book focuses on microstrip as it is by far the most commonly used planar transmission line.

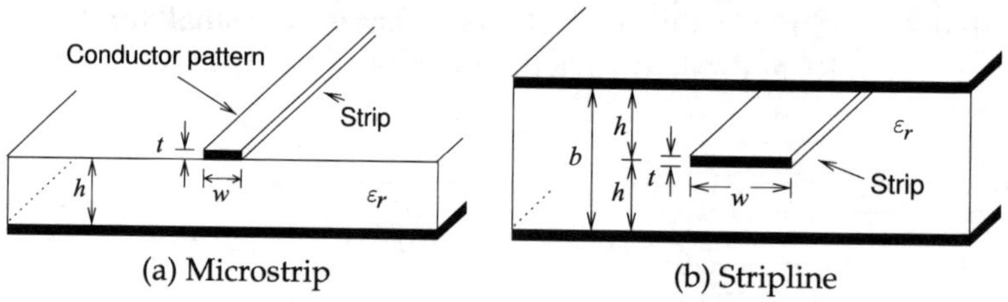

Figure 4-4: Planar transmission lines.

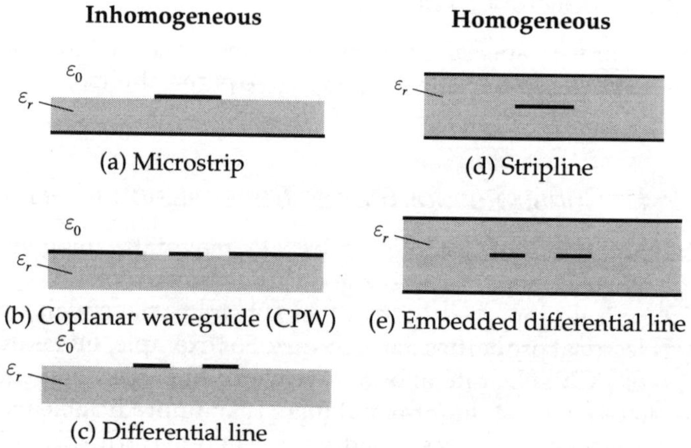

Figure 4-5: Cross sections of several homogeneous and inhomogeneous planar transmission line structures.

4.4 Microstrip Transmission Lines

Microstrip has conductors embedded in two dielectric mediums and cannot support a pure TEM mode. In most practical cases, the dielectric substrate is electrically thin, that is, $h \ll \lambda$. Then the transverse field is dominant and the fields are called quasi-TEM.

4.4.1 Microstrip Line in the Quasi-TEM Approximation

In this section relations are developed based on the principle that the phase velocity of an EM wave in an air-only homogeneous transmission with a TEM field line is just c. As a first step, the potential of the conductor strip is set to V_0 and Laplace's equation is solved using an EM simulator for the electrostatic potential everywhere in the dielectric. Then the per unit length (p.u.l.) electric charge, Q, on the conductor is determined. Using this in the following relation gives the line capacitance:

$$C = \frac{Q}{V_0}.$$

In the next step, the process is repeated with $\varepsilon_r = 1$ to determine C_{air} (the capacitance of the line without a dielectric).

If the microstrip line is now an air-filled lossless TEM structure,

$$v_{p,\text{air}} = c = \frac{1}{LC_{\text{air}}} \quad (4.7) \qquad \text{and so} \quad L = \frac{1}{c^2 C_{\text{air}}}. \quad (4.8)$$

L is not affected by the dielectric properties of the medium. L calculated above is the desired p.u.l. inductance of the line with the dielectric as well as in free space. Once L and C have been found, the characteristic impedance can be found using

$$Z_0 = \sqrt{\frac{L}{C}}, \quad (4.9) \qquad \text{rewritten as} \quad Z_0 = \frac{1}{c}\frac{1}{\sqrt{C\,C_{\text{air}}}}, \quad (4.10)$$

and the phase velocity is

$$v_p = \frac{1}{\sqrt{LC}} = c\sqrt{\frac{C_{\text{air}}}{C}}. \quad (4.11)$$

Now the field is distributed in the inhomogeneous medium and in free space, as shown in Figure 4-6(a). So the effective relative permittivity, ε_e, of the equivalent homogeneous microstrip line (see Figure 4-6(b)) is defined by

$$\sqrt{\varepsilon_e} = \frac{c}{v_p}. \quad (4.12)$$

Combining Equations (4.11) and (4.12), the effective relative permittivity (usually just the term effective permittivity is used) is obtained:

$$\varepsilon_e = \frac{C}{C_{\text{air}}}. \quad (4.13)$$

The effective permittivity can be interpreted as the permittivity of a homogeneous medium that replaces the air and the dielectric regions of the

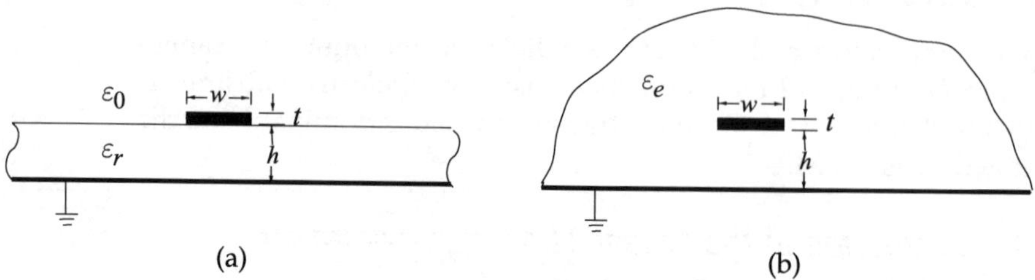

Figure 4-6: Microstrip line: (a) cross section; and (b) eps .. equivalent structure where the strip is embedded in a dielectric of semi-infinite extent with effective relative permittivity ε_e.

microstrip, as shown in Figure 4-6. Since some of the field is in the dielectric and some is in air, the effective relative permittivity must satisfy

$$1 < \varepsilon_e < \varepsilon_r. \tag{4.14}$$

However, the minimum ε_e will be greater than 1 as electrical energy will be distributed in air and dielectric. The wavelength on a transmission line, the guide wavelength λ_g, is related to the free space wavelength by $\lambda_g = \lambda_0/\sqrt{\varepsilon_e}$.

EXAMPLE 4.1 Microstrip Calculations

A microstrip line has a characteristic impedance Z_0 of 50 Ω derived from reflection coefficient measurements and an effective permittivity, ε_e, of 7 derived from measurement of phase velocity. What is the line's per-unit-length inductance, L, and capacitance, C?

Solution: The key equations are $Z_0 = \sqrt{L/C}$, $\varepsilon_e = C/C_{\text{air}}$, and in air $v_p = 1/\sqrt{LC_{\text{air}}} = c$. Also assume that $\mu_r = 1$ which is the default if not specified otherwise and also that L does not change if only the dielectric is changed. Thus

$$C_{\text{air}} = \frac{C}{\varepsilon_e} \quad \text{and then} \quad L = \frac{\varepsilon_e}{c^2 C} \quad \text{so that} \quad Z_0 = \sqrt{\frac{L}{C}} = \frac{\sqrt{\varepsilon_e}}{cC}, \quad \text{that is} \quad C = \frac{\sqrt{\varepsilon_e}}{cZ_0}.$$

So $C = \sqrt{7}/(2.998 \cdot 10^8 \times 50) = 1.765 \cdot 10^{-10} = 176.5\text{pF/m}$ and $L = Z_0^2 C = 44.13\ \mu\text{H/m}$.

4.4.2 Effective Permittivity and Characteristic Impedance

This section presents formulas for the effective permittivity and characteristic impedance of a microstrip line. These formulas are fits to the results of detailed EM simulations. Also, the form of the equations is based on good physical understanding. First, assume that the thickness, t, is zero. This is not a bad approximation, as $t \ll w, h$ for most microwave circuits.

Hammerstad and others provide well-accepted formulas for calculating the effective permittivity and characteristic impedance of microstrip lines [1–3]. Given ε_r, w, and h, the effective relative permittivity is

$$\varepsilon_e = \frac{\varepsilon_r + 1}{2} + \frac{\varepsilon_r - 1}{2}\left(1 + \frac{10h}{w}\right)^{-a \cdot b}, \tag{4.15}$$

where

$$a(u)|_{u=w/h} = 1 + \frac{1}{49}\ln\left[\frac{u^4 + \{u/52\}^2}{u^4 + 0.432}\right] + \frac{1}{18.7}\ln\left[1 + \left(\frac{u}{18.1}\right)^3\right] \tag{4.16}$$

and $$(\varepsilon_r) = 0.564\left[\frac{\varepsilon_r - 0.9}{\varepsilon_r + 3}\right]^{0.053}. \tag{4.17}$$

Take some time to interpret Equation (4.15), the formula for **effective relative permittivity.** If $\varepsilon_r = 1$, then $\varepsilon_e = (1+1)/2 + 0 = 1$, as expected. If ε_r is not that of air, then ε_e will be between 1 and ε_r, dependent on the geometry of the line, or more specifically, the ratio w/h. For a very wide line, $w/h \gg 1$, $\varepsilon_e = (\varepsilon_r + 1)/2 + (\varepsilon_r - 1)/2 = \varepsilon_r$, corresponding to the EM energy being confined to the dielectric. For a thin line $w/h \ll 1$, $\varepsilon_e = (\varepsilon_r + 1)/2$, the average of the dielectric and air permittivities.

Mostly the term **"effective permittivity"** is used to mean effective relative permittivity (check the magnitude).

The characteristic impedance is given by

$$Z_0 = \frac{Z_{01}}{\sqrt{\varepsilon_e}}, \tag{4.18}$$

where the characteristic impedance of the microstrip line in free space is

$$Z_{01} = Z_0|_{(\varepsilon_r=1)} = 60\ln\left[\frac{F_1 h}{w} + \sqrt{1 + \left(\frac{2h}{w}\right)^2}\right] \tag{4.19}$$

and $$F_1 = 6 + (2\pi - 6)\exp\left\{-(30.666h/w)^{0.7528}\right\}. \tag{4.20}$$

The accuracy of Equation (4.15) is better than 0.2% for $0.01 \le w/h \le 100$ and $1 \le \varepsilon_r \le 128$. Also, the accuracy of Equation (4.19) is better than 0.1% for $w/h < 1000$. Note that Z_0 has a maximum value when w is small and a minimum value when w is large.

Now consider the special case where w is vanishingly small. Then ε_e has its minimum value:

$$\varepsilon_e = \tfrac{1}{2}(\varepsilon_r + 1). \tag{4.21}$$

This leads to an approximate (and convenient) form of Equation (4.15):

$$\varepsilon_e = \frac{(\varepsilon_r + 1)}{2} + \frac{(\varepsilon_r - 1)}{2}\frac{1}{\sqrt{1 + 12h/w}}. \tag{4.22}$$

This approximation has its greatest error for low and high ε_r and narrow lines, $w/h \ll 1$, where the maximum error is 1%. Again, Equation (4.18) is used to calculate the characteristic impedance. The more exact analysis, represented by Equation (4.15), was used to develop Table 4-2, which can be used in the design of microstrip.

EXAMPLE 4.2 Microstrip Characteristic Impedance Calculation

The strip of a microstrip line has a width of 600 μm and is fabricated on a lossless substrate that is 635 μm thick and has a relative permittivity of 4.1.

(a) What is the effective relative permittivity?
(b) What is the characteristic impedance?
(c) What is the propagation constant at 5 GHz ignoring any losses?

Solution:

Use the formulas for effective permittivity, characteristic impedance, and attenuation constant from Section 4.4.2 with $w = 600$ μm ; $h = 635$ μm ; $\varepsilon_r = 4.1$; $w/h = 600/635 = 0.945$.

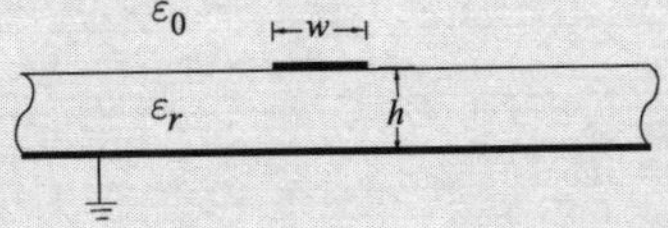

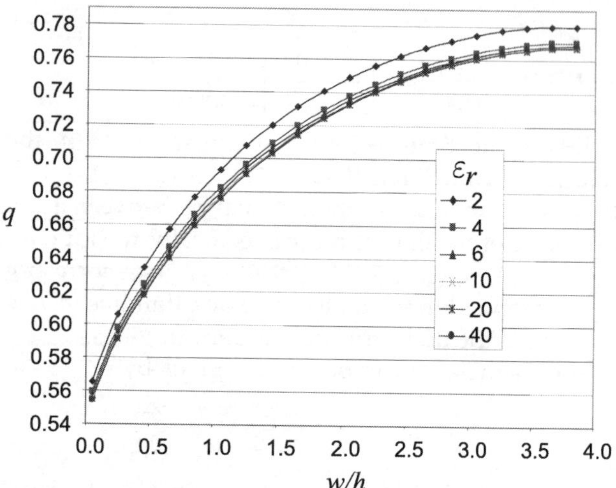

Figure 4-7: Dependence of the q factor of a microstrip line at 1 GHz for various permittivities and aspect (w/h) ratios. (Data obtained from EM field simulations using Sonnet.)

(a)
$$\varepsilon_e = \frac{\varepsilon_r + 1}{2} + \frac{\varepsilon_r - 1}{2}\left(1 + \frac{10h}{w}\right)^{-a \cdot b}$$

From Equations (4.16) and (4.17),

$$a = 1 + \frac{1}{49}\ln\left[\frac{(w/h)^4 + \{w/(52h)\}^2}{(w/h)^4 + 0.432}\right] + \frac{1}{18.7}\ln\left[1 + \left(\frac{w}{18.1h}\right)^3\right] = 0.991,$$

$$b = 0.564\left[\frac{\varepsilon_r - 0.9}{\varepsilon_r + 3}\right]^{0.053} = 0.541.$$

From Equation (4.15), $\varepsilon_e = 2.967$.

(b) In free space,
$$Z_0|_{\text{air}} = 60 \ln\left[\frac{F_1 \cdot h}{w} + \sqrt{1 + \left(\frac{2h}{w}\right)^2}\right],$$

where $F_1 = 6 + (2\pi - 6)\exp\{-(30.666h/w)^{0.7528}\}$, $Z_0 = Z_0|_{\text{air}}/\sqrt{\varepsilon_e}$

$$Z_0|_{\text{air}} = 129.7\,\Omega \quad \text{and} \quad Z_0 = Z_0|_{\text{air}}/\sqrt{\varepsilon_e} = 75.3\,\Omega.$$

(c) $f = 5$ GHz, $\omega = 2\pi f$, $\gamma = \jmath\omega\sqrt{\mu_0\varepsilon_0\varepsilon_e} = \jmath 180.5/\text{ m}$.

4.4.3 Filling Factor

Defining a filling factor, q, provides useful insight into the distribution of energy in an inhomogeneous transmission line. The effective microstrip permittivity is

$$\varepsilon_e = 1 + q(\varepsilon_r - 1), \tag{4.23}$$

where for a microstrip line q has the bounds $\frac{1}{2} \leq q \leq 1$ and is almost independent of ε_r. A q of 1 indicates that all of the fields are in the dielectric region. The dependence of the q of a microstrip line at 1 GHz for various permittivities and aspect (w/h) ratios is shown in Figure 4-7. Fitting yields:

$$q = \tfrac{1}{2}\left(1 + \frac{1}{\sqrt{1 + 12h/w}}\right). \tag{4.24}$$

Table 4-2: Microstrip line normalized width u ($= w/h$) and effective permittivity, ε_e, for specified characteristic impedance Z_0. Data derived from the analysis in Section 4.4.2.

Z_0	$\varepsilon_r = 4$ (SiO_2, FR4)		$\varepsilon_r = 10$ (Alumina)		$\varepsilon_r = 11.9$ (Si)		Z_0	$\varepsilon_r = 4$ (SiO_2, FR4)		$\varepsilon_r = 10$ (Alumina)		$\varepsilon_r = 11.9$ (Si)	
(Ω)	u	ε_e	u	ε_e	u	ε_e	(Ω)	u	ε_e	u	ε_e	u	ε_e
140	0.171	2.718	0.028	5.914	0.017	6.907	110	0.378	2.781	0.089	6.025	0.062	7.032
139	0.176	2.720	0.029	5.917	0.018	6.910	109	0.389	2.783	0.093	6.031	0.065	7.038
138	0.181	2.722	0.030	5.919	0.019	6.914	108	0.399	2.786	0.097	6.036	0.068	7.044
137	0.185	2.723	0.031	5.922	0.020	6.919	107	0.410	2.789	0.100	6.040	0.071	7.050
136	0.190	2.725	0.032	5.924	0.021	6.923	106	0.421	2.791	0.104	6.046	0.074	7.055
135	0.195	2.727	0.033	5.927	0.022	6.925	105	0.432	2.794	0.109	6.052	0.077	7.061
134	0.201	2.729	0.035	5.931	0.022	6.927	104	0.444	2.797	0.113	6.057	0.080	7.066
133	0.206	2.731	0.036	5.933	0.023	6.930	103	0.456	2.800	0.117	6.062	0.084	7.073
132	0.212	2.733	0.037	5.936	0.024	6.934	102	0.468	2.803	0.122	6.069	0.087	7.079
131	0.217	2.734	0.038	5.939	0.025	6.937	101	0.481	2.806	0.127	6.075	0.091	7.085
130	0.223	2.736	0.040	5.942	0.026	6.941	100	0.494	2.809	0.132	6.081	0.095	7.092
129	0.229	2.738	0.043	5.949	0.028	6.948	99	0.507	2.812	0.137	6.087	0.099	7.099
128	0.235	2.740	0.044	5.951	0.029	6.951	98	0.521	2.815	0.143	6.094	0.103	7.105
127	0.241	2.742	0.046	5.955	0.030	6.954	97	0.535	2.819	0.148	6.100	0.108	7.113
126	0.248	2.744	0.048	5.958	0.031	6.957	96	0.550	2.822	0.154	6.106	0.112	7.120
125	0.254	2.746	0.050	5.962	0.033	6.963	95	0.565	2.825	0.160	6.113	0.117	7.127
124	0.261	2.748	0.052	5.966	0.034	6.966	94	0.580	2.829	0.167	6.121	0.122	7.135
123	0.268	2.750	0.054	5.970	0.035	6.969	93	0.596	2.832	0.173	6.127	0.128	7.144
122	0.275	2.752	0.056	5.973	0.038	6.977	92	0.612	2.836	0.180	6.134	0.133	7.151
121	0.283	2.755	0.058	5.977	0.039	6.980	91	0.629	2.839	0.187	6.142	0.139	7.159
120	0.290	2.757	0.061	5.982	0.041	6.985	90	0.646	2.843	0.195	6.150	0.145	7.168
119	0.298	2.759	0.063	5.985	0.043	6.990	89	0.664	2.847	0.202	6.157	0.151	7.176
118	0.306	2.761	0.066	5.990	0.045	6.995	87	0.701	2.855	0.219	6.173	0.164	7.193
117	0.314	2.763	0.068	5.993	0.047	6.999	86	0.721	2.859	0.228	6.182	0.171	7.203
116	0.323	2.766	0.071	5.998	0.049	7.004	85	0.740	2.863	0.237	6.190	0.179	7.213
115	0.331	2.768	0.074	6.003	0.051	7.008	84	0.761	2.867	0.246	6.198	0.187	7.223
114	0.340	2.771	0.077	6.007	0.053	7.013	83	0.782	2.872	0.256	6.208	0.195	7.233
113	0.349	2.773	0.080	6.012	0.055	7.017	82	0.804	2.876	0.266	6.216	0.203	7.242
112	0.359	2.776	0.083	6.016	0.057	7.022	81	0.826	2.881	0.277	6.226	0.212	7.253
111	0.368	2.778	0.086	6.021	0.060	7.028	80	0.849	2.885	0.288	6.235	0.221	7.263

Table 4-2 continued.

Z_0	$\varepsilon_r = 4$ (SiO_2, FR4)		$\varepsilon_r = 10$ (Alumina)		$\varepsilon_r = 11.9$ (Si)	
(Ω)	u	ε_e	u	ε_e	u	ε_e
79	0.873	2.890	0.299	6.245	0.230	7.274
78	0.898	2.895	0.311	6.255	0.240	7.285
77	0.923	2.900	0.324	6.265	0.251	7.297
76	0.949	2.905	0.337	6.276	0.262	7.309
75	0.976	2.910	0.350	6.286	0.273	7.321
74	1.003	2.915	0.364	6.297	0.285	7.333
73	1.032	2.921	0.379	6.309	0.297	7.345
72	1.062	2.926	0.394	6.320	0.310	7.359
71	1.092	2.932	0.410	6.332	0.323	7.371
70	1.123	2.937	0.426	6.344	0.338	7.386
69	1.156	2.943	0.444	6.357	0.352	7.399
68	1.190	2.949	0.462	6.369	0.368	7.414
67	1.224	2.955	0.480	6.382	0.384	7.429
66	1.260	2.961	0.500	6.396	0.400	7.444
65	1.298	2.968	0.520	6.410	0.418	7.460
64	1.336	2.974	0.541	6.424	0.436	7.476
63	1.376	2.980	0.563	6.439	0.455	7.492
62	1.417	2.987	0.586	6.454	0.475	7.509
61	1.460	2.994	0.610	6.470	0.496	7.527
60	1.504	3.001	0.635	6.486	0.518	7.545
59	1.551	3.008	0.661	6.502	0.541	7.564
58	1.598	3.015	0.688	6.519	0.564	7.583
57	1.648	3.022	0.717	6.538	0.589	7.603
56	1.700	3.030	0.746	6.556	0.616	7.624
55	1.753	3.037	0.777	6.575	0.643	7.645
54	1.809	3.045	0.809	6.594	0.672	7.667
53	1.867	3.053	0.843	6.614	0.702	7.690
52	1.927	3.061	0.878	6.635	0.733	7.713
51	1.991	3.069	0.915	6.657	0.766	7.738
50	2.056	3.077	0.954	6.679	0.800	7.763
49	2.125	3.086	0.995	6.702	0.837	7.790
48	2.197	3.094	1.037	6.726	0.875	7.817
47	2.272	3.103	1.081	6.750	0.914	7.845
46	2.350	3.112	1.128	6.775	0.956	7.874
45	2.432	3.121	1.177	6.801	1.000	7.904
44	2.518	3.131	1.229	6.828	1.047	7.936
43	2.609	3.140	1.283	6.856	1.096	7.968
42	2.703	3.150	1.340	6.884	1.147	8.002
41	2.803	3.160	1.400	6.913	1.201	8.036
40	2.908	3.171	1.464	6.944	1.259	8.072
39	3.019	3.181	1.531	6.974	1.319	8.108
38	3.136	3.192	1.602	7.006	1.384	8.147
37	3.259	3.203	1.677	7.039	1.452	8.186
36	3.390	3.214	1.757	7.073	1.524	8.226
35	3.528	3.226	1.841	7.108	1.600	8.268
34	3.675	3.237	1.931	7.143	1.682	8.311
33	3.831	3.250	2.027	7.180	1.769	8.355
32	3.997	3.262	2.129	7.218	1.862	8.402
31	4.174	3.275	2.238	7.258	1.961	8.449
30	4.364	3.288	2.355	7.298	2.067	8.498
29	4.567	3.301	2.480	7.340	2.181	8.549
28	4.785	3.315	2.615	7.384	2.304	8.601
27	5.020	3.329	2.760	7.428	2.436	8.655
26	5.273	3.344	2.917	7.475	2.579	8.712
25	5.547	3.359	3.087	7.523	2.734	8.770
24	5.845	3.374	3.272	7.573	2.902	8.831
23	6.169	3.390	3.474	7.625	3.086	8.894
22	6.523	3.407	3.694	7.679	3.287	8.960
21	6.912	3.424	3.936	7.734	3.508	9.028
20	7.341	3.441	4.203	7.793	3.752	9.100
19	7.815	3.459	4.499	7.854	4.022	9.174
18	8.344	3.478	4.829	7.917	4.323	9.252
17	8.936	3.497	5.199	7.983	4.661	9.334
16	9.603	3.517	5.616	8.053	5.043	9.419
15	10.361	3.538	6.090	8.126	5.476	9.509
14	11.229	3.559	6.633	8.202	5.972	9.604
13	12.233	3.581	7.262	8.282	6.547	9.704
12	13.407	3.604	7.997	8.367	7.219	9.809
11	14.798	3.628	8.868	8.456	8.016	9.920
10	16.471	3.652	9.916	8.550	8.975	10.038

4.5 Microstrip Design Formulas

The formulas developed in Section 4.4.2 enable the electrical characteristics to be determined given the material properties and the physical dimensions of a microstrip line. In design, the physical dimensions must be determined given the desired electrical properties. Several people have developed procedures that can be used to synthesize microstrip lines. This subject is considered in much more depth in [4], and here just one approach is reported. The formulas are useful outside the range indicated, but with reduced accuracy. Again, these formulas are the result of curve fits, but starting with physically based equation forms.

4.5.1 High Impedance

For narrow strips, that is, when $Z_0 > (44 - 2\varepsilon_r)\ \Omega$,

$$\frac{w}{h} = \left(\frac{\exp H'}{8} - \frac{1}{4\exp H'}\right)^{-1}, \tag{4.25}$$

where $$H' = \frac{Z_0\sqrt{2(\varepsilon_r+1)}}{119.9} + \frac{1}{2}\left(\frac{\varepsilon_r - 1}{\varepsilon_r + 1}\right)\left(\ln\frac{\pi}{2} + \frac{1}{\varepsilon_r}\ln\frac{4}{\pi}\right). \tag{4.26}$$

For $Z_0 > (63 - 2\varepsilon_r)\ \Omega$,

$$\varepsilon_e = \frac{\varepsilon_r + 1}{2}\left[1 + \frac{29.98}{Z_0}\left(\frac{2}{\varepsilon_r + 1}\right)^{1/2}\left(\frac{\varepsilon_r - 1}{\varepsilon_r + 1}\right)\left(\ln\frac{\pi}{2} + \frac{1}{\varepsilon_r}\ln\frac{4}{\pi}\right)\right]^2. \tag{4.27}$$

The formula for ε_e is accurate to better than 1% for $Z_0 > (44 - 2\varepsilon_r)\ \Omega$ (i.e $w/h < 1.3$) for $8 < \varepsilon_r < 12$. Overall the synthesis of w/h has an accuracy of better than 1%.

4.5.2 Low Impedance

Strips with low Z_0 are relatively wide and the formulas below can be used when $Z_0 < (44 - 2\varepsilon_r)\ \Omega$. The cross-sectional geometry is given by

$$\frac{w}{h} = \frac{2}{\pi}\left[(d_{\varepsilon_r} - 1) - \ln(2d_{\varepsilon_r} - 1)\right] + \frac{(\varepsilon_r - 1)}{\pi\varepsilon_r}\left[\ln(d_{\varepsilon_r} - 1) + 0.293 - \frac{0.517}{\varepsilon_r}\right], \tag{4.28}$$

where $$d_{\varepsilon_r} = \frac{59.95\pi^2}{Z_0\sqrt{\varepsilon_r}}. \tag{4.29}$$

For $Z_0 < (63 - 2\varepsilon_r)\ \Omega$

$$\varepsilon_e = \frac{\varepsilon_r}{0.96 + \varepsilon_r(0.109 - 0.004\varepsilon_r)[\log(10 + Z_0) - 1]}. \tag{4.30}$$

The expression for ε_e is accurate to better than 1% for $8 < \varepsilon_r < 12$ and $8 \leq Z_0 \leq (63 - 2\varepsilon_r)\ \Omega$.

EXAMPLE 4.3 Microstrip Design

Design a microstrip line to have a characteristic impedance of 75 Ω at 10 GHz. The microstrip a substrate that is 500 μm thick with a relative permittivity of 5.6. (a) What is the width of the line? (b) What is the effective permittivity of the line?

Solution:

(a) The high-impedance (or narrow-strip) formula (Equation (4.25)) is to be used for $Z_0 > (44 - \varepsilon_r)$ $[= (44 - 5.6) = 38.4]$ Ω.
With $\varepsilon_r = 5.6$ and $Z_0 = 75$ Ω, Equation (4.26) yields $H' = 2.445$. From Equation (4.25), $w/h = 0.704$, thus $w = w/h \times h = 0.704 \times 500$ μm $= 352$ μm.

(b) The effective permittivity formula is Equation (4.27), and so $\varepsilon_e = 3.82$.

4.6 Summary

This chapter considered micostrip, the most important planar transmission line. At frequencies below 1 GHz, economics require that standard FR4 circuit boards be used. The weave of a conventional FR4 circuit board can be a significant fraction of critical transmission line dimensions and so affect electrical performance. Nonwoven substrates and sometimes hard substrates such as alumina ceramic, sapphire, or silicon crystal wafers are often required. These provide higher-dimensional tolerance than can be achieved using conventional woven FR4 substrates. In general, once a design has been optimized in fabrication so that the desired electrical characteristics are obtained, microwave circuits using planar transmission lines can be cheaply and repeatably manufactured in volume.

4.7 References

[1] E. Hammerstad and O. Jensen, "Accurate models for microstrip computer-aided design," in *1980 IEEE MTT-S Int. Microwave Symp. Digest*, May 1980, pp. 407–409.

[2] E. Hammerstad and F. Bekkadal, "A microstrip handbook, ELAB Report STF44 A74169," University of Trondheim, Norway, Tech. Rep., Feb. 1975.

[3] E. O. Hammerstad, "Equations for microstrip circuit design," in *5th European Microwave Conf.*, Sep. 1975, pp. 268–272.

[4] T. Edwards and M. Steer, *Foundations for Microstrip Circuit Design*. John Wiley & Sons, 2016.

4.8 Exercises

1. A microstrip line on 250 μm thick GaAs has a minimum and maximum strip widths of 50 μm and 250 μm respectively. What is the range of characteristic impedances that can be used in design?

2. A microstrip line with a substrate having a relative permittivity of 10 has an effective permittivity of 8. What is the wavelength of a 10 GHz signal propagating on the microstrip?

3. A microstrip line has a width of 500 μm and a substrate that is 635 μm thick with a relative permittivity of 20. What is the effective permittivity of the line?

4. The strip of a microstrip has a width of 250 μm and is fabricated on a lossless substrate that is 500 μm thick and has a relative permittivity of 2.3. [Parallels Example 4.2]
   (a) What is the effective relative permittivity of the line?
   (b) What is the characteristic impedance of the line?
   (c) What is the propagation constant at 3 GHz ignoring any losses?
   (d) If the strip has a resistance of 0.5 Ω/cm and the ground plane resistance can be ignored, what is the attenuation constant of the line at 3 GHz?

5. A microstrip line on a 250 μm-thick silicon substrate has a width of 200 μm. Use Table 4-2.
   (a) What is line's effective permittivity.
   (b) What is its characteristic impedance?

6. A 600 μm-wide microstrip line on a 500 μm-thick alumina substrate. Use Table 4-2.
   (a) What is line's effective permittivity.
   (b) What is its characteristic impedance?

7. A microstrip line on a 1 mm-thick FR4 substrate has a width of 0.497 mm. Use Table 4-2.
   (a) What is line's effective permittivity.
   (b) What is its characteristic impedance?

8. Consider a microstrip line on a substrate with a relative permittivity of 12 and thickness of 1 mm.
   (a) What is the minimum effective permittivity of the microstrip line if there is no limit on the minimum or maximum width of the strip?
   (b) What is the maximum effective permittivity of the microstrip line if there is no limit on the minimum or maximum width of the strip?

9. A microstrip line has a width of 1 mm and a substrate that is 1 mm thick with a relative permittivity of 20. What is the geometric filling factor of the line?

10. The substrate of a microstrip line has a relative permittivity of 16 but the calculated effective permittivity is 12. What is the filling factor?

11. A microstrip line has a strip width of 250 μm and a substrate with a relative permittivity of 10 and a thickness of 125 μm. What is the filling factor?

12. A microstrip line has a strip width of 250 μm and a substrate with a relative permittivity of 4 and thickness of 250 μm. Determine the line's filling factor and thus its effective relative permittivity.

13. A microstrip line has a strip with a width of 100 μm and the substrate which is 250 μm thick and a relative permittivity of 8.
    (a) What is the filling factor, q, of the line?
    (b) What is the line's effective relative permittivity?
    (c) What is the characteristic impedance of the line?

14. An inhomogeneous transmission line is fabricated using a medium with a relative permittivity of 10 and has an effective permittivity of 7. What is the fill factor q?

15. A microstrip technology uses a substrate with a relative permittivity of 10 and thickness of 400 μm. The minimum strip width is 20 μm. What is the highest characteristic impedance that can be achieved?

16. A microstrip transmission line has a characteristic impedance of 75 Ω, a strip resistance of 5 Ω/m, and a ground plane resistance of 5 Ω/m. The dielectric of the line is lossless.
    (a) What is the total resistance of the line in Ω/m?
    (b) What is the attenuation constant in Np/m?
    (c) What is the attenuation constant in dB/cm?

17. A microstrip line has a characteristic impedance of 50 Ω, a strip resistance of 10 Ω/m, and a ground plane resistance of 3 Ω/m.
    (a) What is the total resistance of the line in Ω/m?
    (b) What is the attenuation constant in Np/m?
    (c) What is the attenuation constant in dB/cm?

18. A microstrip line has 10 μm-thick gold metallization for both the strip and ground plane. The strip has a width of 125 μm and the substrate is 125 μm thick.
    (a) What is the low frequency resistance (in Ω/m) of the strip?
    (b) What is the low frequency resistance of the ground plane?
    (c) What is the total low frequency resistance of the microstrip line?

19. A 50 Ω microstrip line has 10 μm-thick gold metallization for both the strip and ground plane. The strip has a width of 250 μm and the lossless substrate is 250 μm thick.
    (a) What is the low frequency resistance (in Ω/m) of the strip?
    (b) What is the low frequency resistance of the ground plane?
    (c) What is the total low frequency resistance of the microstrip line?
    (d) What is the attenuation in dB/m of the line at low frequencies?

20. A 50 Ω microstrip line with a lossless substrate has a 0.5 mm-wide strip with a sheet resistance of 1.5 mΩ and the ground plane resistance can be ignored. What is the attenuation constant at 1 GHz? [Parallels Example 4.0]

21. A microstrip line operating at 10 GHz has a substrate with a relative permittivity of 10 and a loss tangent of 0.005. It has a characteristic impedance of 50 Ω and an effective permittivity of 7.
    (a) What is the conductance of the line in S/m?
    (b) What is the attenuation constant in Np/m?
    (c) What is the attenuation constant in dB/cm?

22. A microstrip line has the per unit length parameters L = 2 nH/m and C = 1 pF/m, also at 10 GHz the substrate has a conductance G of 0.001 S/m. The substrate loss is solely due to dielectric relaxation loss and there is no substrate conductive loss. The resistances of the ground and strip are zero.
    (a) What is G at 1 GHz?
    (b) What is the magnitude of the characteristic impedance at 1 GHz?
    (c) What is the dielectric attenuation constant of the line at 1 GHz in dB/m?

23. A microstrip line has the per unit length parameters L = 1 nH/m and C = 1 pF/m, also at 1 GHz the substrate has a conductance G of 0.001 S/m. The substrate loss is solely due to dielectric relaxation loss and there is no substrate conductive loss. The resistance of the strip is 0.5 Ω/m and the resistance of the ground plane is 0.1 Ω/m.
    (a) What is the per unit length resistance of the microstrip line at 1 GHz?
    (b) What is the magnitude of the characteristic impedance at 1 GHz?
    (c) What is the conductive attenuation constant in Np/m?
    (d) What is the dielectric attenuation constant of the line at 1 GHz in dB/m?

24. A microstrip line operating at 2 GHz has perfect metallization for both the strip and ground plane. The strip has a width of 250 μm and the substrate is 250 μm thick with a relative permittivity of 10 and a loss tangent of 0.001.
    (a) What is the filling factor, q, of the line?
    (b) What is the line's effective relative permittivity?
    (c) What is the line's attenuation in Np/m?
    (d) What is the line's attenuation in dB/m?

25. A 50 Ω microstrip line operating at 1 GHz has perfect metallization for both the strip and ground plane. The substrate has a relative permittivity of 10 and a loss tangent of 0.001. Without the dielectric the line has a capacitance of 100 pF/m.
    (a) What is the line conductance in S/m?
    (b) What is the line's attenuation in Np/m?
    (c) What is the line's attenuation in dB/m?

26. Design a microstrip line having a 50 Ω characteristic impedance. The substrate has a permittivity of 2.3 and is 250 μm thick. The operating frequency is 18 GHz. You need to determine the width of the microstrip line.

27. Design a microstrip line to have a characteristic impedance of 65 Ω at 5 GHz. The substrate is 635 μm thick with a relative permittivity of 9.8. Ignore the thickness of the strip. [Parallels Example 4.3]
    (a) What is the width of the line?
    (b) What is the effective permittivity of the line?

28. Design a microstrip line to have a characteristic impedance of 20 Ω. The microstrip is to be constructed on a substrate that is 1 mm thick with a relative permittivity of 12. [Parallels Example 4.3]
    (a) What is the width of the line? Ignore the thickness of the strip and frequency-dependent effects.
    (b) What is the effective permittivity of the line?

4.8.1 Exercises by Section

†challenging

§4.4 1, 2, 3†, 4†, 5, 6, 7, 8, 9, 10, 11, 12, 13, 14, 15, 16, 17, 18, 19, 20, 21, 22†, 23†, 24, 25

§4.5 26†, 27†, 28†

4.8.2 Answers to Selected Exercises

3 12.75

4(c) $\jmath 84.1\ \text{m}^{-1}$

16(c) 0.579 dB/m

17(a) 13 Ω/m

22(b) 44.72 Ω

28(b) 9.17

CHAPTER 5

Extraordinary Transmission Line Effects

5.1 Introduction

Previous chapters discussed the low-frequency operation of transmission lines. This chapter describes the origins of frequency-dependent behavior. Figure 5-1 shows the typical frequency dependence of a line's $RLGC$ parameters, and, except with semiconductor substrates, G is usually negligible. This is called dispersion and a typical example is the spreading out of a pulse, see Figure 5-2.

The major limitation on the dimensions and maximum operating frequency of a transmission line is determined by the origination of higher-order modes (i.e., orientations of the fields). Different modes on a transmission line travel at different velocities. Thus the problem is that if a signal is split between two modes, then information sent from one end of the line will reach the other end in two packets arriving at different times. Multimoding must always be avoided.

5.2 Frequency-Dependent Characteristics

In this section the origins of the frequency-dependent behavior of a microstrip line are examined. The most important frequency-dependent

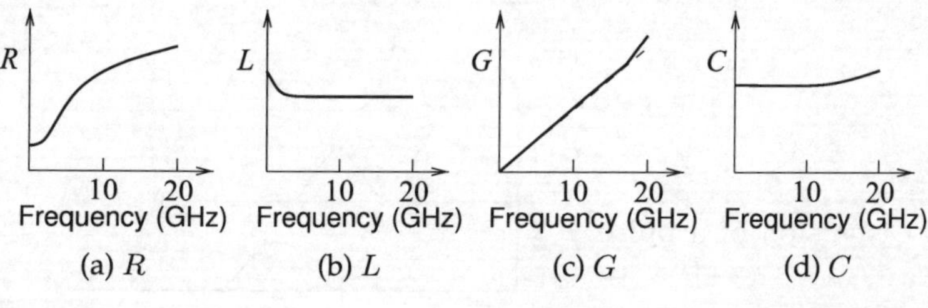

(a) R (b) L (c) G (d) C

Figure 5-1: Frequency dependence of transmission line parameters.

effects are

- changes of material properties (permittivity, permeability, and conductivity) with frequency (Section 5.2.1),
- current bunching (discussed in Section 5.2.3),
- skin effect (Sections 5.2.4 and 5.2.5),
- internal conductor inductance variation (Section 5.2.4),
- dielectric dispersion (Section 5.2.6), and
- multimoding (Section 5.3).

5.2.1 Material Dependency

Changes of permittivity, permeability, and conductivity with frequency are properties of the materials used. Fortunately the characteristics microwave materials are almost independent of frequency, at least up to 300 GHz.

5.2.2 Frequency-Dependent Charge Distribution

Skin effect, current bunching, and internal conductor inductance are all due to the necessary delay in transferring EM information from one location to another. This information cannot travel faster than the speed of light in the medium. In a dielectric material the speed of an EM wave will be slower than that in free space by a factor of $\sqrt{\varepsilon_r}$, e.g. $c/3.2$ for $\varepsilon_r = 10$. The velocity in a conductor is extremely low, around $c/1000$, because of high conductivity. In brief, current bunching is due to changes related to the finite velocity of information transfer through the dielectric, and skin effect is due to the very slow speed of information transfer inside a conductor. As frequency increases, only limited information to rearrange charges can be sent before the polarity of the signal reverses and information is 'sent' to reverse the changes. The skin and charge-bunching effects on a microstrip line are illustrated in Figure 5-3.

5.2.3 Current Bunching

Consider the microstrip charge distribution shown in Figure 5-3. The thickness of the microstrip is often a significant fraction of its width, although this is exaggerated here.

The charge distribution shown in Figure 5-3(a) applies when there is a positive DC voltage on the strip and the positive charges on the top conductor arranged with a fairly uniform distribution. The individual positive charges tend to repel each other, but this has little effect on the

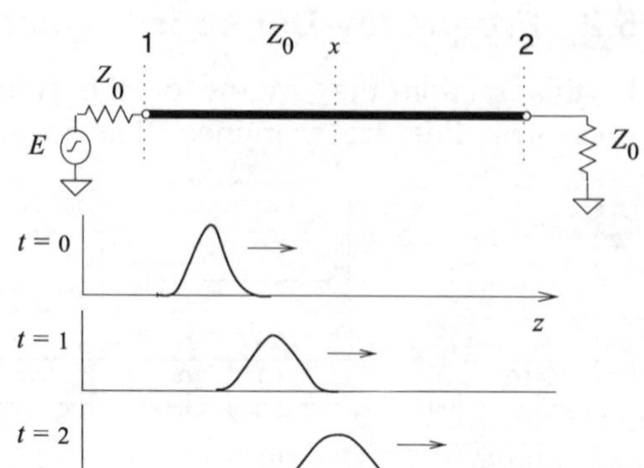

Figure 5-2: Dispersion of a pulse along a line.

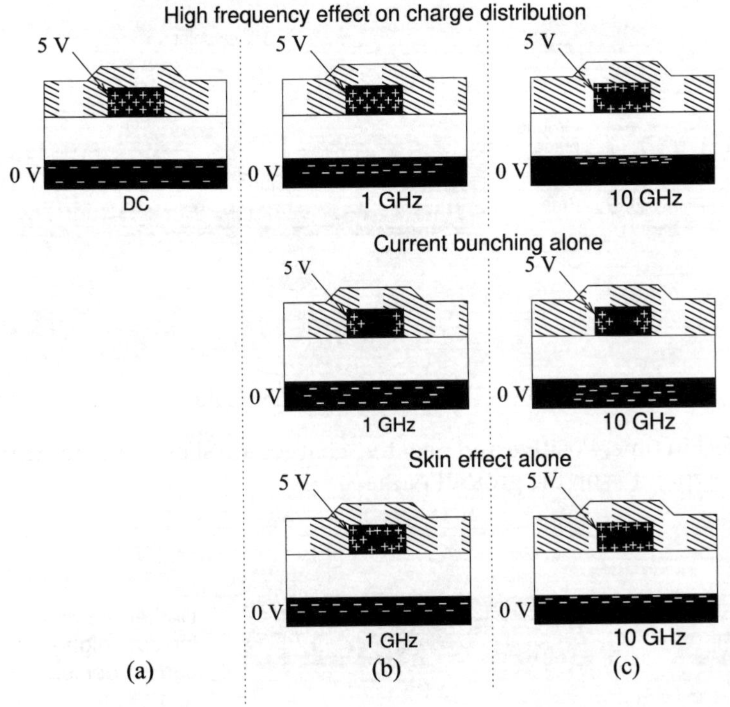

Figure 5-3: Cross-sectional view of the charge distribution on an interconnect at different frequencies. The + and − indicate charge concentrations of different polarity and corresponding current densities. There is no current bunching or skin effect at DC.

charge distribution at low frequencies for practical conductors with finite conductivity. On the ground plane there are balancing negative charges which are uniformly distributed across the whole of the ground plane. The charge distribution at DC, indicates that current would flow uniformly throughout the strip and the return current in the ground plane would be distributed over the whole of the ground plane.

The charge distribution becomes less uniform as frequency increases and eventually the signal changes so quickly that information to rearrange charges on the ground plane is soon (half a period latter) countered by reverse instructions. Thus the charge distribution depends on how fast the signaling changes. One way of looking at this effect is to view the charges on the strip of the microstrip line at one time. This is shown in Figure 5-4 for a DC signal on the line and for a high-frequency signal.

The electric field lines, which must originate and terminate on charges, will concentrate in the substrate more closely under the strip as frequency increases. The two major effects are that the effective permittivity of the microstrip line increases with frequency, and resistive loss increases as the current density in the ground, which corresponds to the net charge density, increases. Thus the line resistance and capacitance increase with frequency, see Figures 5-1(a and d). For the majority of substrates G is due to dielectric relaxation and so increases linearly with frequency with a superlinear increase at very high frequencies when the electric field is more concentrated in the dielectric, see Figure 5-1(c).

In the frequency domain the current bunching effects are seen in the higher-frequency views shown in Figures 5-3(b and c). (The concentration of charges near the metal surface is a separate effect known as the skin effect.)

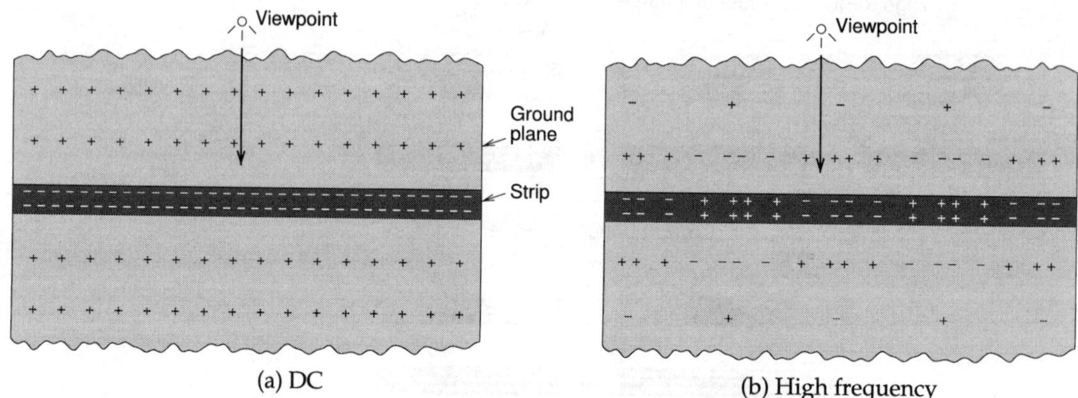

Figure 5-4: Current bunching effect in time. Positive and negative charges are shown on the strip and on the ground plane. The viewpoint is on the ground plane.

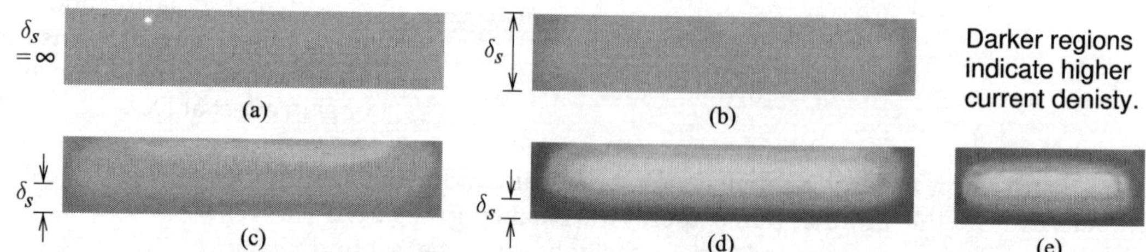

Figure 5-5: Cross-sections of the strip of a microstrip line showing the impact of skin effect and current bunching on current density: (a) at dc (uniform current density); (b) the strip thickness, t, is equal to the skin depth, δ_s (i.e. at a low microwave frequency); (c) $t = 3\delta_s$; (d) $t = 5\delta_s$ (i.e. at a high microwave frequency); and (e) $t = 5\delta_s$ for a narrow strip. The plots are the result of 3D simulations of a microstrip line using internal conductor gridding.

5.2.4 Skin Effect and Internal Conductor Inductance

At low frequencies, currents are distributed uniformly throughout a conductor. Thus there are magnetic fields inside the conductor and hence magnetic energy storage. As a result, there is additional inductance. Transferring charge to the interior of conductors is particularly slow, and as the frequency of the signal increases, charges are confined closer to the surface of the metal, see Figure 5-5. This phenomenon is known as the skin effect. With lower internal currents, the internal conductor inductance reduces and the total inductance of the line drops, see Figure 5-1(b). Only above a few gigahertz or so can the line inductance be approximated as being constant for a typical microstrip line.

The skin depth, δ_s, is the distance at which the electric field, or the charge density, reduces to $1/\mathrm{e}$ of its value at the surface. The skin depth is

$$\delta_s = 1/\sqrt{\pi f \mu_0 \sigma_2}. \tag{5.1}$$

Here f is frequency and σ_2 is the conductivity of the conductor. The (real part of the) permittivity and permeability of metals typically used for interconnects (e.g., gold, silver, copper, and aluminum) are that of free space as there is no mechanism to store additional electric or magnetic energy.

5.2.5 Skin Effect and Line Resistance

The skin effect is illustrated in Figure 5-3(b) at 1 GHz. The situation is more extreme as the frequency continues to increase. There are several important consequences of this. On the top conductor, as frequency increases, current flow is mostly concentrated near the surface of the conductors and the effective cross-sectional area of the conductor, as far as the current is concerned, is less. Thus the resistance of the top conductor increases. A more dramatic situation exists for the charge in the ground plane which becomes more concentrated under the strip. In addition to this, charges and current are confined to the skin of the ground conductor so that the frequency-dependent relative change of ground plane resistance with increasing frequency is greater than that of the strip.

The skin effect and current bunching result in frequency dependence of the line resistance, R, with

$$R(f) = \begin{cases} R(0) & f \text{ such that } t \leq 3\delta_s \\ R(0) + R_{\text{skin}}(f) & f \text{ such that } t > 3\delta_s, \end{cases} \tag{5.2}$$

where $R(0) = R_{\text{strip}}(0) + R_{\text{ground}}(0)$ is the resistance of the line at low frequencies. $R(f)$ describes the frequency-dependent line resistance that is due to both the skin effect and current bunching. Approximately,

$$R_{\text{skin}}(f) = R(0)k\sqrt{f}. \tag{5.3}$$

where k is a geometry-dependent constant. Note that it is pointless to make a strip or of ground thicker than three times the skin depth.

EXAMPLE 5.1 Skin Depth

Determine the skin depth for copper (Cu), silver (Ag), aluminum (Al), gold (Au), and titanium (Ti) at 100 MHz, 1 GHz, 10 GHz, and 100 GHz.

Solution:

The skin depth is calculated using Equation (5.1).

Metal	Resistivity (nΩ · m)	Conductivity (MS/m)	Skin depth, δ_s (μm) 100 MHz	1 GHz	10 GHz	100 GHz
Copper (Cu)	16.78	59.60	6.52	2.06	0.652	0.206
Silver (Ag)	15.87	63.01	6.34	2.01	0.634	0.201
Aluminum (Al)	26.50	37.74	8.19	2.59	0.819	0.259
Gold (Ag)	22.14	45.17	7.489	2.37	0.749	0.237
Titanium (Ti)	4200	0.2381	103.1	32.6	10.3	3.26

5.2.6 Dielectric Dispersion

Dispersion is principally the result of the propagation velocity of a sinusoidal signal being dependent on frequency since the effective permittivity is frequency-dependent. The electric field lines shift as a result with more of the electric energy being in the dielectric as frequency increases. At high frequencies the rearrangement results in the capacitance of the line increases—typically less than 10% from DC to 100 GHz.

The limits of $\varepsilon_e(f)$ are readily established; at the low-frequency extreme it reduces to the static TEM value ε_e (or $\varepsilon_e(0)$), while as frequency is increased indefinitely, $\varepsilon_e(f)$ approaches the substrate permittivity itself, ε_r. That is,

$$\varepsilon_e(f) \rightarrow \begin{cases} \varepsilon_e(0) & \text{as } f \rightarrow 0 \\ \varepsilon_r & \text{as } f \rightarrow \infty, \end{cases} \tag{5.4}$$

where $\varepsilon_e(0)$ is given by Equation 4.18. Detailed analysis yields

$$\varepsilon_e(f) = \varepsilon_r - \frac{\varepsilon_r - \varepsilon_e(0)}{1 + (f/f_a)^m}, \tag{5.5}$$

where m and f_a depend on w, h, and ε_r, see Section 24.3 of [1]. Very approximately $m \approx 2$ and $f_a \approx 50\ GHz$ for a 50 Ω microstrip line with/ a 500 μm substrate. The **frequency-dependent characteristic impedance** is

$$Z_0(f) = Z_0(0)\frac{\sqrt{\varepsilon_e(0)}}{\sqrt{\varepsilon_e(f)}}. \tag{5.6}$$

where $Z_0(0)$ is the low-frequency value given by Equation (4.15).

5.2.7 Summary

For microstrip, with increasing frequency the proportion of signal energy in the air region reduces and the proportion in the dielectric increases. The overall trend is for the fields to be more concentrated in the dielectric as frequency increases and thus effective relative permittivity increases.

5.3 Multimoding on Transmission Lines

Multimoding on a transmission line occurs when there are two or more EM field configurations that can support a propagating wave. Different field configurations travel at different speeds so that the information traveling in two modes will combine randomly and it will be impossible to discern the intended signal. It is critical that the dimensions of a transmission line be small enough to avoid multimoding. Larger dimensions however are more easily manufactured. With microstrip the quasi-TEM mode is supported from DC. Other modes can propagate above a cut-off frequency when transverse dimensions are larger than a quarter or half of a wavelength.

The important concept from EM throry is that electric and magnetic walls in the transverse direction (perpendicular to the direction of propagation) impose boundary conditions on the fields.

5.3.1 Multimoding and Electric and Magnetic Walls

The introduction of a magnetic wall greatly aids in the intuitive understanding of multimoding and in identifying problem situations. A magnetic wall can only be approximated since magnetic charges do not exist. An electric wall requires that the E field be perpendicular and the H field be parallel to the wall, see Figure 5-6. Similarly a magnetic wall requires that the H field be perpendicular and the E field be parallel to the wall, see Figure 5-7. A magnetic wall is approximated at the interface of the dielectric and air, see Figure 5-8(a), and near the edges of the strip, see Figure 5-8(b). The electric and magnetic walls establish boundary conditions for Maxwell's equations with the lowest frequency solutions shown in Table 5-9. With two electric or two magnetic walls, a TEM mode (having no field variations in the transverse plane) can be supported. Modes (other than TEM) with geometric variations in the transverse plane have a critical wavelength, λ_c, and hence a critical frequency, f_c, below which the mode cannot propagate. In Figure 5-9 the distance between the walls is d. For the case of two like walls (Figures 5-9(a and c)), $\lambda_c = 2h$, as one-half sinusoidal variation is required. With unlike walls (see Figure 5-9(b)), the varying modes are supported with just one-quarter sinusoidal variation, and so $\lambda_c = 4h$.

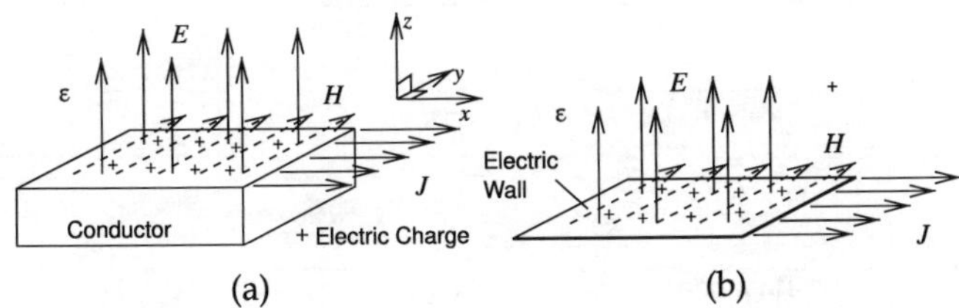

Figure 5-6: Properties of an electric wall: (a) the electrical field is perpendicular to a conductor and the magnetic field is parallel to it; and (b) a conductor can be approximated by an electric wall.

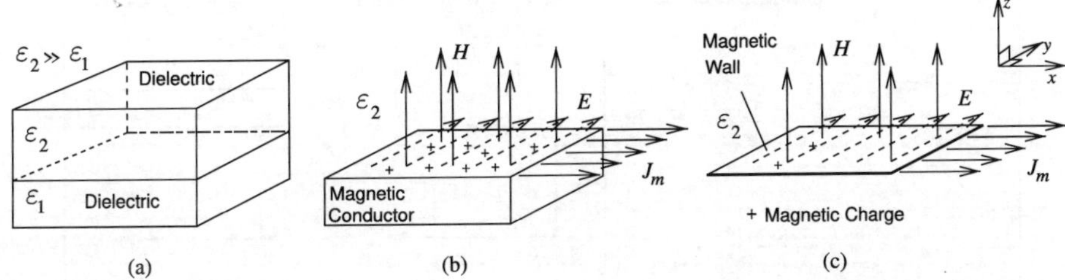

Figure 5-7: Properties of the EM field at a magnetic wall: (a) the interface of two dielectrics of contrasting permittivities approximates a magnetic wall; (b) the dielectric with lower permittivity can be approximated as a magnetic conductor; and (c) a magnetic wall.

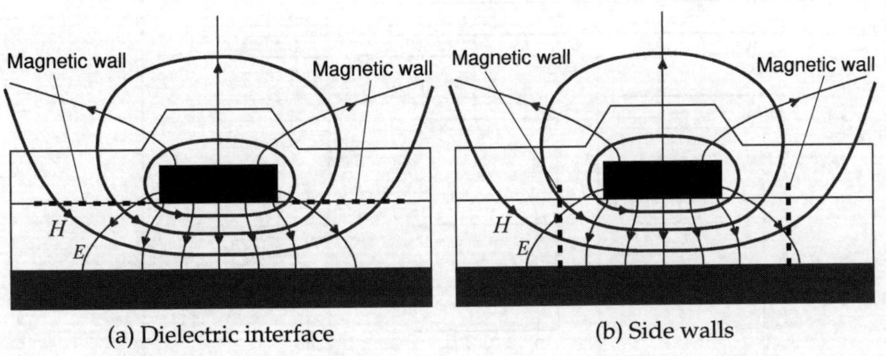

Figure 5-8: Approximate magnetic walls in microstrip where the H field is almost normal and the E field is almost parallel to the wall.

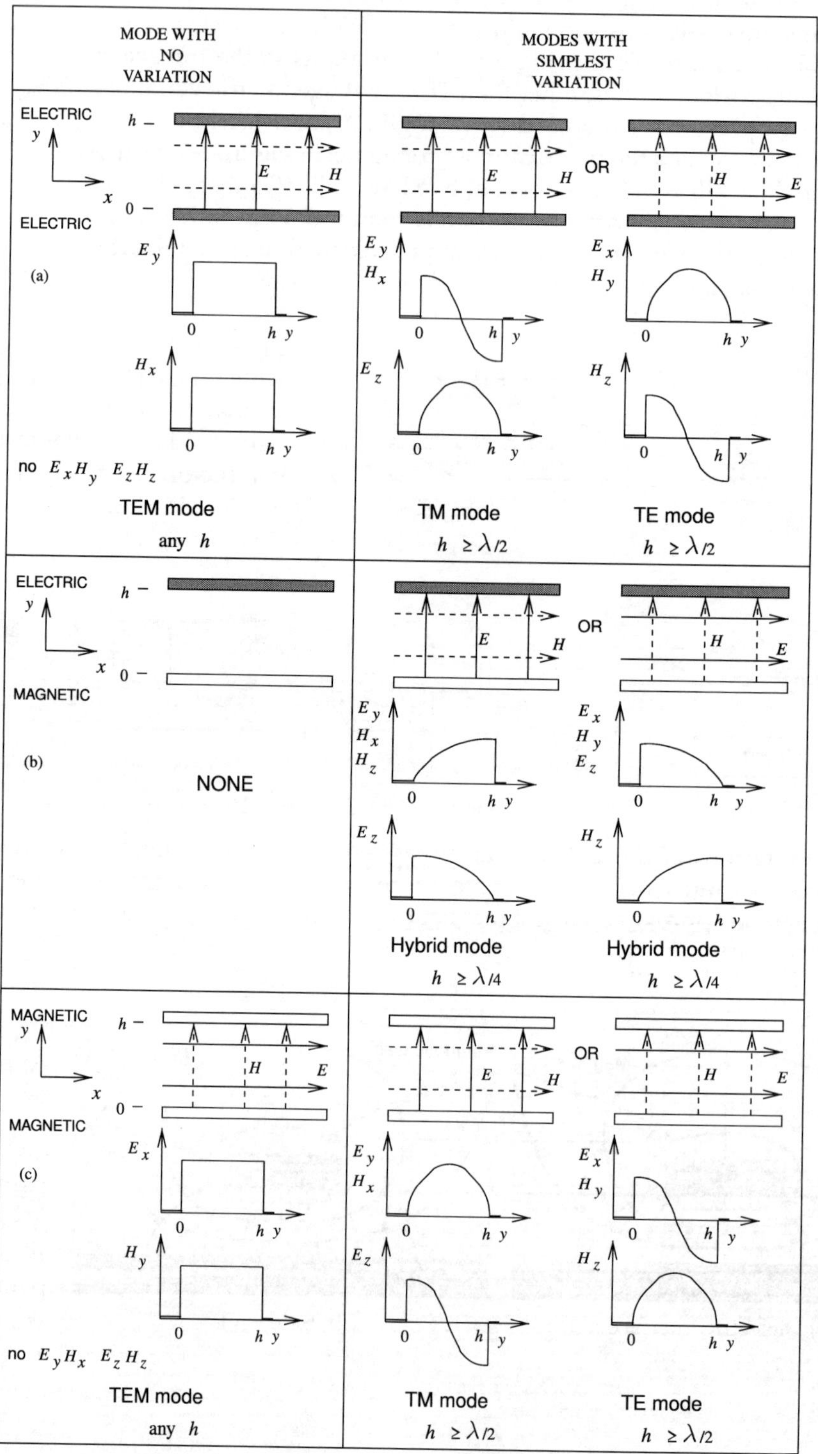

Figure 5-9: Lowest-order modes supported by combinations of electric and magnetic walls.

EXAMPLE 5.2 Modes and Electric and Magnetic Walls

A magnetic wall and an electric wall are 1 cm apart and are separated by a lossless material having $\varepsilon_r = 9$. What is the cut-off frequency of the lowest-order mode in this system?

Solution:

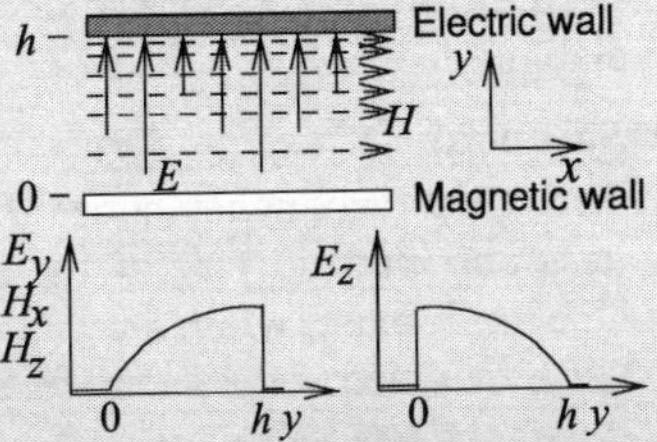

The EM field established by the electric and magnetic walls is described in Figure 5-9(b). There is no solution to the Maxwell's equations that has no variation of the EM fields since it is not possible to have a spatially uniform electric field which is perpendicular to an electric wall while also being perpendicular to a parallel magnetic wall. The other solutions of Maxwell's equations require that the fields vary spatially, i.e. curl. Without electric and magnetic walls the minimum distance over which the EM fields will fold back on to themselves is a wavelength. With parallel electric and magnetic wall separated by h the minimum distance for a solution of Maxwell's equations is a quarter wavelength, λ, of the walls as shown on the right in Figure 5-9(b). That is

$$h = \lambda/4 \quad \text{or} \quad \lambda = 4h = 4\text{ cm} = \lambda_0/(\sqrt{\varepsilon_t \mu_r}). \tag{5.7}$$

Since the relative permeability has not been specified we assume that $\mu_r = 1$ so

$$\lambda_0 = \lambda\sqrt{9} = 12\text{ cm} = c/f$$

The cut-off frequency is

$$\begin{aligned} f &= (2.998 \times 10^8\text{ m/s})/(0.12\text{ m}) \\ &= 2.498\text{ GHz} \end{aligned} \tag{5.8}$$

5.4 Microstrip Operating Frequency Limitations

The higher-order modes, i.e. other than the quasi-TEM mode, with the lowest cut-off or critical frequency, f_c are usually (a) the lowest-order TM mode and (b) the lowest-order transverse microstrip resonance mode.

5.4.1 *Microstrip Dielectric Mode (Slab Mode)*

A dielectric on a ground plane with an air region (of a wavelength or more above it) can support a TM mode, variously called a microstrip dielectric mode, **substrate mode, microstrip TM mode**, or slab mode. Referring to Figure 5-9(b), the cut-off or critical frequency, $f_{c,\text{SLAB}}$ above which the slab mode exists is when the distance between the electric and magnetic walls $h > \lambda/4$, where $\lambda = \lambda_0/\sqrt{\varepsilon_r}$ is the wavelength in the dielectric, so

$$f_{c,\text{SLAB}} = \frac{c}{4h\sqrt{\varepsilon_r}}. \tag{5.9}$$

Experience is that the the critical frequency is less than this, e.g. 20% less, which is consistent with the distance between the electric and magnetic walls being slightly larger so that the magnetic wall is slightly into the air region. The critical frequency is also affected by discontinuities. This stresses the importance of measurements of a design in development and the design engineer aware of effects such as multimoding that are not fully accounted for in modeling. No software can properly account for all effects in part because EM simulation tools rely on symmetry, smooth surface, and dielectrics and conductors with uniform properties.

EXAMPLE 5.3 Dielectric Mode

The strip of a microstrip has a width of 1 mm and the substrate is 2.5 mm thick with a relative permittivity of 9. What is the lowest frequency that slab mode exists?

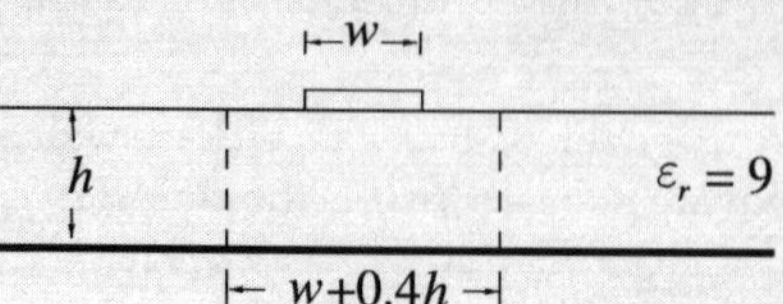

Solution:

The critical frequency comes from the thickness and permittivity of the substrate. The slab mode exists when a variation of the magnetic or electric field can be supported between the ground plane and the approximate magnetic wall at the substrate/air interface. This is when $h = \frac{1}{4}\lambda = \lambda_0/(4\sqrt{9}) = 2.5\text{ mm} \Rightarrow \lambda_0 = 3\text{ cm}$. Thus the critical slab mode critical frequency is

$$f_{c,\text{SLAB}} = 10\text{ GHz}.$$

5.4.2 Higher-Order Microstrip Mode

The next highest microstrip mode (or parallel-plate TE mode) above the quasi-TEM mode occurs when there is a half-sinusoidal variation of the electric field between the strip and the ground plane. This corresponds to Figure 5-9(a). However the first higher-order microstrip mode has been found be experiemntally found to exist at a lower frequency than derived using the parallel electrical wall model. This is because the microstrip fields do not follow the shortest distance between the strip and the ground plane. Thus the fields along the longer paths to the sides of the strip can vary at a lower frequency than on the direct path. With experimental support it has been established that the first higher-order microstrip mode can exist at frequencies above [2]

$$f_{\text{Higher-Microstrip}} = \frac{c}{4h\sqrt{\varepsilon_r - 1}}. \tag{5.10}$$

EXAMPLE 5.4 Higher-Order Microstrip Mode

The strip of a microstrip has a width of 1 mm and is fabricated on a lossless substrate that is 2.5 mm thick and has a relative permittivity of 9. At what frequency does the first higher microstrip mode first propagate?

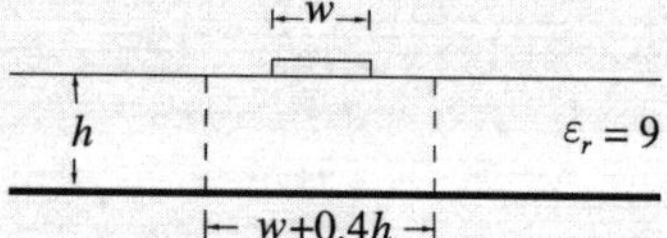

Shorter half-wavelength path
Higher frequency estimate

Longer half-wavelength path
Lower frequency estimate

Solution:

The higher-order microstrip mode occurs when a half-wavelength variation of the electric field between the strip and the ground plane can be supported. When $h = \lambda/2 = \lambda_0/(3 \cdot 2) = 2.5$ mm; that is, the mode will occur when $\lambda_0 = 15$ mm. So

$$f_{\text{Higher-Microstrip}} = 20\text{ GHz}.$$

A better estimate of the frequency where the higher-order microstrip mode becomes a problem is given by Equation (5.10):

$$f_{\text{Higher-Microstrip}} = c/(4h\sqrt{\varepsilon_r - 1}) = 10.6\text{ GHz}.$$

So two estimates have been calculated for the frequency at which the first higher-order microstrip mode can first exist. The first estimate is approximate and is based on a half-wavelength variation of the electric field confined to the direct path between the strip and the ground plane. The second estimate is more accurate as it considers that on the edge of the strip the fields follow a longer path to the ground plane. It is the half-wavelength variation on this longer path that determines if the higher-order microstrip mode will exist.

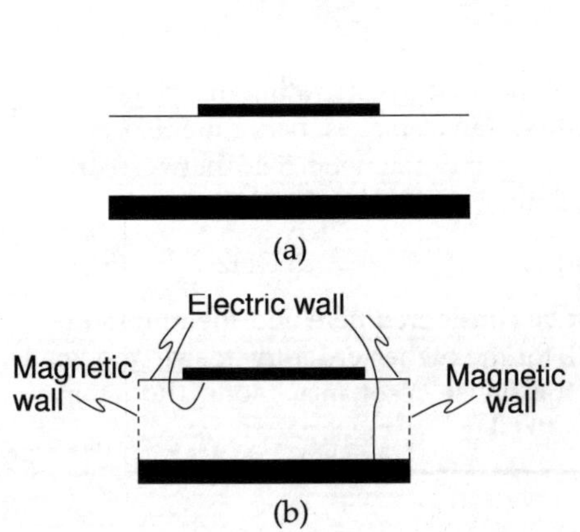

Figure 5-10: Approximation of a microstrip line as a waveguide: (a) cross section of microstrip; and (b) **microstrip waveguide model** having effective width $w + 0.4h$ with magnetic and electric walls.

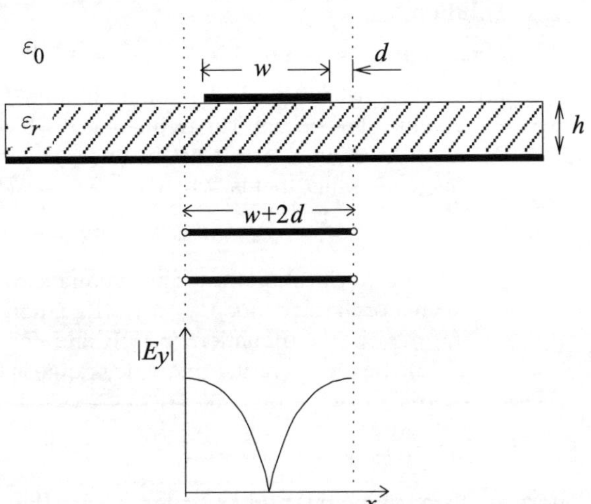

Figure 5-11: Transverse resonance: standing wave ($|E_y|$) and equivalent transmission line of length $w + 2d$, where $d = 0.2h$.

5.4.3 Transverse Microstrip Resonance

For a wide microstrip line, a transverse resonance mode can exist. This is the mode that occurs when EM energy bounces between the edges of the strip, see Figure 5-10. Here the microstrip is modelled by magnetic walls on the sides and an extended electrical wall on the top surface of the dielectric. The transverse resonance mode corresponds to the lowest-order H field variation between the magnetic walls. The equivalent circuit for the transverse-resonant mode is a resonant transmission line of length $w + 2d$, as shown in Figure 5-11, where $d = 0.2h$ accounts for the microstrip side fringing. A half-wavelength must be supported by the length $w + 2d$. Therefore the cutoff half-wavelength is

$$\frac{\lambda_c}{2} = w + 2d = w + 0.4h, \quad \text{that is,} \quad \frac{c}{2f_c\sqrt{\varepsilon_r}} = w + 0.4h. \tag{5.11}$$

Hence the critical frequency for transverse resonance is

$$f_{c,\text{TRAN}} = \frac{c}{\sqrt{\varepsilon_r}\,(2w + 0.8h)}. \tag{5.12}$$

EXAMPLE 5.5 Transverse Resonance Mode

The strip of a microstrip has a width of 1 mm and is fabricated on a lossless substrate that is 2.5 mm thick and has a relative permittivity of 9.

(a) At what frequency does the transverse resonance first occur?

(b) What is the operating frequency range of the microstrip line?

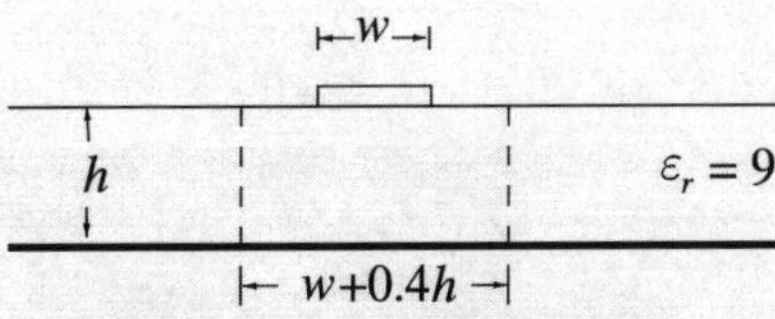

Solution:

$h = 2.5$ mm, $w = 1$ mm , $\lambda = \lambda_0/\sqrt{\varepsilon_r} = \lambda_0/3$

(a) The magnetic waveguide model of Figure 5-10 can be used in estimating the frequency at which this occurs. The frequency at which the first transverse resonance mode occurs is when there is a full half-wavelength variation of the magnetic field between the magnetic walls, that is, when $w + 0.4h = \lambda/2 = 2$ mm:

$$\frac{\lambda_0}{3 \cdot 2} = 2 \text{ mm} \Rightarrow \lambda_0 = 12 \text{ mm}, \qquad \text{and so} \qquad f_{c,\text{TRAN}} = 25 \text{ GHz}. \tag{5.13}$$

(b) All of the critical multimoding frequencies must be considered here and the minimum taken: for the slab mode, $f_{c,\text{SLAB}}$ (Equation (5.9)); for the higher-order microstrip mode, $f_{\text{High-Microstrip}}$ (Equation (5.10)); and for the transverse resonance mode (Equation (5.13)). So the operating frequency range is DC to 10 GHz.

5.4.4 Summary

There are three principal higher-order modes that need to be considered with microstrip transmission lines:

Mode	Critical frequency
Dielectric (or substrate) mode with discontinuity	Equation (5.9)
Higher order microstrip mode	Equation (5.10)
Transverse resonance mode	Equation (5.12)

The lowest frequency determines the upper frequency of transmission line operation with a single mode.

5.5 Summary

Microstrip can support multiple modes, but most are cut off by keeping transverse dimensions small with respect to a wavelength. When it is possible for a second (or higher-order) mode to exist, whether that mode is generated depends on the coupling mechanism between modes. This coupling mechanism is a discontinuity, which of course is common if circuit structures are to be incorporated. A microwave designer must always be aware of frequency dependence and multimoding and choose dimensions to avoid their occurrence except when the use is intentional and controlled.

5.6 References

[1] M. Steer, *Microwave and RF Design, Transmission Lines*, 3rd ed. North Carolina State University, 2019.

[2] T. Edwards and M. Steer, *Foundations for Microstrip Circuit Design*. John Wiley & Sons, 2016.

5.7 Exercises

1. What is the skin depth on a copper microstrip line at 10 GHz? Assume that the conductivity of the deposited copper forming the strip is half that of bulk single-crystal copper. Use the data in Example 5.1.

2. What is the skin depth on a silver microstrip line at 1 GHz? Assume that the conductivity of the fabricated silver conductor is 75% that of bulk single-crystal silver. Use the data in Example 5.1.

3. A magnetic wall and an electric wall are 2 cm apart and are separated by a lossless material having a relative permittivity of 10 and a relative permeability of 23. What is the cut-off frequency of the lowest-order mode in this system?

4. The strip of a microstrip has a width of 600 μm and is fabricated on a lossless substrate that is 1 mm thick and has a relative permittivity of 10.
   (a) Draw the microstrip waveguide model of the microstrip line. Put dimensions on your drawing.
   (b) Sketch the electric field distribution of the first transverse resonance mode and calculate the frequency at which the transverse resonance mode occurs.
   (c) Sketch the electric field distribution of the first higher-order microstrip mode and calculate the frequency at which it occurs.
   (d) Sketch the electric field distribution of the slab mode and calculate the frequency at which it occurs.

5. A microstrip line has a width of 352 μm and is constructed on a substrate that is 500 μm thick with a relative permittivity of 5.6.
   (a) Determine the frequency at which transverse resonance would first occur.
   (b) When the dielectric is slightly less than one-quarter wavelength in thickness the dielectric slab mode can be supported. Some of the fields will appear in the air region as well as in the dielectric, extending the effective thickness of the dielectric. Ignoring the fields in the air (use a one-quarter wavelength criterion), at what frequency will the dielectric slab mode first occur?

6. The strip of a microstrip has a width of 600 μm and uses a lossless substrate that is 635 μm thick and has a relative permittivity of 4.1.
   (a) At what frequency will the first transverse resonance occur?
   (b) At what frequency will the first higher-order microstrip mode occur?
   (c) At what frequency will the slab mode occur?
   (d) Identify the useful operating frequency range of the microstrip.

7. The strip of a microstrip has a width of 500 μm and is fabricated on a lossless substrate that is 635 μm thick and has a relative permittivity of 12. [Parallels Examples 5.3, 5.4, and 5.5]
   (a) At what frequency does the transverse resonance first occur?
   (b) At what frequency does the first higher-order microstrip mode first propagate?
   (c) At what frequency does the substrate (or slab) mode first occur?

8. The strip of a microstrip has a width of 250 μm and uses a lossless substrate that is 300 μm thick and has a relative permittivity of 15.
   (a) At what frequency does the transverse resonance first occur?
   (b) At what frequency does the first higher-order microstrip mode propagate?
   (c) At what frequency does the substrate (or slab) mode first occur?
   (d) What is the highest operating frequency of the microstrip?

9. A microstrip line has a strip width of 100 μm and is fabricated on a substrate that is 150 μm thick and has a relative permittivity of 9.
   (a) Draw the microstrip waveguide model and indicate and calculate the dimensions of the model.
   (b) Based only on the microstrip waveguide model, determine the frequency at which the first transverse resonance occurs?
   (c) Based on the microstrip waveguide model, determine the frequency at which the first higher-order microstrip mode occurs?
   (d) At what frequency will the slab mode occur? For this you cannot use the microstrip waveguide model.

10. A microstrip line has a strip width of 100 μm and is fabricated on a substrate that is 150 μm thick and has a relative permittivity of 9.
   (a) Define the properties of a magnetic wall.
   (b) Identify two situations where a magnetic wall can be used in the analysis of a microstrip line; that is, give two situations where a magnetic wall approximation can be used.
   (c) Draw the microstrip waveguide model and indicate and calculate the dimensions of the model.

11. The strip of a microstrip line has a width of 0.5 mm, and the microstrip substrate is 1 mm thick and has a relative permittivity of 9 and relative permeability of 1.
    (a) Draw the microstrip waveguide model and calculate the dimensions of the model. Clearly show the electric and magnetic walls in the model.
    (b) Use the microstrip waveguide model to calculate the cut off frequency of the transverse resonance mode?
    (c) A substrate mode can also be excited but the cut off frequency of this mode cannot be calculated using the microstrip waveguide model. Provide a brief description of the substrate mode and calculate the lowest frequency at which it can exist.
12. A microstrip technology uses a substrate with a relative permittivity of 10 and thickness of 400 μm. If the operating frequency is 10 GHz, what is the maximum width of the strip from higher-order mode considerations.
13. A microstrip line operating at 18 GHz has a 200 μm thick substrate with a relative permittivity of 20.
    (a) Determine the maximum width of the strip from higher-order mode considerations. Consider the transverse resonance mode, the higher-order microstrip mode, and the slab mode.
    (b) Thus determine the minimum achievable characteristic impedance.
14. A microstrip line has a strip width of 100 μm and is fabricated on a 150 μm-thick lossless substrate with a relative permittivity of 9.
    (a) Define the properties of a magnetic wall.
    (b) Identify two situations where a magnetic wall can be used in determining multimoding on a microstrip line. That is, give two locations where a magnetic wall approximation can be used.
15. Two magnetic walls are separated by 1 mm in a lossless material having a relative permittivity of 9 and a relative permeability of 1.
    (a) What is the wavelength of a 10 GHz signal in this material?
    (b) Now consider a field variation, i.e. a mode and not constant, established by the magnetic walls. Describe this lowest order field variation. That is how does the H field vary or how does the E field vary (one is sufficient)?
    (c) What is the lowest frequency at which a field variation can be supported by those walls in the specified medium?
16. A microstrip line has a strip width of 250 μm and a 300 μm thick substrate with a relative permittivity of 15. At what frequency can the substrate mode first occur?
17. A microstrip line has a strip width of 250 μm and a 300 μm thick substrate with a relative permittivity of 15. At what frequency can a higher-order microstrip mode first propagate?
18. A microstrip line has a strip width of 250 μm and a 300 μm thick substrate with a relative permittivity of 15. At what frequency can transverse resonance first occur?
19. The strip of a microstrip line has a width of 200 μm and the substrate is 400 μm thick and has a relative permittivity of 4.
    (a) Draw the effective waveguide model of a microstrip line with magnetic walls and an effective strip width, w_{eff}.
    (b) What is the effective relative permittivity of the microstrip waveguide model?
    (c) What is w_{eff}?
    (d) Can the lowest frequency at which the transverse resonance mode first occurs be determined from the microstrip waveguide model?

5.7.1 Exercises by Section

†challenging

§5.2 1, 2

§5.3 3

§5.4 $4^{\dagger}, 5^{\dagger}, 6^{\dagger}, 7^{\dagger}, 8^{\dagger}, 9^{\dagger}, 10^{\dagger}, 11^{\dagger}$, 12, 13, 14, 15, 16, 17, 18, 19

5.7.2 Answers to Selected Exercises

2 2.315 μm

3 247 MHz

4(c) 25 GHz

5(c) DC to 63.4 GHz

6(d) DC $\leq f \leq$ 48.6 GHz

8 104.6 GHz

11(b) 55.6 GHz

CHAPTER 6

Coupled Lines and Applications

6.1 Introduction

This chapter describes what happens when two transmission lines are so close together that the fields produced by one line interfere with the other line. Then a portion of the signal energy on one line is transferred to the other resulting in coupling. For microstrip lines the coupling reduces as the lines separate and usually the coupling is small enough to be ignored if the separation is at least three times the height of the strips. Transmission line coupling may be undesirable in many situations, but the phenomenon is exploited to realize many novel RF and microwave elements such as filters. A coupled pair of transmissions lines also enables the forward- and backward-traveling waves on a line to be separately measured.

The chapter begins with a discussion of the physics of coupling which leads to the preferred description of propagation on a pair of **parallel coupled lines** (**PCLs**) as supporting even and odd modes.

6.2 Physics of Coupling

If the fields of one transmission line intersect the region around another transmission line, then some of the energy propagating on the first line appears on the second. This coupling discriminates in terms of forward- and backward-traveling waves. In particular, consider the coupled microstrips shown in Figure 6-1. The direction of propagation comes from $\bar{\mathcal{E}} \times \bar{\mathcal{H}}$. Using the right-hand rule, the direction of propagation of the signal on the left-hand strip is out of the page. This is called the forward-traveling direction. Now consider the fields around the right-hand strip, the inactive line, and note the direction of the E field. The H field immediately to the left of the

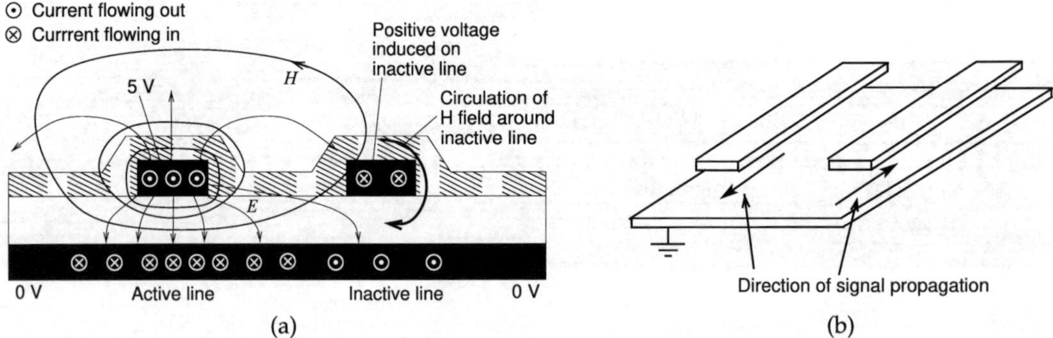

Figure 6-1: Edge-coupled microstrip lines with the left line driven with the forward-traveling field coming out of the page: (a) cross section of microstrip lines as found on a printed circuit board showing a conformal top passivation or solder resist layer; and (b) in perspective showing the direction of propagation of the driven line (left) and the direction of propagation of the main coupled signal on the victim line (right).

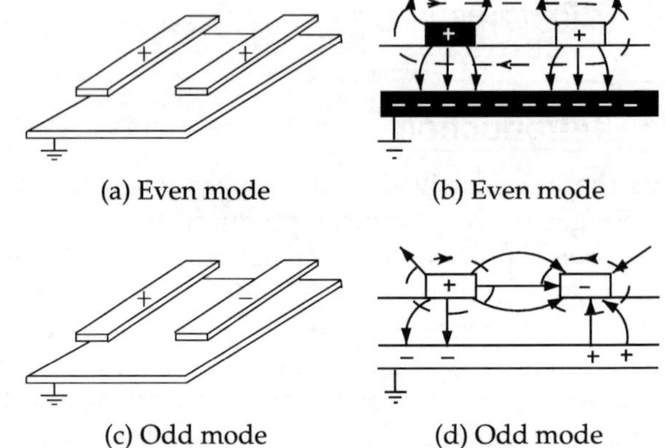

Figure 6-2: Modes on parallel-coupled microstrip lines. In (b) and (d) the electric fields are indicated as arrowed lines and the magnetic field lines are dashed.

right-hand strip is stronger than on the right of the strip. So effectively there is a clockwise circulation of the H field around the right hand strip. By applying $\bar{\mathcal{E}} \times \bar{\mathcal{H}}$ to the signal induced on the right-hand strip, it is seen that the induced signal travels into the page, or in the backward-traveling direction. The coupling with parallel-coupled microstrip lines is called backward-wave coupling. Both the even mode and the odd mode, and both have forward- and backward-traveling components (see Figure 6-2).

In the even mode (Figures 6-2(a and b)), the amplitude and polarity of the voltages on the two signal conductors are the same. In the odd mode (Figures 6-2(c and d)), the voltages on the two signal conductors are equal but have opposite polarity. Any EM field orientation on the coupled lines can be represented as a weighted addition of the even-mode and odd-mode field configurations. The actual fields will be a superposition of the even and odd modes. Also the voltages on the lines are composed of even and odd components.

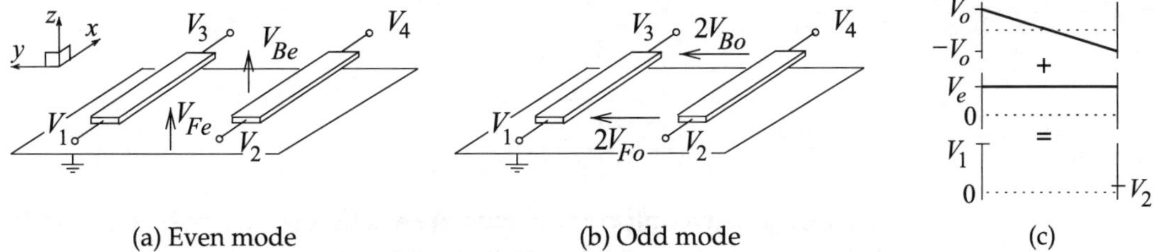

Figure 6-3: Definitions of total voltages and even- and odd-mode voltage phasors on a pair of coupled microstrip lines: (a) even-mode voltage definition; (b) odd-mode voltage definition; (c) depiction of how even- and odd-mode voltages combine to yield the total voltages on individual lines. F indicates front and B indicates back.

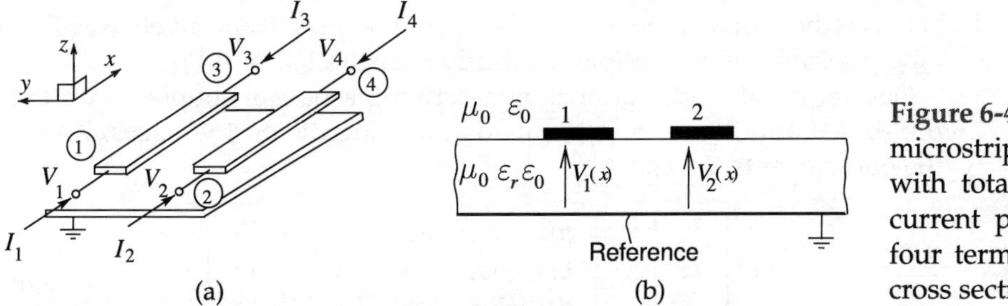

Figure 6-4: Coupled microstrip lines: (a) with total voltage and current phasors at the four terminals; and (b) cross section.

The characteristic impedances of the even and odd modes will differ because of the different field orientations. These are termed the even-mode and odd-mode characteristic impedances, denoted by Z_{0e} and Z_{0o}, respectively. Here the e subscript identifies the even mode and the o subscript identifies the odd mode. With coupled microstrip lines, the phase velocities of the two modes will differ since the two modes have different amounts of energy in the air and in the dielectric, resulting in different effective permittivities of the two modes.

Coupled lines are modeled by determining the propagation characteristics of the even and odd modes. Using the definitions shown in Figures 6-3 and 6-4, the total voltage and current phasors on the original structure are a superposition of the even- and odd-mode solutions:

$$\left.\begin{array}{ll} V_1 = V_{Fe} + V_{Fo} & I_1 = I_{Fe} + I_{Fo} \\ V_2 = V_{Fe} - V_{Fo} & I_2 = I_{Fe} - I_{Fo} \\ V_3 = V_{Be} + V_{Bo} & I_3 = I_{Be} + I_{Bo} \\ V_4 = V_{Be} - V_{Bo} & I_4 = I_{Be} - I_{Bo} \end{array}\right\}, \tag{6.1}$$

with the subscripts F and B indicating the front and back, respectively, of the even- and odd-mode components. Also

$$\left.\begin{array}{ll} V_{Fe} = (V_1 + V_2)/2 & I_{Fe} = (I_1 + I_2)/2 \\ V_{Fo} = (V_1 - V_2)/2 & I_{Fo} = (I_1 - I_2)/2 \\ V_{Be} = (V_3 + V_4)/2 & I_{Be} = (I_3 + I_4)/2 \\ V_{Bo} = (V_3 - V_4)/2 & I_{Bo} = (I_3 - I_4)/2 \end{array}\right\}. \tag{6.2}$$

The forward-traveling components can be written as

$$\left.\begin{array}{ll} V_1^+ = V_{Fe}^+ + V_{Fo}^+ & I_1^+ = I_{Fe}^+ + I_{Fo}^- \\ V_2^+ = V_{Fe}^+ - V_{Fo}^+ & I_2^+ = I_{Fe}^+ - I_{Fo}^- \\ V_3^+ = V_{Be}^+ + V_{Bo}^+ & I_3^+ = I_{Be}^+ + I_{Bo}^- \\ V_4^+ = V_{Be}^+ - V_{Bo}^+ & I_4^+ = I_{Be}^+ - I_{Bo}^- \end{array}\right\} . \tag{6.3}$$

The backward-traveling components are written similarly so that the total front and back even- and odd-mode voltages are

$$\left.\begin{array}{ll} V_{Fe} = V_{Fe}^+ + V_{Fe}^- & V_{Be} = V_{Be}^+ + V_{Be}^- \\ V_{Fo} = V_{Fo}^+ + V_{Fo}^- & V_{Bo} = V_{Bo}^+ + V_{Bo}^- \end{array}\right\} . \tag{6.4}$$

With an ideal voltmeter the V_1, V_2, V_3, and V_4 voltages would be measured from a point on the strip to a point on the ground plane immediately below. The voltages $2V_{Fo}$ and $2V_{Bo}$ are the voltages that would be measured between the strips at the front and back of the lines, respectively (see Figure 6-2). It would not be possible to directly measure V_{Fe} and V_{Be}.

One set of network parameters describing a pair of coupled lines is the port-based admittance matrix equation relating the port voltages, V_1–V_4, to the port currents, I_1–I_4:

$$\begin{bmatrix} I_1 \\ I_2 \\ I_3 \\ I_4 \end{bmatrix} = \begin{bmatrix} y_{11} & y_{12} & y_{13} & y_{14} \\ y_{21} & y_{22} & y_{23} & y_{24} \\ y_{31} & y_{32} & y_{33} & y_{34} \\ y_{41} & y_{42} & y_{43} & y_{44} \end{bmatrix} \begin{bmatrix} V_1 \\ V_2 \\ V_3 \\ V_4 \end{bmatrix} . \tag{6.5}$$

These are port-based y parameters as the currents and voltages are defined for ports. Since microstrip lines are fabricated (most of the time) on non-magnetic dielectrics, then they are reciprocal so that $y_{ij} = y_{ji}$. Coupling from one line to another is described by the terms y_{12} (=y_{21}) and y_{34} (=y_{43}).

6.2.1 Summary

The important concept introduced in this section is that fields and the propagating waves on a pair of parallel coupled lines can be described as a combination of odd and even modes each of which has forward- and backward-wave versions.

6.3 Low-Frequency Capacitance Model of Coupled Lines

The low-frequency model of a pair of lossless coupled lines comprises only capacitances. A pair of coupled lines, as shown in Figure 6-5(a), has four terminals. At very low frequencies V_1 and V_3 are identical as are voltages V_2 and V_4. So the low-frequency model of the pair of coupled lines has just two terminals in addition to ground, as shown in Figure 6-5(b).

The capacitances in Figure 6-5(b) are the shunt capacitance C_1 and C_2 and the mutual capacitance C_g. In the even mode, the voltages at terminals 1 and 2 are the same so that C_g vanishes, see Figure 6-5(c). In the odd mode, the voltage at terminal 2 is the negative of the voltage at terminal 1. The result is that there is a virtual ground between the terminals. Now a better circuit model is that shown in Figure 6-5(d). This is where the restriction that the lines are of equal width is used. This assumption places the virtual ground

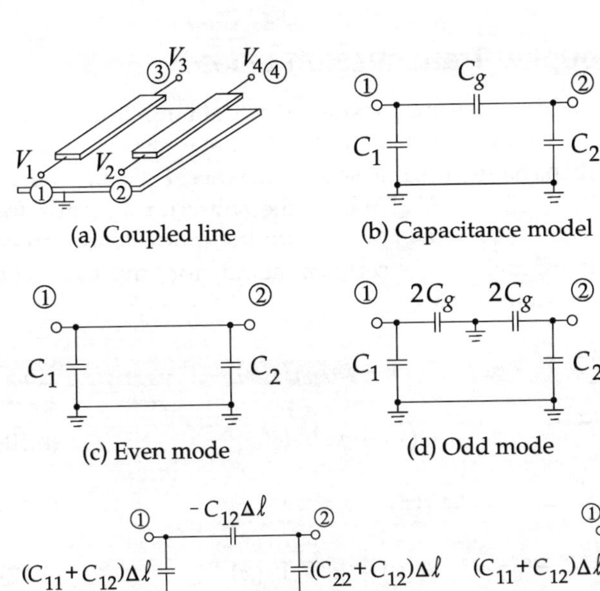

Figure 6-5: Very low frequency models of a pair of coupled lines.

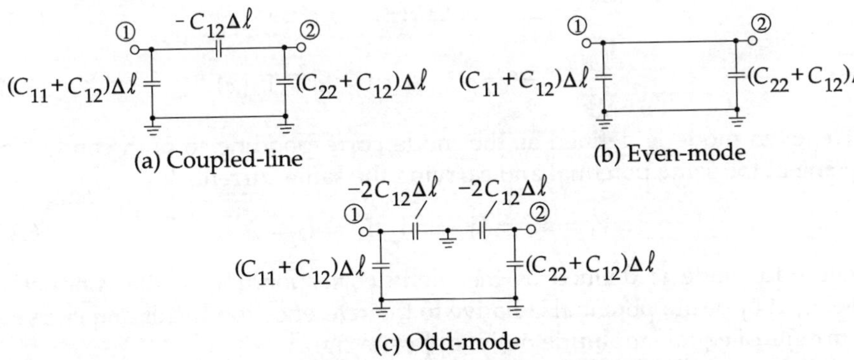

Figure 6-6: Low frequency capacitance models of a pair of coupled lines of length $\Delta\ell$. C_{12} is negative.

between equal-value capacitances. The symmetrical case is the one of most interest.

To proceed, the capacitance model must be put in the form of per unit length capacitances and put in terms of the elements of a capacitance matrix. The indefinite nodal admittance matrix of the low-frequency coupled-line model of Figure 6-5(b) is

$$\mathbf{Y} = \jmath\omega \begin{bmatrix} C_1 + C_g & -C_g \\ -C_g & C_2 - C_g \end{bmatrix} = \jmath\omega\mathbf{C}\Delta\ell, \tag{6.6}$$

where $\Delta\ell$ is the length of the coupled lines, and $\mathbf{C}$ is the per unit length capacitance matrix. Thus the low-frequency capacitance model of a pair of coupled lines of length $\Delta\ell$ and equal width is as shown in Figure 6-6(a). It is found in analysis that C_{12} is negative.

For symmetrical coupled lines (the strips having the same width) the per unit length even- and odd-mode capacitances, as defined in the definition of odd and even modes in Section 6.2, are

$$C_e = C_{11} + C_{12} \quad \text{and} \quad C_o = C_{11} - C_{12}. \tag{6.7}$$

That is, $$\mathbf{C} = \begin{bmatrix} C_{11} & C_{12} \\ C_{12} & C_{22} \end{bmatrix} = \begin{bmatrix} \frac{1}{2}(C_e + C_o) & \frac{1}{2}(C_e - C_o) \\ \frac{1}{2}(C_e - C_o) & \frac{1}{2}(C_e + C_o) \end{bmatrix}. \tag{6.8}$$

6.4 Symmetric Coupled Transmission Lines

In this section, even and odd modes are considered as defining independent transmission lines. The development is restricted to a symmetrical pair of coupled lines. Thus the strips have the same self-inductance, $L_s = L_{11} = L_{22}$, and self-capacitance, $C_s = C_{11} = C_{22}$, where the subscript s stands for "self." $L_m = L_{12} = L_{21}$ and $C_m = C_{12} = C_{21}$ are the mutual inductance and capacitance of the lines, and the subscript m stands for "mutual." The coupled transmission line equations are

$$\frac{dV_1(x)}{dx} = -\jmath\omega L_s I_1(x) - \jmath\omega L_m I_2(x) \tag{6.9}$$

$$\frac{dV_2(x)}{dx} = -\jmath\omega L_m I_1(x) - \jmath\omega L_s I_2(x) \tag{6.10}$$

$$\frac{dI_1(x)}{dx} = -\jmath\omega C_s V_1(x) - \jmath\omega C_m V_2(x) \tag{6.11}$$

$$\frac{dI_2(x)}{dx} = -\jmath\omega C_m V_1(x) - \jmath\omega C_s V_2(x). \tag{6.12}$$

The even mode is defined as the mode corresponding to both conductors being at the same potential and carrying the same currents:[1]

$$V_1 = V_2 = V_e \quad \text{and} \quad I_1 = I_2 = I_e. \tag{6.13}$$

The odd mode is defined as the mode corresponding to the conductors being at opposite potentials relative to the reference conductor and carrying currents of equal amplitude but of opposite sign:[2]

$$V_1 = -V_2 = V_o \quad \text{and} \quad I_1 = -I_2 = I_o. \tag{6.14}$$

The characteristics of the two possible modes of the coupled transmission lines are now described. For the even mode, from Equations (6.9) and (6.10),

$$\frac{d}{dx}\left[V_1(x) + V_2(x)\right] = -\jmath\omega\left[L_m + L_s\right]\left[I_1(x) + I_2(x)\right], \tag{6.15}$$

which becomes

$$\frac{dV_e(x)}{dx} = -\jmath\omega\left(L_s + L_m\right) I_e(x). \tag{6.16}$$

Similarly, using Equations (6.11) and (6.12),

$$\frac{d}{dx}\left[I_1(x) + I_2(x)\right] = -\jmath\omega\left(C_s + C_m\right)\left[V_1(x) + V_2(x)\right], \tag{6.17}$$

which in turn becomes

$$\frac{dI_e(x)}{dx} = -\jmath\omega\left(C_s + C_m\right) V_e(x). \tag{6.18}$$

Defining the even-mode inductance and capacitance, L_e and C_e, respectively, as

$$L_e = L_s + L_m = L_{11} + L_{12} \quad \text{and} \quad C_e = C_s + C_m = C_{11} + C_{12} \tag{6.19}$$

[1] Here $I_e = (I_1 + I_2)/2$ and $V_e = (V_1 + V_2)/2$.
[2] Here $I_o = (I_1 - I_2)/2$ and $V_o = (V_1 - V_2)/2$.

leads to the **even-mode telegrapher's equations:**

$$\frac{dV_e(x)}{dx} = -\jmath\omega\, L_e I_e(x) \quad (6.20) \quad \text{and} \quad \frac{dI_e(x)}{dx} = -\jmath\omega\, C_e V_e(x). \quad (6.21)$$

From these, the even-mode characteristic impedance can be found,

$$Z_{0e} = \sqrt{\frac{L_e}{C_e}} = \sqrt{\frac{L_s + L_m}{C_s + C_m}}, \tag{6.22}$$

and also the even-mode phase velocity,

$$v_{pe} = \frac{1}{\sqrt{L_e C_e}}. \tag{6.23}$$

The characteristics of the odd-mode operation of the coupled transmission line can be determined in a similar procedure to that used for the even mode. Using Equations (6.9)–(6.12), the **odd-mode telegrapher's equations** become

$$\frac{dV_o(x)}{dx} = -\jmath\omega\,(L_s - L_m)\, I_o(x) \quad (6.24) \quad \text{and} \quad \frac{dI_o(x)}{dx} = -\jmath\omega\,(C_s - C_m)\, V_o(x). \quad (6.25)$$

Defining L_o and C_o for the odd mode such that

$$L_o = L_s - L_m = L_{11} - L_{12} \quad \text{and} \quad C_o = C_s - C_m = C_{11} - C_{12}, \tag{6.26}$$

then the odd-mode characteristic impedance is

$$Z_{0o} = \sqrt{\frac{L_o}{C_o}} = \sqrt{\frac{L_s - L_m}{C_s - C_m}} \tag{6.27}$$

and the odd-mode phase velocity is

$$v_{po} = \frac{1}{\sqrt{L_o C_o}}. \tag{6.28}$$

Now for a sanity check. If the individual strips are widely separated, L_m and C_m will become very small and Z_{0e} and Z_{0o} will be almost equal. As the strips become closer, L_m and C_m will become larger and Z_{0e} and Z_{0o} will diverge. This is as expected.

6.5 Directional Coupler

Coupling can be exploited to realize a new type of element called a directional coupler. The schematic of a directional coupler is shown in Figure 6-7(a) and a microstrip realization is shown in Figure 6-7(b). A usable directional coupler has a coupled line length of at least one-quarter

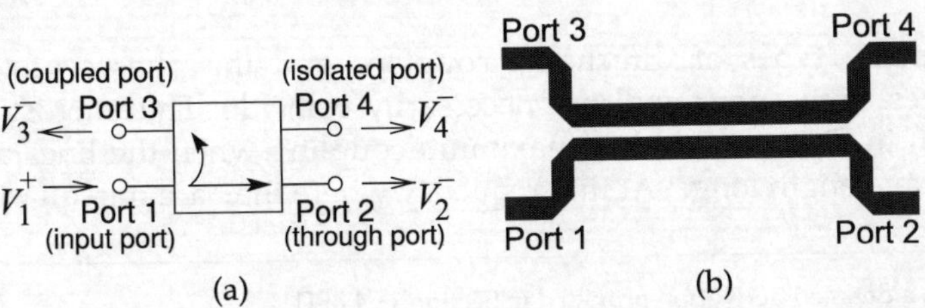

Figure 6-7: Directional couplers: (a) schematic; and (b) backward-coupled microstrip coupler. (Note that not all couplers are backward-wave couplers as shown in (a).)

Table 6-1: Ideal and typical parameters of a directional coupler.

Parameter	Ideal	Ideal (dB)	Typical
Coupling, C	-	-	3–40 dB
Transmission, T	$\lvert\sqrt{1-1/C^2}\rvert$	$20\log\lvert\sqrt{1-1/C^2}\rvert$	-0.5 dB
Directivity, D	∞	∞	40 dB
Isolation, I	∞	∞	40 dB

wavelength, with longer lengths of line resulting in broader bandwidth operation. Directional couplers are used to sample a traveling wave on one line and to induce a usually much smaller image of the wave on another line. That is, the forward- and backward-traveling waves are separated. Here a prescribed amount of the incident power is coupled out of the system. Thus, for example, a 20 dB microstrip coupler is a pair of coupled microstrip lines in which 1/100 of the power input is coupled from one microstrip line onto the another.

Referring to Figure 6-7, a coupler is specified in terms of the following parameters (always check the magnitude of the factors, as some papers and books on couplers use the inverse of the C used here):

- Coupling factor:
  $C = V_1^+/V_3^-$ = inverse of the voltage fraction "transferred" (coupled) across to the opposite arm ($C > 1$).
- Transmission factor (inverse of insertion loss):
  $T = V_2^-/V_1^+$ = transmission directly through the "primary" arm of the structure ($T < 1$).
- Directivity factor:
  $D = V_3^-/V_4^-$ = measure of the undesired coupling from Port 1 to Port 4 relative to the signal level at Port 3 ($D > 1$).
- Isolation factor:
  $I = V_1^+/V_4^-$ = isolation between Port 4 and Port 1 ($I > 1$).

It is usual to quote these quantities in decibels. For example, the coupling factor in decibels is $C|_{\text{dB}} = 20\log C$. So 20 dB coupling indicates that the coupling factor is 10. An ideal quarter-wave coupler has $D = \infty$ (i.e., infinite directivity) and

$$C = \frac{Z_{0e} + Z_{0o}}{Z_{0e} - Z_{0o}}. \tag{6.29}$$

In decibels the coupling is

$$C|_{\text{dB}} = 20\log\left(\frac{Z_{0e} + Z_{0o}}{Z_{0e} - Z_{0o}}\right). \tag{6.30}$$

Typical and ideal parameters of a directional coupler are given in Table 6-1. Since an ideal coupler does not dissipate power, the magnitude of the transmission coefficient is

$$|T| = |\sqrt{1 - 1/C^2}|. \tag{6.31}$$

There are many types of directional couplers, and the phases of the traveling waves at the ports will not necessarily coincide. The microstrip coupler shown in Figure 6-7(b) has maximum coupling when the lines are one-quarter wavelength long.[3] At the frequency where they are one-quarter

[3] This is shown in a detailed derivation provided in Section 11.4 of [1].

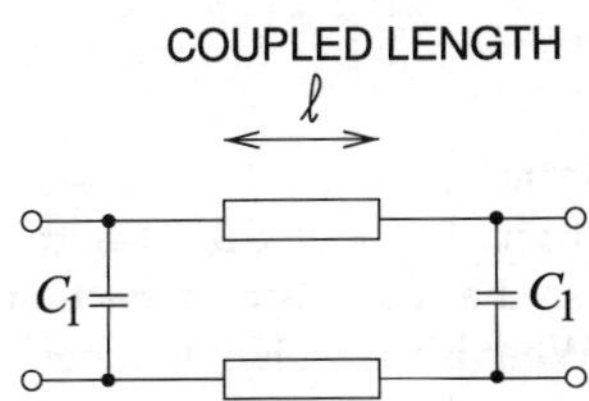

Figure 6-8: Parallel coupled lines with lumped capacitors reducing the length of the coupled lines.

wavelength long, the phase difference between traveling waves entering at Port 1 and leaving at Port 2 will be 90°.

EXAMPLE 6.1 Directional Coupler Isolation

A lossless directional coupler has coupling $C = 20$ dB, transmission factor 0.8, and directivity 20 dB. What is the isolation? Express your answer in decibels.

Solution:

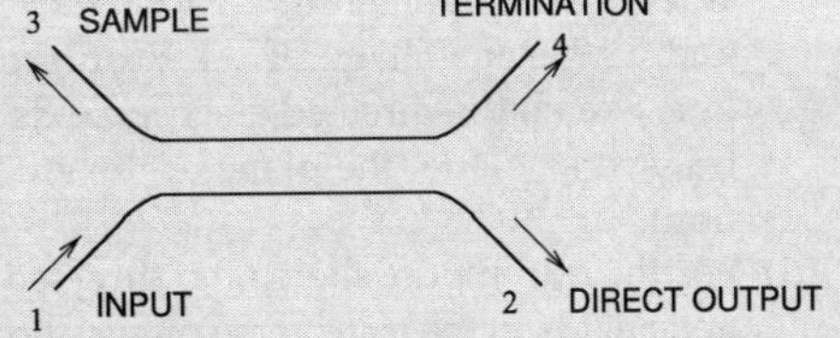

Coupling factor: $C = V_1^+/V_3^-$
Transmission factor: $T = V_2^-/V_1^+$
Directivity factor: $D = V_3^-/V_4^-$
Isolation factor: $I = V_1^+/V_4^-$

$$D = 20\text{ dB} = 10 \quad \text{and} \quad C = 20\text{ dB} = 10,$$

so the isolation is

$$I = \frac{V_1^+}{V_4^-} = \frac{V_3^-}{V_4^-} \cdot \frac{V_1^+}{V_3^-} = D \cdot C = 10 \cdot 10 = 100 = 40\text{ dB}.$$

6.5.1 Directional Coupler With Lumped Capacitors

Directional couplers using only coupled transmission lines can be large at low frequencies, as the minimum length is approximately one-quarter of a wavelength. This can be a problem at RF and low microwave frequencies, say, below 3 GHz. The length of the line can be reduced by incorporating lumped elements, as shown in Figure 6-8(a).

6.6 Summary

Coupling from one transmission line to a nearby neighbor may often be undesirable. However, the coupling can be controlled and coupled lines have become an important circuit component in distributed microwave circuits. The suite of microwave elements that exploit distributed effects available to a microwave designer is surprisingly large. Coupled lines comprise a large proportion of these elements. This chapter concludes with an example of the broadband response of a pair of coupled microstrip lines.

6.7 References

[1] T. Edwards and M. Steer, *Foundations for Microstrip Circuit Design*. John Wiley & Sons, 2016.

6.8 Exercises

1. Consider the cross section of a coupled transmission line, as shown in Figure 6-1, with even and odd modes both traveling out of the page.
   (a) For an even mode on the coupled line, consider a phasor voltage of 1 V on each of the lines above the ground plane at 0 V. Sketch the directed electric field in the transverse plane, i.e. show the direction of the electric field.
   (b) For the even mode, sketch the directed magnetic fields in the transverse plane (the plane of the cross section).
   (c) For an odd mode on the coupled line, consider a phasor voltage of +1 V on the left line and a phasor voltage of −1 V on the right line. Sketch the directed electric fields in the transverse plane (the plane of the cross section).
   (d) For the odd mode, sketch the directed magnetic fields in the transverse plane (the plane of the cross section).
2. An ideal directional coupler is lossless and there are no reflections at the ports. If the coupling factor is 10, what is the the magnitude of the transmission coefficient?
3. A directional coupler has the following characteristics: coupling factor $C = 20$, transmission factor 0.9, and directivity factor 25 dB. Also, the coupler is matched so that there is no reflection at any of the ports. What is the isolation in decibels?
4. A lossy 6 dB directional coupler is matched so that there is no reflection at any of the ports. The insertion loss (considering the through path) is 2 dB. If 1 mW is input to the directional coupler, what is the power in microwatts dissipated in the directional coupler? Ignore power leaving the isolated port.
5. A matched directional coupler has a coupling factor C of 20, transmission factor 0.9, and directivity of 25 dB. What is the power dissipated in the directional coupler if the input power to Port 1 is 1 W.
6. Consider a pair of parallel microstrip lines separated by a spacing, s, of 100 μm.
   (a) What happens to the coupling factor of the lines as s reduces?
   (b) What happens to the system impedance as s reduces and no other dimensions change?
   (c) In terms of wavelengths, what is the optimum length of the coupled lines for maximum coupling?

6.8.1 Exercises by Section

†challenging

§6.2 1† §6.5 2, 3, 4, 5, 6†

6.8.2 Answers to Selected Exercises

2 0.9950 3(b) 187 mW

CHAPTER 7

Microwave Network Analysis

7.1 Introduction

Analog circuits at frequencies up to a few tens of megahertz are characterized by admittances, impedances, voltages, and currents. Above these frequencies it is not possible to measure voltage, current, or impedance directly. It is better to use quantities such as voltage reflection and transmission coefficients that can be quite readily measured and are related to power flow. As well, in RF and microwave circuit design the power of signals and of noise is always of interest. Thus there is a predisposition to focus on measurement parameters that are related to the reflection and transmission of power.

Scattering parameters, S parameters, embody the effects of reflection and transmission. As will be seen, it is easy to convert these to more familiar network parameters such as admittance and impedance parameters. In this chapter S parameters will be defined and related to impedance and admittance parameters, then it will be demonstrated that the use of S parameters helps in the design and interpretation of RF circuits. S parameters have become the most important parameters for RF and microwave engineers and many design methodologies have been developed around them.

This chapter presents microwave circuit theory which is based on the representation of distributed structures such as transmission lines, and other structures that are too large to be considered to be dimensionless, by lumped element circuits.

7.2 Two-Port Networks

In microwave circuits it is generally difficult to do this. Recall that with transmission lines it is not possible to establish a common ground point. However, with transmission lines it was seen that for each signal current there is a signal return current. Thus at radio frequencies, and for circuits that are distributed, ports are used, as shown in Figure 7-1(a), which define the voltages and currents for a two-port network, or just two-port.[1] The network in Figure 7-1(a) has four terminals and two ports. A port voltage is defined as the voltage difference between a pair of terminals with one of the terminals in the pair becoming the reference terminal. The current entering the network at the top terminal of Port 1 is I_1 and there is an equal current leaving the reference terminal. This arrangement clearly makes sense when transmission lines are attached to Ports 1 and 2, as in Figure 7-1(b). With transmission lines at Ports 1 and 2 there will be traveling-wave voltages, and at the ports the traveling-wave components add to give the total port voltage. In dealing with nondistributed circuits it is preferable to use the total port voltages and currents—V_1, I_1, V_2, and I_2, shown in Figure 7-1(a). However, with distributed elements it is preferable to deal with traveling voltages and currents—V_1^+, V_1^-, V_2^+, and V_2^-, shown in Figure 7-1(b). RF and microwave design necessarily requires switching between the two forms.

7.2.1 Reciprocity, Symmetry, Passivity, and Linearity

Reciprocity, symmetry, passivity, and linearity are fundamental properties of networks. A network is linear if the response is linearly dependent on the drive level, and superposition also applies. So if the two-port shown in Figure 7-1(a) is linear, the currents I_1 and I_2 are linear functions of V_1 and V_2. An example of a linear network would be one with resistors and capacitors. A network with a diode would be nonlinear. A passive network has no internal sources of power and a symmetrical two-port has the same characteristics at each of the ports. An example of a symmetrical network is a transmission line with a uniform cross section.

A reciprocal two-port has a response at Port 2 from an excitation at Port 1 that is the same as the response at Port 1 to the same excitation at Port 2. As an example, consider the two-port in Figure 7-1(a) with $V_2 = 0$. If the network is reciprocal, then the ratio I_2/V_1 with $V_2 = 0$ will be the same as the ratio I_1/V_2 with $V_1 = 0$. Networks with resistors, capacitors, and transmission lines are reciprocal. A transistor amplifier is not reciprocal as gain is in one

[1] Even when the term "two-port" is used on its own, the hyphen is used, as it is referring to a two-port network.

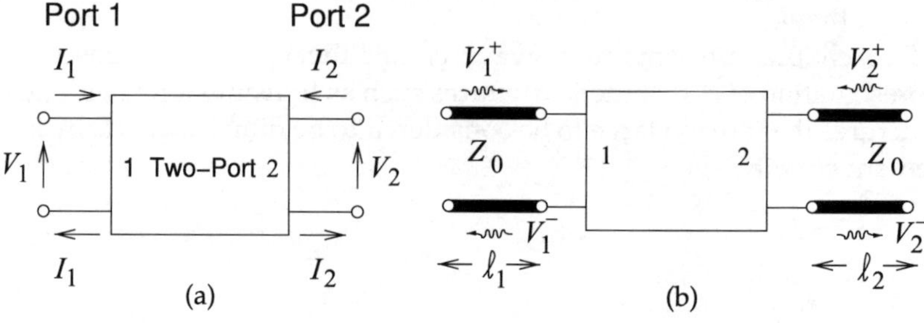

Figure 7-1: A two-port network: (a) port voltages; and (b) with transmission lines at the ports.

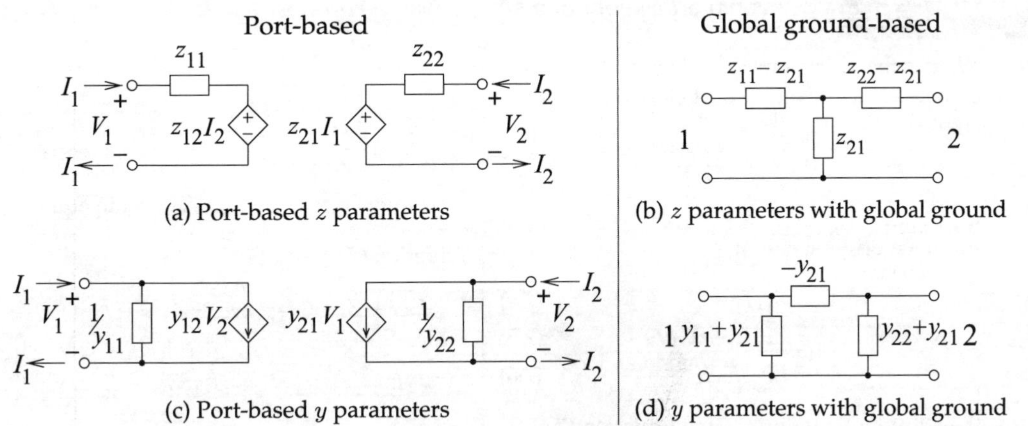

(a) Port-based z parameters (b) z parameters with global ground

(c) Port-based y parameters (d) y parameters with global ground

Figure 7-2: The z- and y-parameter equivalent circuits for port-based and global ground representations. The immittances are shown as impedances.

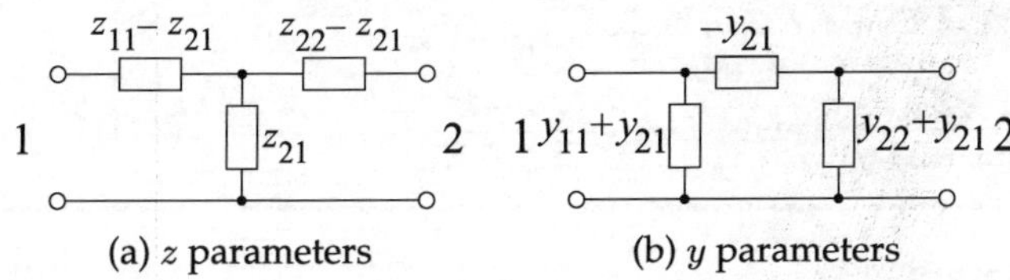

(a) z parameters (b) y parameters

Figure 7-3: Circuit equivalence of the z and y parameters for a reciprocal network (in (b) the elements are admittances).

direction.

7.2.2 Parameters Based on Total Voltage and Current

Here port-based impedance (z), admittance (y), and hybrid (h) parameters will be described. These are similar to the more conventional z, y, and h parameters defined with respect to a common or global ground. The contrasting circuit representations of the parameters are shown in Figure 7-2.

Impedance parameters

With reference to Figure 7-1the port-based impedance parameters, or z parameters, are defined as

$$V_1 = z_{11}I_1 + z_{12}I_2 \quad (7.1) \qquad V_2 = z_{21}I_1 + z_{22}I_2, \quad (7.2)$$

or in matrix form as $\mathbf{V} = \mathbf{ZI}$. (7.3)

The double subscript on a parameter is ordered so that the first refers to the output and the second refers to the input, so z_{ij} relates the voltage output at Port i to the current input at Port j. If the network is reciprocal, then $z_{12} = z_{21}$, but this simple type of relationship does not apply to all network parameters. The reciprocal circuit equivalence of the z parameters is shown in Figure 7-3(a).

EXAMPLE 7.1 Thevenin equivalent of a source with a two-port

What is the Thevenin equivalent circuit of the source-terminated two-port network on the right.

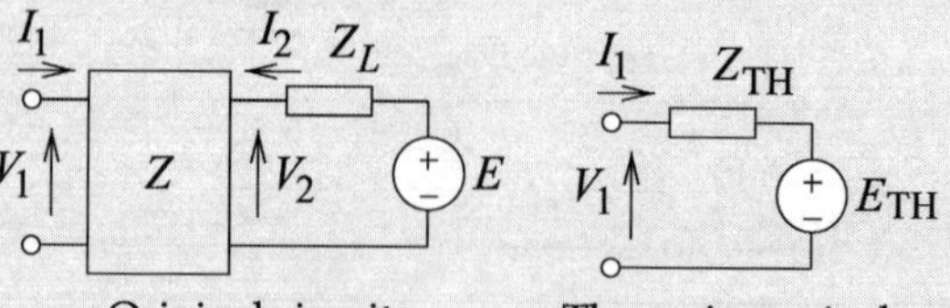

Solution:
From the original circuit

$$V_1 = Z_{11}I_1 + Z_{12}I_2 \quad (7.4)$$
$$V_2 = Z_{21}I_1 + Z_{22}I_2 \quad (7.5)$$
$$V_2 = E - I_2 Z_L \quad (7.6)$$

Substituting Equation (7.6) in Equation (7.5)

$$E = Z_{21}I_1 + (Z_{22} + Z_L)I_2 \quad (7.7)$$

Multiplying Equation (7.4) by $(Z_{22} + Z_L)$ and Equation (7.7) by Z_{12}

$$(Z_{22} + Z_L)V_1 = (Z_{22} + Z_L)Z_{11}I_1 + (Z_{22} + Z_L)Z_{12}I_2 \quad (7.8)$$
$$Z_{12}E = Z_{12}Z_{21}I_1 + Z_{12}(Z_{22} + Z_L)I_2 \quad (7.9)$$

Subtracting Equation (7.9) from Equation (7.8)

$$(Z_{22} + Z_L)V_1 - Z_{12}E = [(Z_{22} + Z_L)Z_{11} -]I_1$$
$$V_1 = \frac{Z_{12}E}{Z_{22} + Z_L} + \left(Z_{11} - \frac{Z_{12}Z_{21}}{Z_{22} + Z_L}\right) I_1 \quad (7.10)$$

For the Thevenin equivalent circuit $V_1 = E_{TH} + I_1 Z_{TH}$ and so

$$Z_{TH} = \left(Z_{11} - \frac{Z_{12}Z_{21}}{Z_{22} + Z_L}\right) \quad \text{and} \quad E_{TH} = \frac{Z_{12}E}{Z_{22} + Z_L} \quad (7.11)$$

Admittance parameters

The port-based admittance parameters, or y parameters, are defined as

$$I_1 = y_{11}V_1 + y_{12}V_2 \quad (7.12) \qquad I_2 = y_{21}V_1 + y_{22}V_2, \quad (7.13)$$

or in matrix form as $\mathbf{I} = \mathbf{YV}$. (7.14)

Now, for reciprocity, $y_{12} = y_{21}$ and the circuit equivalence of the y parameters is shown in Figure 7-3(b).

7.3 Scattering Parameters

Direct measurement of the z and y parameters requires that the ports be terminated in either short or open circuits. For active circuits this could result in undesired behavior, including oscillation or destruction. Also, at RF it is difficult to realize a good open or short. Since RF circuits are designed with close attention to maximum power transfer conditions, resistive terminations are preferred, as these are closer to the actual operating conditions. Thus the effect of measurement errors will have less impact. This is the way scattering parameters are measured.

The discussion of scattering parameters, S parameters[2], begins by considering the reflection coefficient, which is the S parameter of a one-port network.

[2] For historical reasons a capital "S" is used when referring to S parameters. For most other network parameters, lowercase is used (e.g., z parameters for impedance parameters).

7.3.1 Reflection Coefficient

The reflection coefficient, Γ, of a load Z_L can be determined by separately measuring the forward- and backward-traveling voltages on a transmission line terminated by the load:

$$\Gamma(x) = \frac{V^-(x)}{V^+(x)}. \tag{7.15}$$

Imagine between the source and the load that there is a line of characteristic impedance Z_0 and with infinitesimal length, then Γ at the load is related to the impedance Z_L by

$$\Gamma(0) = \frac{Z_L - Z_0}{Z_L + Z_0}, \tag{7.16}$$

where Z_0 is the characteristic impedance of the connecting transmission line. This can also be written as

$$\Gamma(0) = \frac{Y_0 - Y_L}{Y_0 + Y_L}, \tag{7.17}$$

where $Y_0 = 1/Z_0$ and $Y_L = 1/Z_L$. More completely, Γ as defined above is called the voltage reflection coefficient sometimes denoted Γ^V.

Recall that the current reflection coefficient $\Gamma^I = -\Gamma^V$.

7.3.2 Two-Port S Parameters

Two-port S parameters are defined in terms of traveling waves on transmission lines with real characteristic impedance Z_0 attached to each of the ports of a network, see Figure 7-1(b):

$$V_1^- = S_{11}V_1^+ + S_{12}V_2^+ \tag{7.18}$$ $$V_2^- = S_{21}V_1^+ + S_{22}V_2^+, \tag{7.19}$$

where S_{ij} are the individual S parameters. In matrix form

$$\begin{bmatrix} V_1^- \\ V_2^- \end{bmatrix} = \begin{bmatrix} S_{11} & S_{12} \\ S_{21} & S_{22} \end{bmatrix} \begin{bmatrix} V_1^+ \\ V_2^+ \end{bmatrix} = \mathbf{S} \begin{bmatrix} V_1^+ \\ V_2^+ \end{bmatrix}. \tag{7.20}$$

Individual S parameters are determined by measuring the forward- and backward-traveling waves with loads Z_0 at the ports. Since the load is Z_0 it cannot reflect power and so $V_2^+ = 0$, then

$$S_{11} = \left.\frac{V_1^-}{V_1^+}\right|_{V_2^+=0}. \tag{7.21}$$

The remaining parameters are determined similarly and so S_{22} is found as

$$S_{22} = \left.\frac{V_2^-}{V_2^+}\right|_{V_1^+=0} \tag{7.22}$$

and the transmission parameter as

$$S_{21} = \left.\frac{V_2^-}{V_1^+}\right|_{V_2^+=0}. \tag{7.23}$$

Table 7-1: Two-port S parameter conversion chart. The z, and y parameters are normalized to Z_0. Z' and Y' are the actual parameters.

	S	In terms of S
z	$Z'_{11} = z_{11}Z_0 \quad Z'_{12} = z_{12}Z_0$	$Z'_{21} = z_{21}Z_0 \quad Z'_{22} = z_{22}Z_0$
	$\delta_z = (1 + z_{11})(1 + z_{22}) - z_{12}z_{21}$	$\delta_S = (1 - S_{11})(1 - S_{22}) - S_{12}S_{21}$
	$S_{11} = [(z_{11} - 1)(z_{22} + 1) - z_{12}z_{21}]/\delta_z$	$z_{11} = [(1 + S_{11})(1 - S_{22}) + S_{12}S_{21}]/\delta_S$
	$S_{12} = 2z_{12}/\delta_z$	$z_{12} = 2S_{12}/\delta_S$
	$S_{21} = 2z_{21}/\delta_z$	$z_{21} = 2S_{21}/\delta_S$
	$S_{22} = [(z_{11} + 1)(z_{22} - 1) - z_{12}z_{21}]/\delta_z$	$z_{22} = [(1 - S_{11})(1 + S_{22}) + S_{12}S_{21}]/\delta_S$
y	$Y'_{11} = y_{11}/Z_0 \quad Y'_{12} = y_{12}/Z_0$	$Y'_{21} = y_{21}/Z_0 \quad Y'_{22} = y_{22}/Z_0$
	$\delta_y = (1 + y_{11})(1 + y_{22}) - y_{12}y_{21}$	$\delta_S = (1 + S_{11})(1 + S_{22}) - S_{12}S_{21}$
	$S_{11} = [(1 - y_{11})(1 + y_{22}) + y_{12}y_{21}]/\delta_y$	$y_{11} = [(1 - S_{11})(1 + S_{22}) + S_{12}S_{21}]/\delta_S$
	$S_{12} = -2y_{12}/\delta_y$	$y_{12} = -2S_{12}/\delta_S$
	$S_{21} = -2y_{21}/\delta_y$	$y_{21} = -2S_{21}/\delta_S$
	$S_{22} = [(1 + y_{11})(1 - y_{22}) + y_{12}y_{21}]/\delta_y$	$y_{22} = [(1 + S_{11})(1 - S_{22}) + S_{12}S_{21}]/\delta_S$

In the reverse direction,

$$S_{12} = \left.\frac{V_1^-}{V_2^+}\right|_{V_1^+=0}. \tag{7.24}$$

In the above Z_0 is referred to as the **normalization impedance** or equivalently the **reference impedance**. In some circumstances Z_{REF} is used to denote reference impedance to avoid possible confusion with a transmission line impedance that is not the same as the reference impedance. The S parameters here are also called **normalized S parameters**, and the S parameters are normalized to the same real reference impedance at each port.

The relationships between the two-port S parameters and the common network parameters are given in Table 7-1. It is interesting to note that $S_{21}/S_{12} = z_{21}/z_{12} = y_{21}/y_{12}$. That is, the ratio of the forward to reverse parameters (at least for S, z, and y parameters) are the same and this ratio is one for a reciprocal device. An S parameter is a voltage ratio, so when it is expressed in decibels $S_{ij}|_{\text{dB}} = 20\log(S_{ij})$.

A reciprocal network has $S_{12} = S_{21}$. If unit power flows into a two-port, a fraction, $|S_{11}|^2$, is reflected and a further fraction, $|S_{21}|^2$, is transmitted through the network.

EXAMPLE 7.2 Two-Port S Parameters

What are the S parameters of a 30 dB attenuator?

Solution:

An attenuator is shown with a system impedance of Z_0. An ideal attenuator has no reflection at each of the two ports when the attenuator is embedded in its system impedance. Thus $\Gamma_{\text{in}} = 0 = \Gamma_{\text{out}}$. Since there is no reflection from the load or the source, this implies that $S_{11} = 0 = S_{22}$.

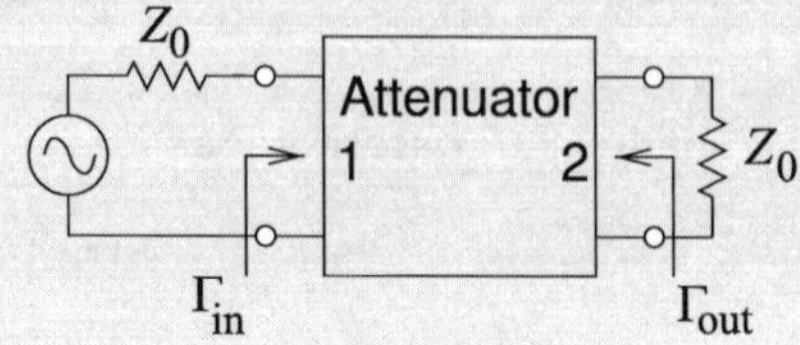

Since this is a 30 dB attenuator, the power delivered to the load impedance Z_0 is 30 dB below the power available from the source, thus $S_{21} = -30$ dB $= 0.0316$. The attenuator is reciprocal and so $S_{12} = S_{21}$. Thus the S parameters of the attenuator are

$$\mathbf{S} = \begin{bmatrix} 0 & 0.0316 \\ 0.0316 & 0 \end{bmatrix}. \tag{7.25}$$

Note that the reference impedance did not need to be known to develop the S parameters.

7.3.3 Input Reflection Coefficient of a Terminated Two-Port Network

A two-port is shown in Figure 7-4 that is terminated at Port 2 in a load with a reflection coefficient Γ_L. The lines at each of the ports are of infinitesimal length (i.e., $\ell_1 \to 0$ and $\ell_2 \to 0$) and are used to make it easier to visualize the separation of the voltage into forward- and backward-traveling components. The aim in this section is to develop a formula for the input reflection coefficient $\Gamma_{\text{in}} = V_1^-/V_1^+$. For the circuit in Figure 7-4 , three equations can be developed:

$$V_1^- = S_{11}V_1^+ + S_{12}V_2^+ \tag{7.26}$$

$$V_2^- = S_{21}V_1^+ + S_{22}V_2^+ \tag{7.27}$$

$$V_2^+ = \Gamma_L V_2^-, \quad \text{i.e.,} \quad V_2^- = V_2^+/\Gamma_L. \tag{7.28}$$

Note that V_2^- is the voltage wave that leaves the two-port but is incident on the load Γ_L. The aim here is to eliminate V_2^+ and V_2^-. Substituting Equation (7.28) into Equation (7.27) leads to

$$V_2^+/\Gamma_L = S_{21}V_1^+ + S_{22}V_2^+ \tag{7.29}$$

$$V_2^+\left(\frac{1 - S_{22}\Gamma_L}{\Gamma_L}\right) = S_{21}V_1^+ \tag{7.30}$$

$$V_2^+ = \left(\frac{S_{21}\Gamma_L}{1 - S_{22}\Gamma_L}\right)V_1^+. \tag{7.31}$$

Now substituting Equation (7.31) in Equation (7.26) yields

$$V_1^- = S_{11}V_1^+ + S_{12}\left(\frac{S_{21}\Gamma_L}{1 - S_{22}\Gamma_L}\right)V_1^+ \tag{7.32}$$

and so $$\Gamma_{\text{in}} = S_{11} + \frac{S_{12}S_{21}\Gamma_L}{1 - S_{22}\Gamma_L}. \tag{7.33}$$

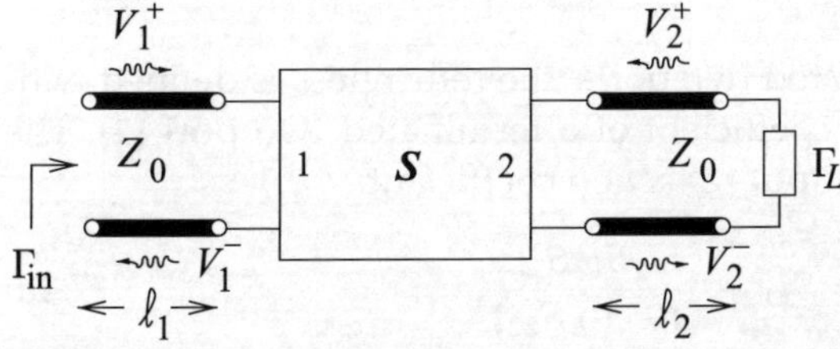

Figure 7-4: A terminated two-port network with transmission lines of infinitesimal length at the ports.

7.3.4 Properties of a Two-Port in Terms of S Parameters

The properties of most interest are whether the two-port network is lossless, passive, or reciprocal.

If a network is lossless, all of the power input to the network must leave the network. The power incident on Port 1 of a network is

$$P_1^+ = \left| \frac{\frac{1}{2} V_1^+}{Z_0} \right|^2 \tag{7.34}$$

and the power leaving Port 1 is

$$P_1^- = \left| \frac{\frac{1}{2} V_1^-}{Z_0} \right|^2 . \tag{7.35}$$

This can be repeated for Port 2 and the factor $\frac{1}{2}/Z_0$ appears in all expressions. So cancelling this factor, the condition for the network to be lossless is

$$|S_{11}|^2 + |S_{21}|^2 = 1 \quad \text{and} \quad |S_{12}|^2 + |S_{22}|^2 = 1. \tag{7.36}$$

For a network to be passive, no more power can leave the network than enters it. So the condition for passivity is

$$|S_{11}|^2 + |S_{21}|^2 \leq 1 \quad \text{and} \quad |S_{12}|^2 + |S_{22}|^2 \leq 1. \tag{7.37}$$

Reciprocity requires that $S_{21} = S_{12}$.

7.4 Return Loss, Substitution Loss, and Insertion Loss

7.4.1 Return Loss

Return loss, also known as **reflection loss**, is a measure of the fraction of available power that is not delivered by a source to a load. If the power incident on a load is P_i and the power reflected by the load is P_r, then the return loss in decibels is [1, 2]

$$\text{RL}_{\text{dB}} = 10 \log \frac{P_i}{P_r}. \tag{7.38}$$

The better the load is matched to the source, the lower the reflected power and hence the higher the return loss. RL is a positive quantity if the reflected power is less than the incident power. If the load has a complex reflection coefficient ρ, then

$$\text{RL}_{\text{dB}} = 10 \log \left| \frac{1}{\rho^2} \right| = -20 \log |\rho| . \tag{7.39}$$

That is, the return loss is the negative of the reflection coefficient expressed in decibels [3].

When generalized to terminated two ports, the return loss is defined with respect to the input reflection coefficient of a terminated two port [4]. The two port in Figure 7-5 has the input reflection coefficient

$$\Gamma_{\text{in}} = S_{11} + \frac{\Gamma_L S_{12} S_{21}}{(1 - \Gamma_L S_{22})}, \tag{7.40}$$

where Γ_L is the reflection coefficient of the load. Thus the return loss of a terminated two-port is

$$\mathrm{RL}_{\mathrm{dB}} = -20\log|\Gamma_{\mathrm{in}}| = -20\log\left|S_{11} + \frac{\Gamma_L S_{12} S_{21}}{(1-\Gamma_L S_{22})}\right| . \quad (7.41)$$

If the load is matched, i.e. $Z_L = Z_0^*$ (the system reference impedance), then

$$\mathrm{RL}_{\mathrm{dB}} = -20\log|S_{11}| . \quad (7.42)$$

This return loss is also called the input return loss since the reflection coefficient is calculated at Port 1. The output return loss is calculated looking into Port 2 of the two-port, where now the termination at Port 1 is just the source impedance.

7.4.2 Substitution Loss and Insertion Loss

The substitution loss is the ratio of the power, ${}^iP_L$, delivered to the load by an initial two-port identified by the leading superscript 'i', and the power delivered to the load, ${}^fP_L$, with a substituted final two-port identified by the leading superscript 'f'. In terms of scattering parameters with reference impedances Z_{REF}, the substitution loss in decibels is (using the results in [5] and noting that Γ_S and Γ_L are referred to Z_{REF})

$$L_S|\,\mathrm{dB} = \frac{{}^iP_L}{{}^fP_L} = 10\log\left|\frac{{}^iS_{21}\left[(1-{}^fS_{11}\Gamma_S)(1-{}^fS_{22}\Gamma_L) - {}^fS_{12}{}^fS_{21}\Gamma_S\Gamma_L\right]}{{}^fS_{21}\left[(1-{}^iS_{11}\Gamma_S)(1-{}^iS_{22}\Gamma_L) - {}^iS_{12}{}^iS_{21}\Gamma_S\Gamma_L\right]}\right|^2 . \quad (7.43)$$

Insertion loss is a special case of substitution loss with a particular type of initial two-port networks.

Insertion Loss with an Ideal Adaptor

Insertion loss is the substitution loss when the initial two-poort is a direct connection so ${}^iS_{11} = 0 = {}^iS_{22}$ (for no reflection), ${}^iS_{12}{}^iS_{21} = 1$ (for no loss in the adaptor and there is no phase shift), and ${}^iS_{12} = {}^iS_{21}$ (for reciprocity), insertion loss in decibels is, using Equation (7.43):

$$\mathrm{IL}|_{\mathrm{dB}} = 10\log\left[\left|\frac{(1-{}^fS_{11}\Gamma_S)(1-{}^fS_{22}\Gamma_L) - {}^fS_{12}{}^fS_{21}\Gamma_S\Gamma_L}{{}^fS_{12}\,(1-\Gamma_S\Gamma)}\right|^2\right] \quad (7.44)$$

Attenuation is defined as the insertion loss without source and load

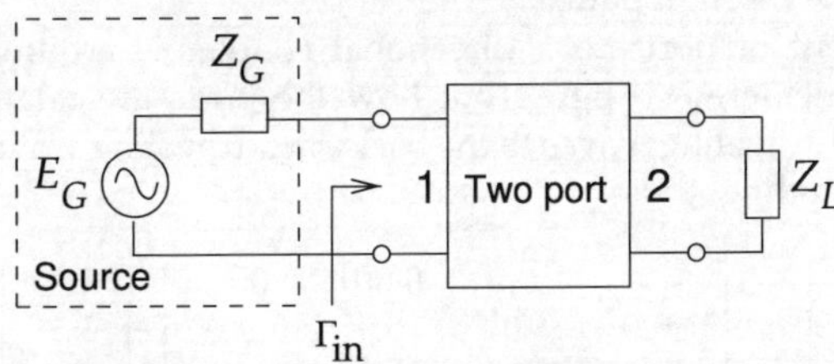

Figure 7-5: Terminated two-port used to define return loss.

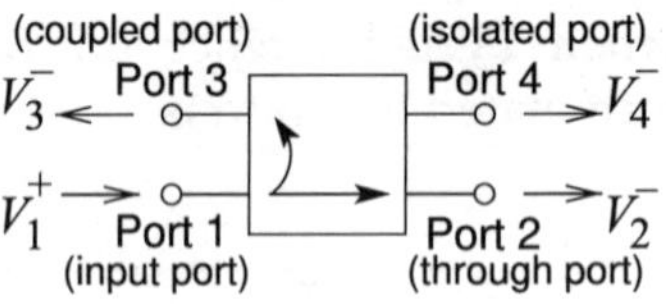

Figure 7-6: Schematic of a directional coupler.

reflections ($\Gamma_S = 0 = \Gamma_L$) [6], and Equation (7.44) becomes

$$A|\,\text{dB} = 10\log\left(\frac{Z_{02}}{Z_{01}}\frac{1}{|S_{21}|^2}\right) \quad (= \text{IL with } \Gamma_S = 0 = \Gamma_L) \tag{7.45}$$

7.5 Scattering Parameters and Directional Couplers

Directional couplers were described in Section 6.5, but without the use of S parameters. A directional coupler with ports defined as in Figure 7-6, and with the ports matched (so that $S_{11} = 0 = S_{22} = S_{33} = S_{44}$), has the following scattering parameter matrix:

$$\mathbf{S} = \begin{bmatrix} 0 & T & 1/C & 1/I \\ T & 0 & 1/I & 1/C \\ 1/C & 1/I & 0 & T \\ 1/I & 1/C & T & 0 \end{bmatrix}. \tag{7.46}$$

There are many types of directional couplers, and the phases of the traveling waves at the ports will not necessarily be in phase as Equation (7.46) implies. When the phase difference between traveling waves entering at Port 1 and leaving at Port 2 is 90°, Equation (7.46) becomes

$$\mathbf{S} = \begin{bmatrix} 0 & -\jmath T & 1/C & 1/I \\ -\jmath T & 0 & 1/I & 1/C \\ 1/C & 1/I & 0 & -\jmath T \\ 1/I & 1/C & -\jmath T & 0 \end{bmatrix}. \tag{7.47}$$

EXAMPLE 7.3 Identifying Ports of a Directional Coupler

A directional coupler has the following S parameters:

$$S = \begin{bmatrix} 0 & 0.9 & 0.001 & 0.1 \\ 0.9 & 0 & 0.1 & 0.001 \\ 0.001 & 0.1 & 0 & 0.9 \\ 0.1 & 0.001 & 0.9 & 0 \end{bmatrix}.$$

(a) What are the through (i.e., transmission) paths? Identify two paths. That is, identify the pairs of ports at the ends of the through paths.
First note that the assignment of ports to a directional coupler is arbitrary. So the S parameters need to be considered to figure out how the ports are related. The S parameters relate the forward-traveling waves to the backward-traveling waves and this leads to the required understanding, thus

$$\begin{bmatrix} V_1^- \\ V_2^- \\ V_3^- \\ V_4^- \end{bmatrix} = S \begin{bmatrix} V_1^+ \\ V_2^+ \\ V_3^+ \\ V_4^+ \end{bmatrix} = \begin{bmatrix} 0 & 0.9 & 0.001 & 0.1 \\ 0.9 & 0 & 0.1 & 0.001 \\ 0.001 & 0.1 & 0 & 0.9 \\ 0.1 & 0.001 & 0.9 & 0 \end{bmatrix} \begin{bmatrix} V_1^+ \\ V_2^+ \\ V_3^+ \\ V_4^+ \end{bmatrix}.$$

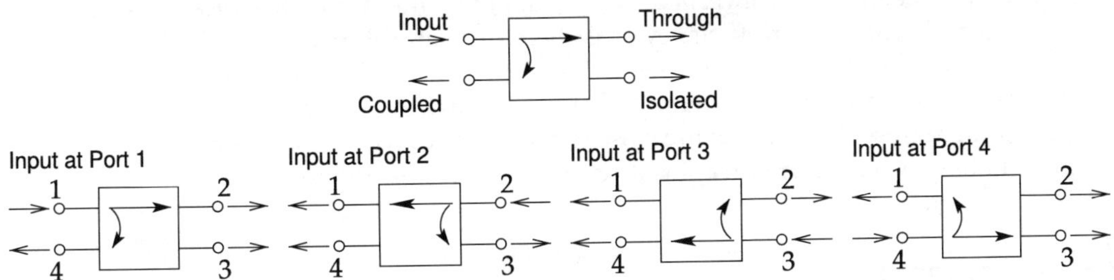

Figure 7-7: Directional coupler schematic drawn with each of the four possible input ports.

Writing the S parameters out this way makes it easier to identify the largest backward-traveling waves for each of the inputs at Ports 1, 2, 3, and 4. The backward-traveling wave will leave the directional coupler and the inputs will be forward-traveling waves. Consider Port 1, the largest backward-traveling wave is at Port 2, and so Ports 1 and 2 define one of the through paths. The other through path is between Ports 3 and 4. So the through paths are 1–2 and 3–4.

(b) What is the coupled port for the signal entering Port 1?
The coupled port is identified by the port with the largest backward-traveling signal not including the port at the other end of the through path. For Port 1 the coupled port is Port 4.

(c) What is the coupling factor?

$$C = \frac{V_1^+}{V_4^-} = \frac{1}{0.1} = 10 = 20 \text{ dB}.$$

(d) What is the isolated port for the signal entering Port 1?
The isolated port is Port 3. The backward-traveling wave at this port is the smallest given an input at Port 1.

(e) What is the isolation factor?

$$I = \frac{V_1^+}{V_3^-} = \frac{1}{0.001} = 1000 = 60 \text{ dB}.$$

(f) What is the directivity factor?
The directivity factor indicates how much stronger the signal is at the coupled port compared to the isolated port for a signal at the input. For an input at Port 1, the directivity factor is

$$D = \frac{V_4^-}{V_3^-} = \frac{0.1}{0.001} = 100 = 40 \text{ dB}.$$

As a check $D = I/C = 1000/10 = 100$.

(g) Draw a schematic of the directional coupler.
There are four ways to draw it depending on the input port chosen, see Figure 7-7.

7.6 Summary

There are several network parameters used with RF and microwave circuits. Which is used depends on which makes the task of visualizing circuit operation more clear, which makes analyzing circuits more convenient, and which enables different circuits to be made electrically equivalent.

Scattering parameters are parameters that are almost exclusively used by RF and microwave engineers. They describe power flow and traveling waves and are essential to describing distributed circuits. Much of RF and microwave engineering is concerned with managing the signal-to-noise power ratio and with power efficiency. It is therefore natural to work with parameters that directly relate to power flow. RF and microwave design is characterized by conceptual insight and it is essential to use parameters and graphical representations that are close to the physical world. Scattering parameters have very natural graphical representations, as will be seen in the next chapter.

7.7 References

[1] "Telecommunications: Glossary of Telecommunication Terms," Federal Standard 1037C, U.S. General Services Administration Information Technology Service, Aug. 7, 1996.

[2] T. S. Bird, "Definition and misuse of return loss [report of the trans. editor-in-chief]," *IEEE Antennas and Propagation Magazine*, vol. 51, no. 2, pp. 166–167, Apr. 2009.

[3] IEEE, *IEEE Standard Dictionary of Electrical and Electronic Engineering Terms*, 4th ed. IEEE Press, 1988.

[4] T. Otoshi, "Maximum and minimum return losses from a passive two-port network terminated with a mismatched load," *IEEE Trans. on Microwave Theory and Techniques*, vol. 42, no. 5, pp. 787–792, May 1994.

[5] M. Steer, *Microwave and RF Design, Networks*, 3rd ed. North Carolina State University, 2019.

[6] R. Beatty, "Insertion loss concepts," *Proc. of the IEEE*, vol. 52, no. 6, pp. 663–671, Jun. 1964.

7.8 Exercises

1. A load has a reflection coefficient of $0.5 - \jmath 0.1$ in a 75 Ω reference system. What is the reflection coefficient in a 50 Ω reference system?

2. The 50 Ω S parameters of a two-port are $S_{11} = 0.5 + \jmath 0.5$, $S_{12} = 0.95 + \jmath 0.25$, $S_{21} = 0.15 - \jmath 0.05$, and $S_{22} = 0.5 - \jmath 0.5$. Port 1 is connected to a 50 Ω source with an available power of 1 W and Port 2 is terminated in 50 Ω. What is the power reflected from Port 1?

3. The scattering parameters of a certain two-port are $S_{11} = 0.5 + \jmath 0.5$, $S_{12} = 0.95 + \jmath 0.25$, $S_{21} = 0.15 - \jmath 0.05$, and $S_{22} = 0.5 - \jmath 0.5$. The system reference impedance is 50 Ω.
   (a) Is the two-port reciprocal? Explain.
   (b) Consider that Port 1 is connected to a 50 Ω source with an available power of 1 W. What is the power delivered to a 50 Ω load placed at Port 2?
   (c) What is the reflection coefficient of the load required for maximum power transfer at Port 2?

4. In characterizing a two-port, power could only be applied at Port 1. The signal reflected was measured and the signal at a 50 Ω load at Port 2 was also measured. This yielded two S parameters referenced to 50 Ω: $S_{11} = 0.3 - \jmath 0.4$ and $S_{21} = 0.5$.
   (a) If the network is reciprocal, what is S_{12}?
   (b) Is the two-port lossless?
   (c) What is the power delivered into the 50 Ω load at Port 2 when the available power at Port 1 is 0 dBm?

5. A matched lossless transmission line has a length of one-quarter wavelength. What are the scattering parameters of the two-port?

6. A connector has the scattering parameters $S_{11} = 0.05$, $S_{21} = 0.9$, $S_{12} = 0.9$, and $S_{22} = 0.04$ and the reference impedance is 50 Ω. What is the return loss in dB of the connector at Port 1 in a 50 Ω system?

7. The scattering parameters of an amplifier are $S_{11} = 0.5$, $S_{21} = 2.$, $S_{12} = 0.1$, and $S_{22} = -0.2$ and the reference impedance is 50 Ω. If the amplifier is terminated at Port 2 in a resistance of 25 Ω, what is the return loss in dB at Port 1?

8. A two-port network has the scattering parameters $S_{11} = -0.5$, $S_{21} = 0.9$, $S_{12} = 0.8$, and $S_{22} = 0.04$ and the reference impedance is 50 Ω.
   (a) What is the return loss in dB of the connector at Port 1 in a 50 Ω system?
   (b) Is the two-port reciprocal and why?

9. A two-port network has the scattering parameters $S_{11} = -0.2$, $S_{21} = 0.8$, $S_{12} = 0.7$, and $S_{22} = 0.5$ and the reference impedance is 75 Ω.
   (a) What is the return loss in dB of the connector at Port 1 in a 75 Ω system?
   (b) Is the two-port reciprocal and why?

10. A cable has the scattering parameters $S_{11} = 0.1$, $S_{21} = 0.7$, $S_{12} = 0.7$, and $S_{22} = 0.1$. At Port 2 is a 55 Ω load and the S parameters and reflection coefficients are referred to 50 Ω.
    (a) What is the load's reflection coefficient?
    (b) What is the input reflection coefficient of the terminated cable?
    (c) What is the return loss, at Port 1 and in dB, of the cable terminated in the load?

11. A cable has the 50 Ω scattering parameters $S_{11} = 0.05$, $S_{21} = 0.5$, $S_{12} = 0.5$, and $S_{22} = 0.05$. What is the insertion loss of the cable if the source at Port 1 has a 50 Ω Thevenin impedance and the termination at Port 2 is 50 Ω? Express your answer in decibels.

12. A 1 m long cable has the 50 Ω scattering parameters $S_{11} = 0.1$, $S_{21} = 0.7$, $S_{12} = 0.7$, and $S_{22} = 0.1$. The cable is used in a 55 Ω system. Express your answers in decibels.
    (a) What is the return loss of the cable in the 55 Ω system? (Hint see Section sec:input:terminated:two:port and consider finding Z_in.]
    (b) What is the insertion loss of the cable in the 55 Ω system? Follow the procedure in Example 7.0
    (c) What is the return loss of the cable in a 50 Ω system?
    (d) What is the insertion loss of the cable in a 50 Ω system?

13. A 1 m long cable has the 50 Ω scattering parameters $S_{11} = 0.05$, $S_{21} = 0.5$, $S_{12} = 0.5$, and $S_{22} = 0.05$. The Thevenin equivalent impedance of the source and terminating load impedances of the cable are 50 Ω. Express your answers in decibels.
    (a) What is the return loss of the cable?
    (b) What is the insertion loss of the cable?

14. A lossy directional coupler has the following 50 Ω S parameters:

$$S = \begin{bmatrix} 0 & -0.95\jmath & 0.005 & 0.1 \\ -0.95\jmath & 0 & 0.1 & 0.005 \\ 0.005 & 0.1 & 0 & -0.95\jmath \\ 0.1 & 0.005 & -0.95\jmath & 0 \end{bmatrix}.$$

    (a) What are the through (transmission) paths (identify two paths)? That is, identify the pairs of ports at the ends of the through paths.
    (b) What is the coupling in decibels?
    (c) What is the isolation in decibels?
    (d) What is the directivity in decibels?

15. A directional coupler has the following characteristics: coupling factor $C = 20$, transmission factor 0.9, and directivity factor 25 dB. Also, the coupler is matched so that $S_{11} = 0 = S_{22} = S_{33} = S_{44}$.
    (a) What is the isolation factor in decibels?
    (b) Determine the power dissipated in the directional coupler if the input power to Port 1 is 1 W.

16. A lossy directional coupler has the following 50 Ω S parameters:

$$S = \begin{bmatrix} 0 & 0.25 & -0.9\jmath & 0.01 \\ 0.25 & 0 & 0.01 & -0.9\jmath \\ -0.9\jmath & 0.01 & 0 & 0.25 \\ 0.01 & -0.9\jmath & 0.25 & 0 \end{bmatrix}.$$

    (a) Which port is the input port (there could be more than one answer)?
    (b) What is the coupling in decibels?
    (c) What is the isolation in decibels?
    (d) What is the directivity factor in decibels?

17. A directional coupler using coupled lines has a coupling factor of 3.38, a transmission factor of $-\jmath 0.955$, and infinite directivity and isolation. The input port is Port 2 and the through port is Port 2. Write down the 4×4 S parameter matrix of the coupler.

7.8.1 Exercises by Section

†challenging

§7.3 1, 2, 3†, 4†, 5†

§7.4 6, 7, 8, 9, 10, 11, 12†, 13

§7.5 14†, 15†, 16†, 17†

7.8.2 Answers to Selected Exercises

1 $0.638 - \jmath 0.079$	6 26 dB	15(b) 187 mW
3(d) $50 + \jmath 100$ Ω	12(b) 2.99 dB	16(d) 28 dB

CHAPTER 8

Graphical Network Analysis

8.1 Introduction

This chapter introduces the Smith chart, most widely used graphical technique for describing and solving problems using S parameters. The Smith chart is an annotated polar plot of S parameters and is one of the most powerful tools in RF and microwave engineering and is used to present measured results, to conceptualize designs, and to intuitively solve problems involving distributed networks.

8.2 Polar Representations of Scattering Parameters

Scattering parameters are most naturally represented in polar form with the square of the magnitude relating to power flow. In this section a greater rationale for representing S parameters on a polar plot is presented and this serves as the basis for a more complicated representation of S parameters on a Smith chart to be described in the next section.

8.2.1 Shift of Reference Planes as a S Parameter Rotation

A polar plot is a natural way to present S parameters graphically. Adding additional lengths of the lines at each port rotates the S parameters. Consider the two-port in Figure 8-1. Here the original two-port with scattering

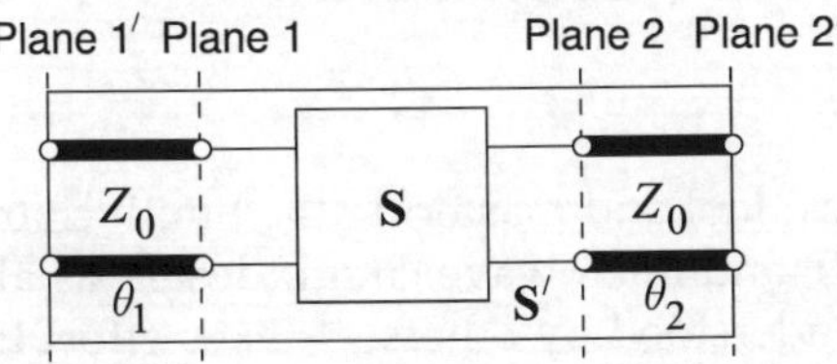

Figure 8-1: A two-port with scattering parameter matrix $\mathbf{S}$ augmented by lines at each port with the lines having the reference characteristic impedance Z_0 and electrical length θ_n. The scattering parameter matrix of the augmented two-port is $\mathbf{S}'$.

parameter matrix **S** is augmented by lines at each port with each having a characteristic impedance equal to the reference impedance. The S parameter matrix of the augmented two-port, $\mathbf{S}'$, are the same as the original S parameter matrix but phase-shifted. That is

$$\mathbf{S}' = \begin{bmatrix} S'_{11} & S'_{12} \\ S'_{21} & S'_{22} \end{bmatrix} = \begin{bmatrix} S_{11}e^{-\jmath 2\Theta 1} & S_{12}e^{-\jmath(\Theta 1+\Theta 2)} \\ S_{21}e^{-\jmath(\Theta 1+\Theta 2)} & S_{22}e^{-\jmath 2\Theta 2} \end{bmatrix}. \tag{8.1}$$

The shift in reference planes simply rotates the S parameters. This is one of the main reasons why S parameters are commonly plotted on a polar plot.

8.2.2 Polar Plot of Reflection Coefficient

The polar plot of reflection coefficient is simply the polar plot of a complex number. Figure 8-2 is used in plotting reflection coefficients and is a polar plot that has a radius of one. So a reflection coefficient with a magnitude of one is on the unit circle. The center of the polar plot is zero so the reflection coefficient of a matched load, which is zero, is plotted at the center of the circle. Plotting a reflection coefficient on the polar plot enables convenient interpretation of the properties of a reflection. The graph has additional notation that enables easy plotting of an S parameter on the graph. Conversely, the magnitude and phase of an S parameter can be easily read from the graph. The horizontal label going from 0 to 1 is used in determining magnitude. The notation arranged on the outer perimeter of the polar plot is used to read off angle information. Notice the additional notation "ANGLE OF REFLECTION COEFFICIENT IN DEGREES" and the scale relates to the actual angle of the polar plot. Verify that the 90° point is just where one would expect it to be.

Figure 8-3 annotates the polar plot of reflection coefficient with real and imaginary axes and shows the location of the short circuit and open circuit points. Note that the reflection coefficient of an inductive impedance is in the top half of the polar plot while the reflection coefficient of a capacitive impedance is in the bottom half of the polar plot.

The nomograph shown in Figure 8-4 aids in interpretation of polar reflection coefficient plots. The nomograph relates the reflection coefficient (RFL. COEFFICIENT), ρ (ρ was originally used instead of Γ and is still used with the Smith chart); the return loss (RTN. LOSS) (in decibels); and the standing wave ratio (SWR); and the standing wave ratio (in decibels) as $20\log(\mathrm{SWR})$. When printed together with the reflection coefficient polar plot (Figures 8-2 and 8-4 combined) the nomograph is scaled properly, but it is expanded here so that it can be read more easily. So with the aid of a compass with one point on the zero point of the polar plot and the other at the reflection coefficient (as plotted on the polar plot), the magnitude of the reflection coefficient is captured. The compass can then be brought down to the nomograph to read ρ, the return loss, and VSWR directly.

8.3 Smith Chart

The Smith chart is a powerful graphical tool and mastering the Smith chart is essential to entering the world of RF and microwave circuit design as all practitioners use this as if it is well understood by others. It takes effort to master but fundamentally it is quite simple combining a polar plot used for

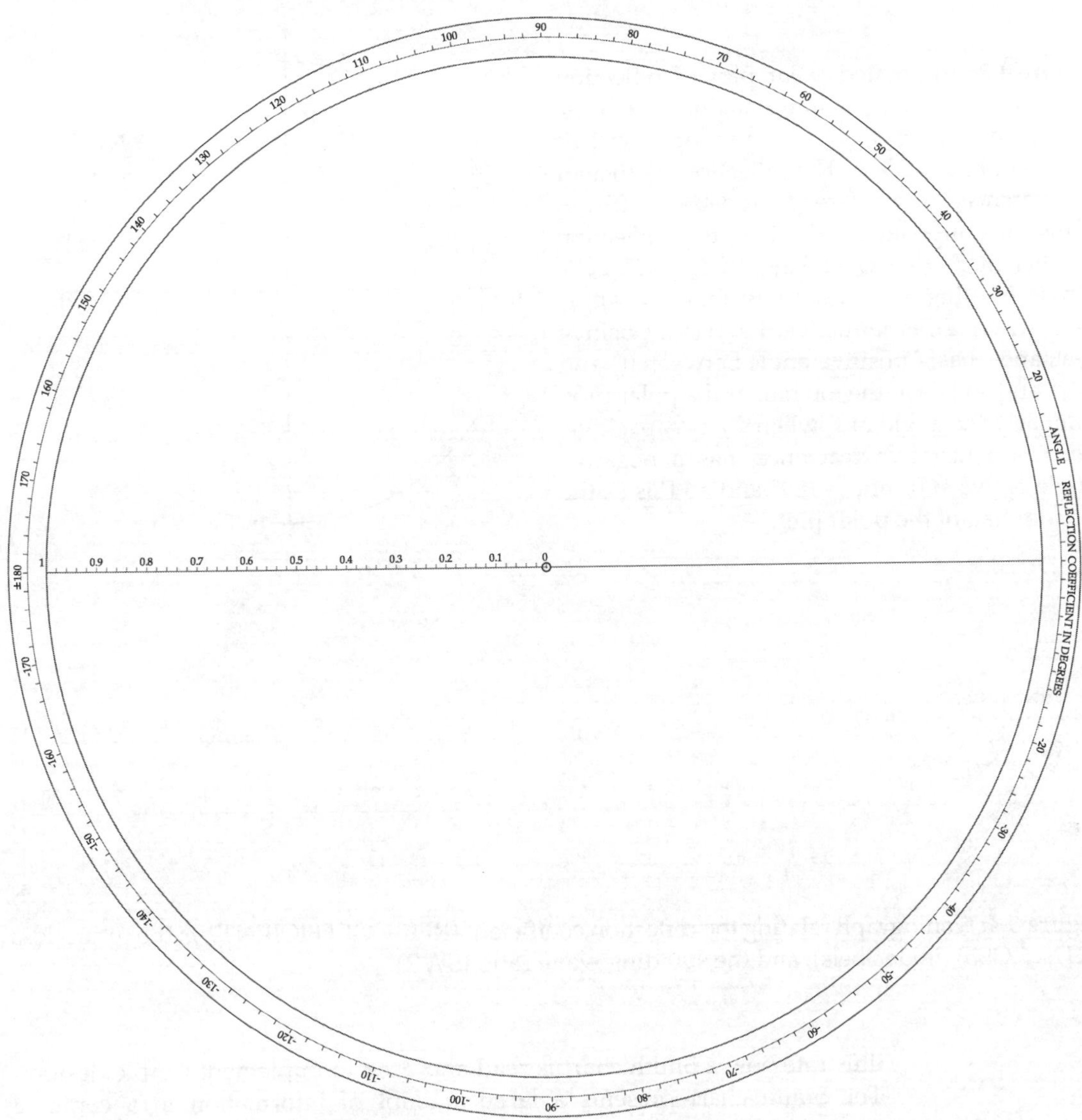

Figure 8-2: Polar chart for plotting reflection coefficient and transmission coefficient.

plotting S parameters directly, curves that enable normalized impedances and admittances to be plotted directly, and scales that enable electrical lengths in terms of wavelengths and degrees to be read off. The chart has many numbers printed in quite small font and with signs dropped off as there is limited room.

The Smith chart was invented by Phillip Smith and presented in close to its current form in 1937, see [1–4]. Once nomographs and graphical calculators were common engineering tools mainly because of limited computing resources. Only a few have survived in electrical engineering usage, with Smith charts being overwhelmingly the most important.

This section first presents the impedance Smith chart and then the admittance Smith chart before introducing a combined Smith chart which is the form needed in design. A number of examples are presented to

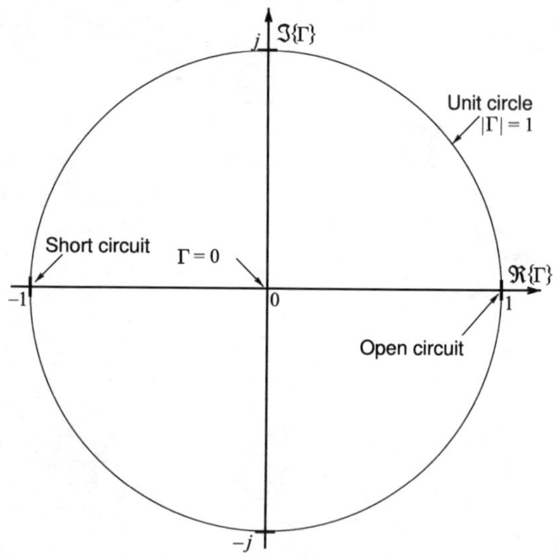

Figure 8-3: Annotated polar plot of reflection coefficient with real, $\Re$, and imaginary, $\Im$, axes. The short circuit $\Gamma = -1$ and open circuit $\Gamma = +1$ are indicated. The reflection coefficient is referenced to a reference impedance Z_{REF}. Thus an impedance Z_L has the reflection coefficient $\Gamma = (Z_L - Z_{\text{REF}})/(Z_L + Z_{\text{REF}})$). An interesting observation is that the angle of Γ when Z_L is inductive, i.e. has a positive reactance, has a positive angle between 0° and 180° and so Γ is in the top half of the polar plot. Similarly the angle of Γ when Z_L is capacitive, i.e. has a negative reactance, has a negative angle between 0° and $-180°$ and so Γ is in the bottom half of the polar plot.

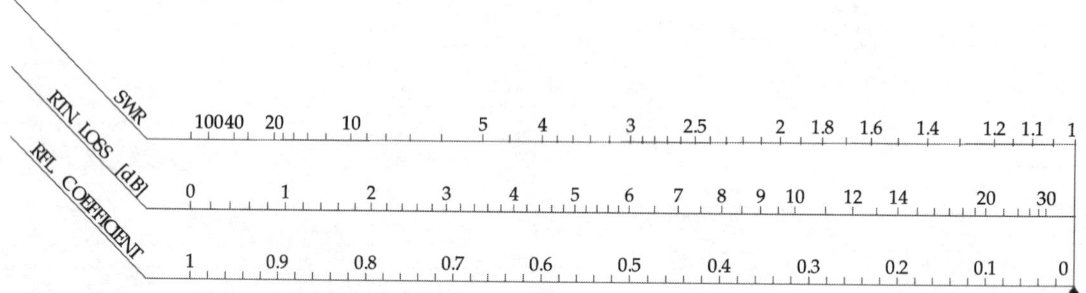

Figure 8-4: Nomograph relating the reflection coefficient (RFL. COEFFICIENT), ρ; the return loss (RTN. LOSS) (in decibels); and the standing wave ratio (SWR).

illustrate how a Smith chart is read and used to implement simple designs. The Smith chart presents a large amount of information in a confined space and interpretation, such as applying appropriate signs, is required to extract values. The Smith chart is a 'back-of-the-envelope' tool that RF and microwave circuit designers use to sketch out designs.

8.3.1 Impedance Smith Chart

The reflection coefficient, Γ, is related to a load, Z_L, by

$$\Gamma = \frac{Z_L - Z_{\text{REF}}}{Z_L + Z_{\text{REF}}}, \tag{8.2}$$

where Z_{REF} is the system reference impedance. With normalized load impedance $z_l = r + \jmath x = Z_L/Z_{\text{REF}}$, this becomes

$$\Gamma = \frac{r + \jmath x - 1}{r + \jmath x + 1}. \tag{8.3}$$

Commonly in network design reactive elements are added either in shunt or in series to an existing network. If a reactive element is added in series then the input reactance, x, is changed while the input resistance, r, is held constant. So superimposing the loci of Γ (on the S parameter polar plot) with fixed values of r, but varying values of x (x varying from $-\infty$ to ∞), proves useful, as will be seen. Also, plotting the loci of Γ with fixed values of x and varying values of r (r varying from 0 to ∞) is also useful. The combination of the reflection/transmission polar plots, the nomographs, and the r and x loci is called the impedance Smith chart, see Figure 8-5. This is still a polar plot of reflection coefficient and the arcs and circles of constant and resistance enable easy conversion between reflection coefficient and impedance.

The full impedance Smith chart shown in Figure 8-5 is daunting so discussion will begin with the less dense form of the impedance Smith chart shown in Figure 8-6(a) which is annotated in Figure 8-6(b). Referring to Figure 8-6(b), the unit circle corresponds to a reflection coefficient magnitude of one and hence a pure reactance. Note that there are lines of constant resistance and arcs of constant reactance. All points in the top half of the Smith chart have positive reactances and so all reflection coefficient points plotted in the top half of the Smith chart indicate inductive impedances. All points in the bottom half of the Smith chart have negative reactances and so all reflection coefficient points plotted in the bottom half of the Smith chart indicate capacitive impedances. The horizontal line across the middle of the Smith chart indicates pure resistance just as the unit circle indicates a pure reactance.

One big difference between the less dense form of the impedance Smith chart, Figure 8-6(a), and the full impedance Smith chart of Figure 8-5 is that the signs of the reactances are missing in the full impedance Smith chart. This is simply because there is not enough room and the user must add the appropriate sign when reading the chart. Thus it is essential that the user keep the annotations in Figure 8-6(b) in mind. Yet another factor that makes it difficult to develop essential Smith chart skills.

EXAMPLE 8.1 Impedance Plotting

Plot the normalized impedances $z_A = 1 - 2\jmath$ and $z_B = 0.3 + 0.6\jmath$ on an impedance Smith chart.

Solution:

The impedance $z_A = 1 - 2\jmath$ is plotted as Point A to the right. To plot this first identify the circle of constant normalized resistance $r = 1$, and then identify the arc of constant normalized reactance $x = -2$. The intersection of the circle and arc locates z_A at point A. The reader is encouraged to do this with the full impedance Smith chart as shown in Figure 8-5. Recall that signs of reactances are missing on the full chart. As an exercise read off the reflection coefficient (the answer is $0.5 - \jmath 0.5 = 0.707\angle -45°$).

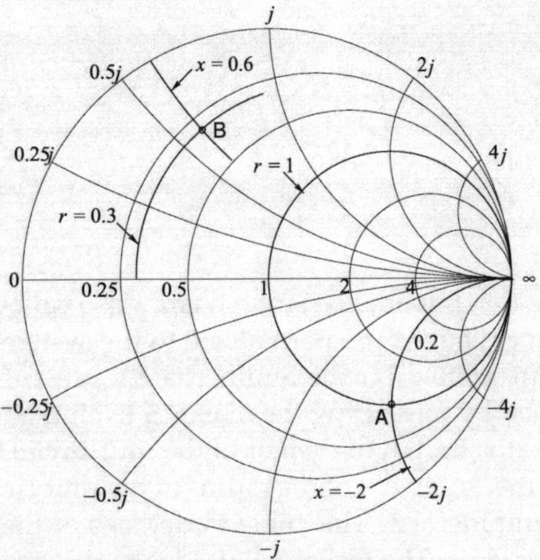

The impedance $z_B = 0.3 + 0.6\jmath$ is plotted as Point B which is at the intersection of the circle $r = 0.3$ and the arc $x = +0.6$. Interpolation is required to identify the required circle and arc. The reader should do this with the full impedance Smith chart. As an exercise read off the reflection coefficient (the answer is $-0.268 + \jmath 0.585 = 0.644\angle 115°$).

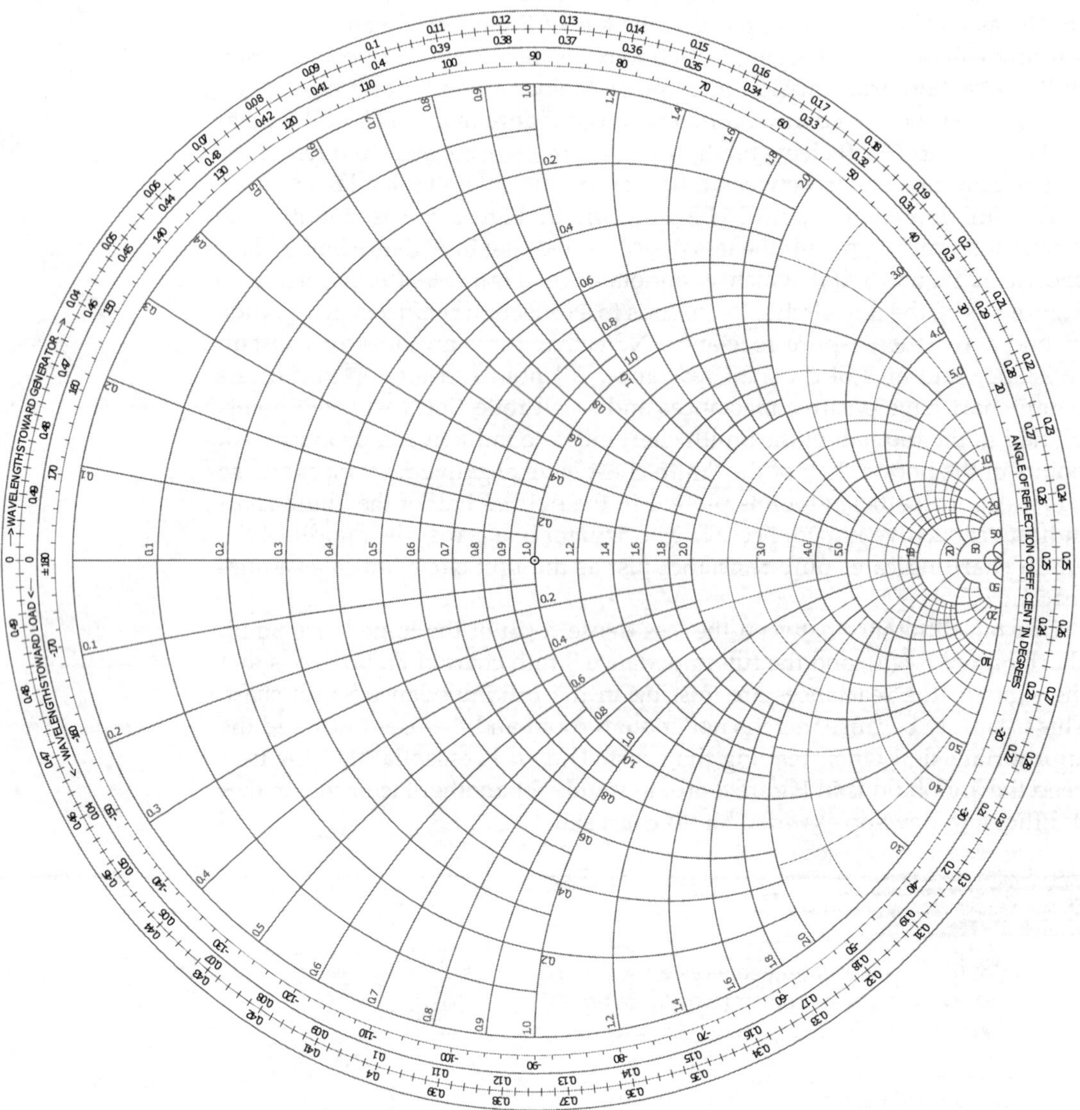

Figure 8-5: Impedance Smith chart. Also called a normalized Smith chart since resistances and reactances have been normalized to the system reference impedance Z_{REF}: $z = Z/Z_{\text{REF}}$.
A point plotted on the Smith chart represents a complex number A. The magnitude of A is obtained by measuring the distance from the origin of the polar plot (the same as the origin of the Smith chart in the center of the unit circle) to point A, say using a ruler, and comparing that to the measurement of the radius of the unit circle which corresponds to a complex number with a magnitude of 1. The angle in degrees of the complex number A is read from the innermost circular scale. The technique used is to draw a straight line from the origin through point A out to the circular scale.

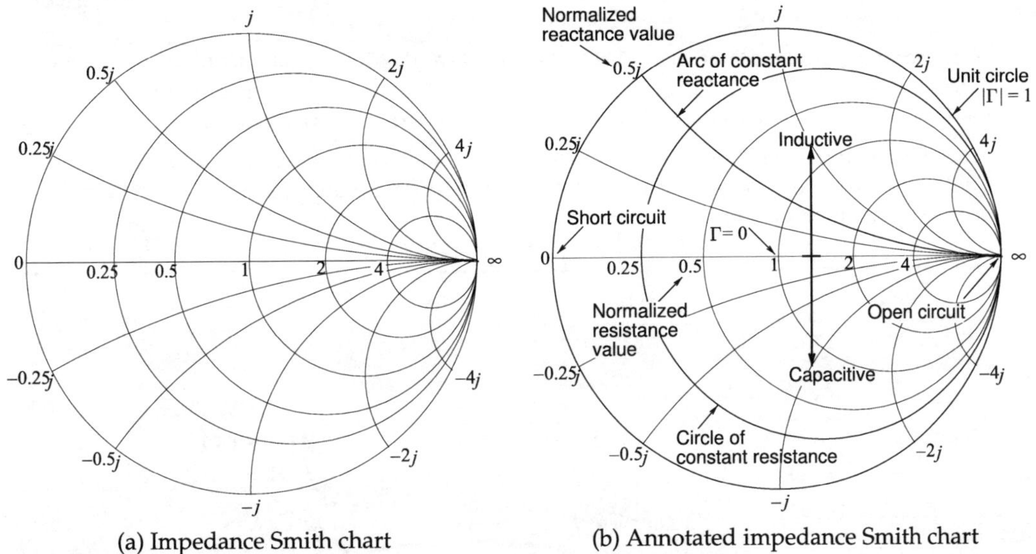

(a) Impedance Smith chart (b) Annotated impedance Smith chart

Figure 8-6: Normalized impedance Smith charts.

EXAMPLE 8.2 Impedance Synthesis

Use a length of terminated transmission line to realize an impedance of $Z_{\text{in}} = \jmath 140\ \Omega$.

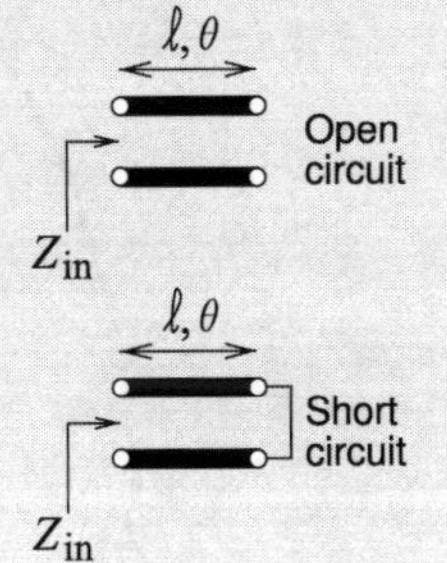

Solution:
The impedance to be synthesized is reactive so the termination must also be lossless. The simplest termination is either a short circuit or an open circuit. Both cases will be considered. Choose a transmission line with a characteristic impedance, Z_0, of 100 Ω so that the desired normalized input impedance is $\jmath 140\ \Omega/Z_0 = 1.4\jmath$, plotted as point B in Figure 8-7.

First, the short-circuit case. In Figure 8-7, consider the path AB. The termination is a short circuit and the impedance of this load is Point A with a reference length of $\ell_A = 0\ \lambda$ (from the outermost circular scale). The corresponding reflection coefficient reference angle from the scale is $\theta_A = 180°$ (from the innermost circular scale labeled 'WAVELENGTHS TOWARDS THE GENERATOR' which is the same as 'wavelengths away from the load'). As the line length increases, the input impedance of the terminated line follows the clockwise path to Point B where the normalized input impedance is $\jmath 1.4$. (To verify your understanding that the locus of the refection coefficient rotates in the clockwise direction, i.e. increasingly negative angle as the line length increases, see Section 3.3.3 .) At Point B the reference line length $\ell_B = 0.1515\ \lambda$ and the corresponding reflection coefficient reference angle from the scale is $\theta_B = 71.2°$. The reflection coefficient angle and length in terms of wavelengths were read directly off the Smith chart and care needs to be taken that the right sign and correct scale are used. A good strategy is to correlate the scales with the easily remembered properties at the open-circuit and short-circuit points. Here the line length is

$$\ell = \ell_B - \ell_A = 0.1515\lambda - 0\lambda = 0.1515\lambda, \tag{8.4}$$

and the electrical length is half of the difference in the reflection coefficient angles,

$$\theta = \tfrac{1}{2}\,|\theta_B - \theta_A| = \tfrac{1}{2}\,|71.2° - 180°| = 54.4°, \tag{8.5}$$

corresponding to a length of $(54.4°/360°)\lambda = 0.1511\lambda$ (the discrepancy with the previously determine line length of 0.1515λ is small). This is as close as could be expected from using the scales. So the length of the stub with a short-circuit termination is 0.1515λ.

For the open-circuited stub, the path begins at the infinite impedance point $\Gamma = +1$ and rotates clockwise to Point A (this is a 90° or 0.25λ rotation) before continuing on to Point B. For the open-circuited stub,

$$\ell = 0.1515\lambda + 0.25\lambda = 0.4015\lambda. \quad (8.6)$$

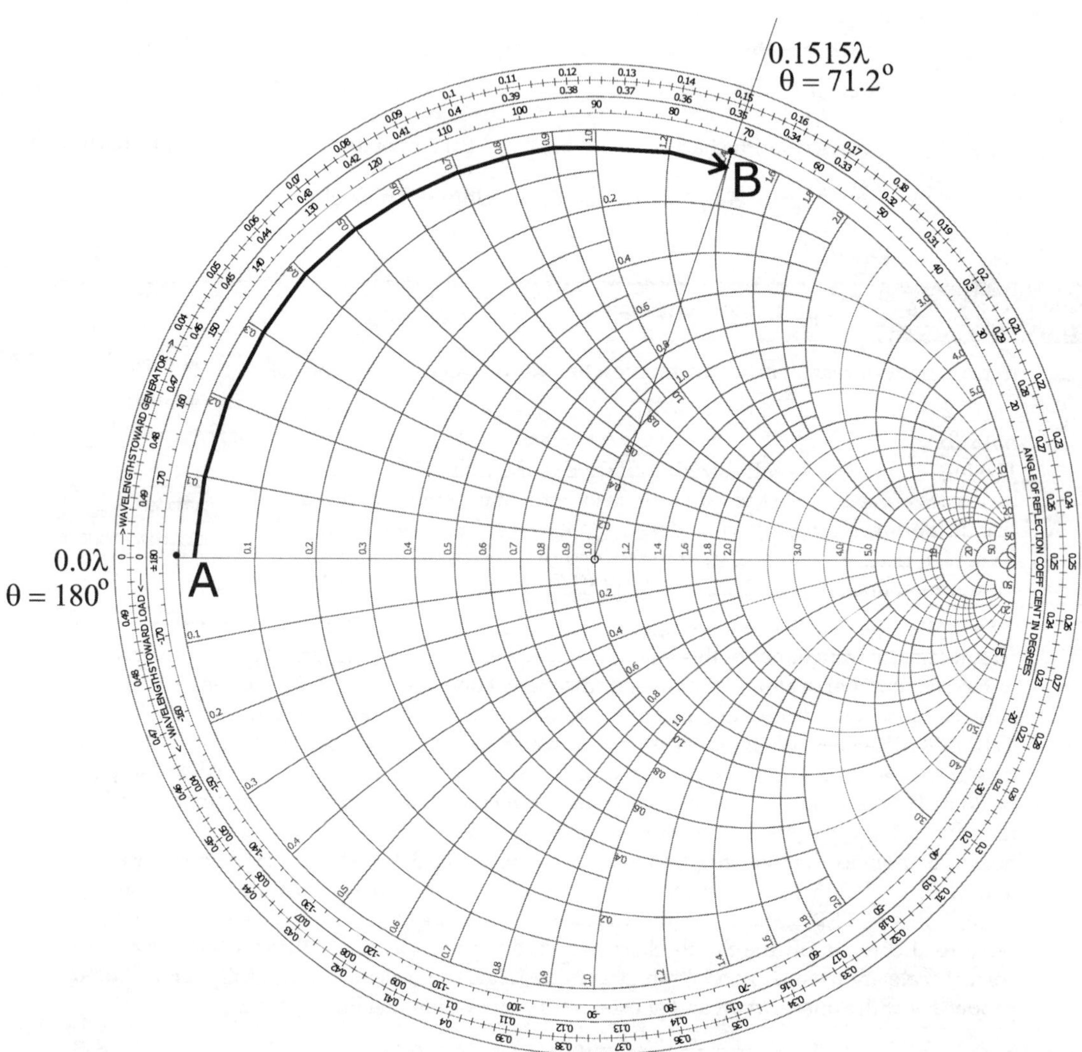

Figure 8-7: Design of a short-circuit stub with a normalized input impedance of $\jmath 1.4$. The path AB is actually on the unit circle but has been displaced here to avoid covering numbers. The electrical length in wavelengths has been read from the outermost circular scale, and the angle, Θ, in degrees refers to the angle of the polar plot (and is twice the electrical length).

8.3.2 Admittance Smith Chart

The admittance Smith chart has loci for discrete constant susceptances ranging from $-\infty$ to ∞, and for discrete constant conductance ranging from 0 to ∞, see Figure 8-8. A less dense form is shown in Figure 8-9(a). This chart looks like the flipped version of the impedance Smith chart but it is the same polar plot of a reflection coefficient so that the positions of the open and short circuit remain the same as do the capacitive and inductive halves of the Smith chart. In the full version of the admittance Smith chart, Figure 8-8, signs have been dropped as there is not room for them. Thus interpreting admittances from the chart requires that the user separately determine the signs of susceptances. The less dense version, Figure 8-9(a),

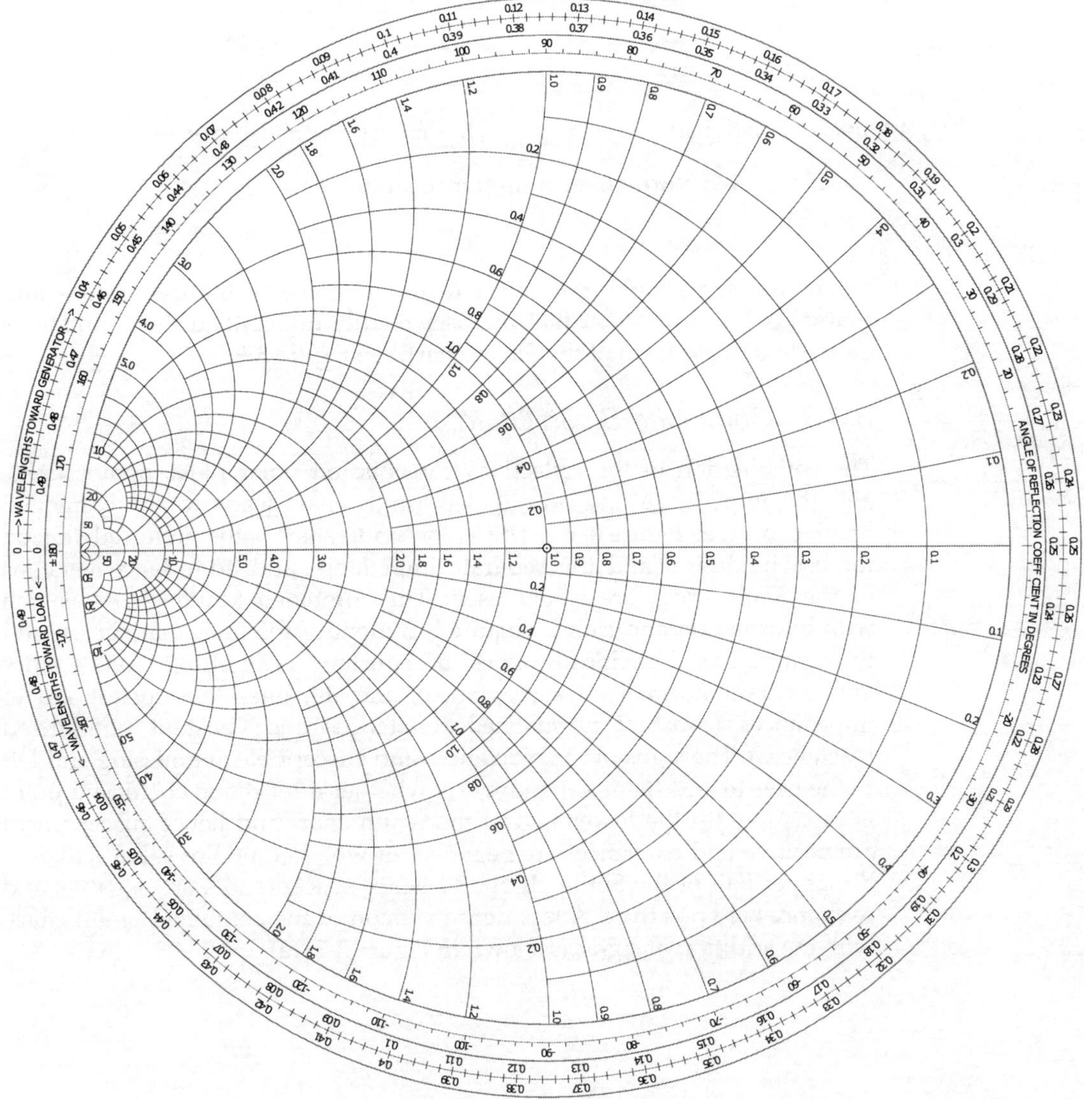

Figure 8-8: Admittance Smith chart.

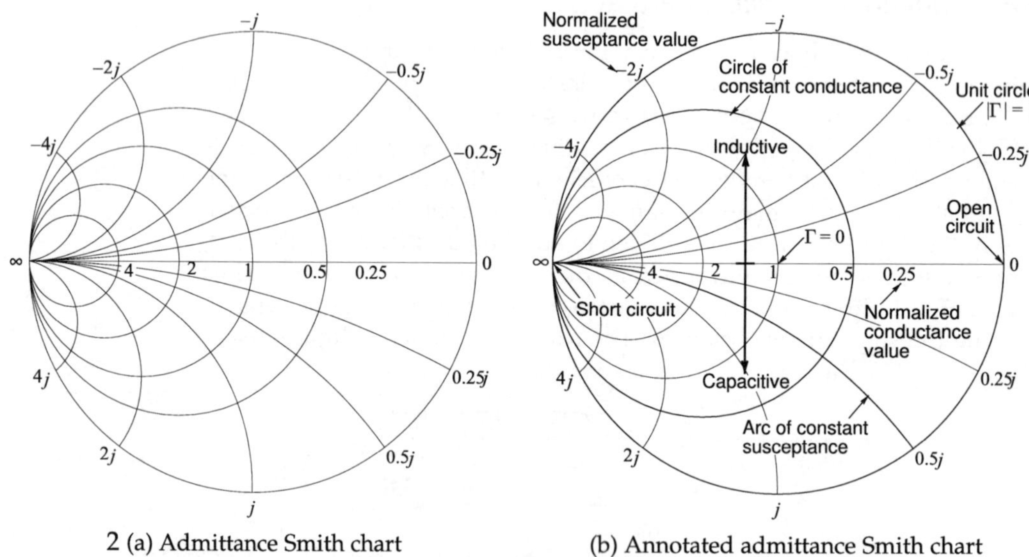

2 (a) Admittance Smith chart (b) Annotated admittance Smith chart

Figure 8-9: Normalized admittance Smith chart.

retains the signs making it easier to follow some of the discussions and examples. It is important that the user readily understand the annotations on the less dense form of the Smith chart, see Figure 8-9(b).

8.3.3 Combined Smith Chart

The combination of the reflection/transmission polar plots, nomographs, and the impedance and admittance Smith chart leads to the combined Smith chart (see Figure 8-10). This color Smith chart is the preferred version for use in design and the separate impedance and admittance versions of the Smith chart are rarely used. The combined Smith charts is rich with information and care is required to identify the lines that correspond to admittances (specifically lines of constant normalized conductance and constant normalized susceptance), and the lines that correspond to impedances (constant normalized resistances and constant normalized reactances). The signs of the reactances and susceptible are missing and left to the user to add them depending on whether a reflection coefficient point is capacitive (in the lower half of the Smith chart and hence susceptances are positive and reactances are negative) or whether a point is inductive (in the upper half of the Smith chart and hence susceptances are negative and reactances are positive). A less dense version of the combined Smith chart, with the addition of signs, is shown in Figure 8-11(a).

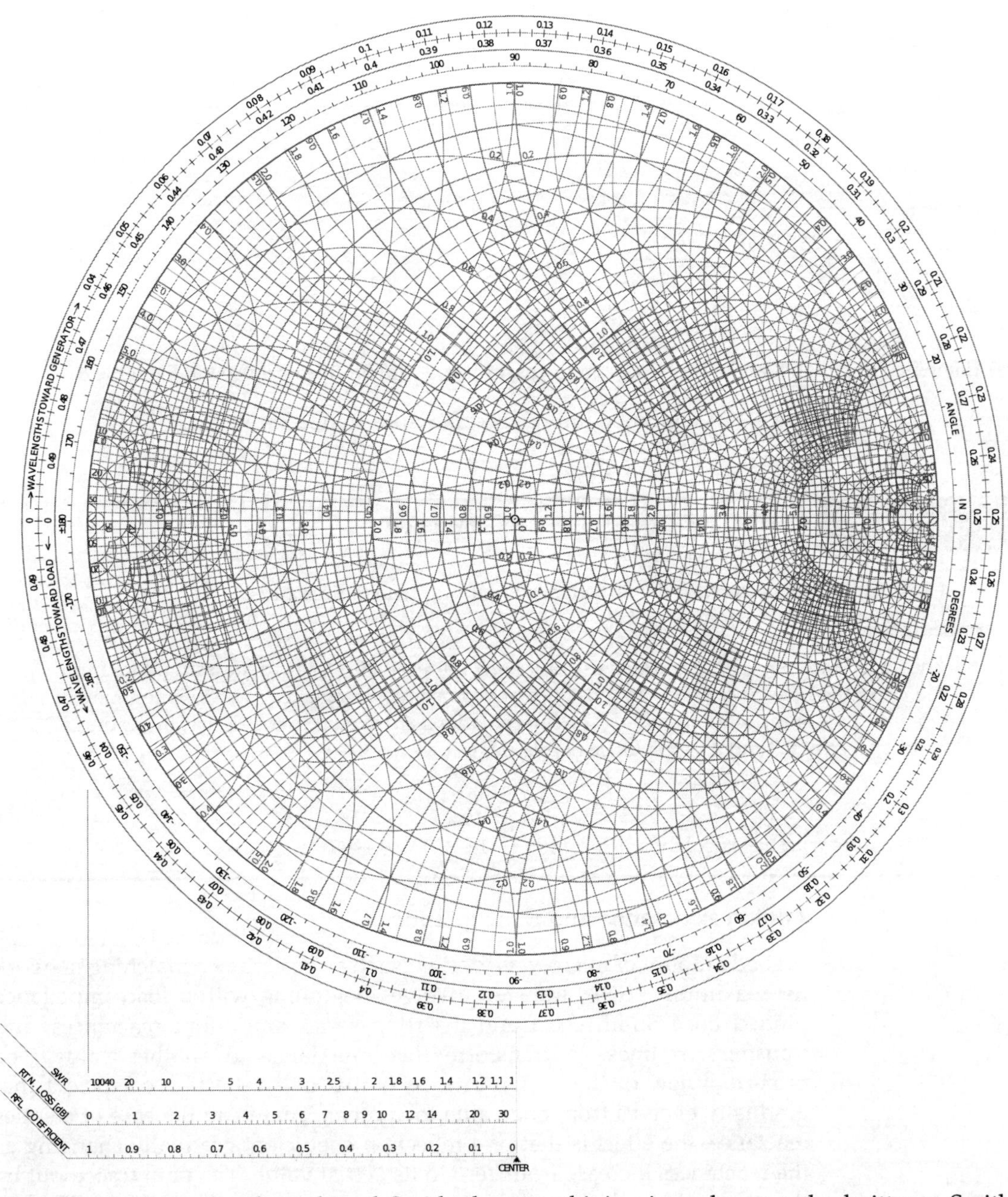

Figure 8-10: Normalized combined Smith chart combining impedance and admittance Smith charts.

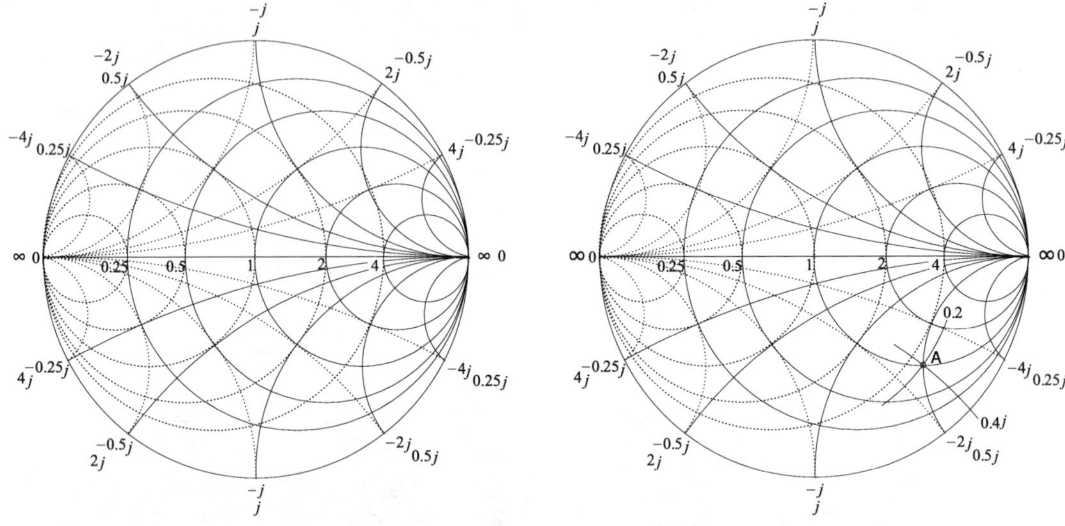

(a) Combined impedance and admittance Smith chart (b) Smith chart used in Example 8.3

Figure 8-11: Combined impedance and admittance Smith chart.

EXAMPLE 8.3 Impedance to Admittance Conversion

Use a Smith chart to convert the impedance $z = 1 - 2\jmath$ to an admittance.

Solution:

The impedance $z = 1-2\jmath$ is plotted as Point A in Figure 8-11(b). To read the admittance from the chart, the lines of constant conductance and constant susceptance must be interpolated from the arcs and circles provided. The interpolations are shown in the figure, indicating a conductance of 0.2 and a susceptance of $0.4\jmath$. Thus

$$y = 0.2 + 0.4\jmath.$$

(This agrees with the calculation: $y = 1/z = 1/(1 - 2\jmath) = 0.2 + 0.4\jmath$.)

Adding Reactance and Susceptance

A good amount of microwave design, such as designing a matching network for maximum power transfer, involves beginning with a load impedance plotted on a Smith chart and inserting series and shunt reactances, and transmission lines, to transform the impedance to another value. The preferred view of the design process is that of a reflection coefficient that gradually evolves from one value to another. That is, in the case of a series reactance, the effect is that of a reflection coefficient gradually changing as the reactances increase from zero to its actual value. The path traced out by the gradually evolving reflection coefficient value is called a locus. The loci of common circuit elements added to various loads are shown in Figure 8-12. For each locus the load is at the start of the arrow with the value of the element increasing from zero to its actual final value at the arrow head.

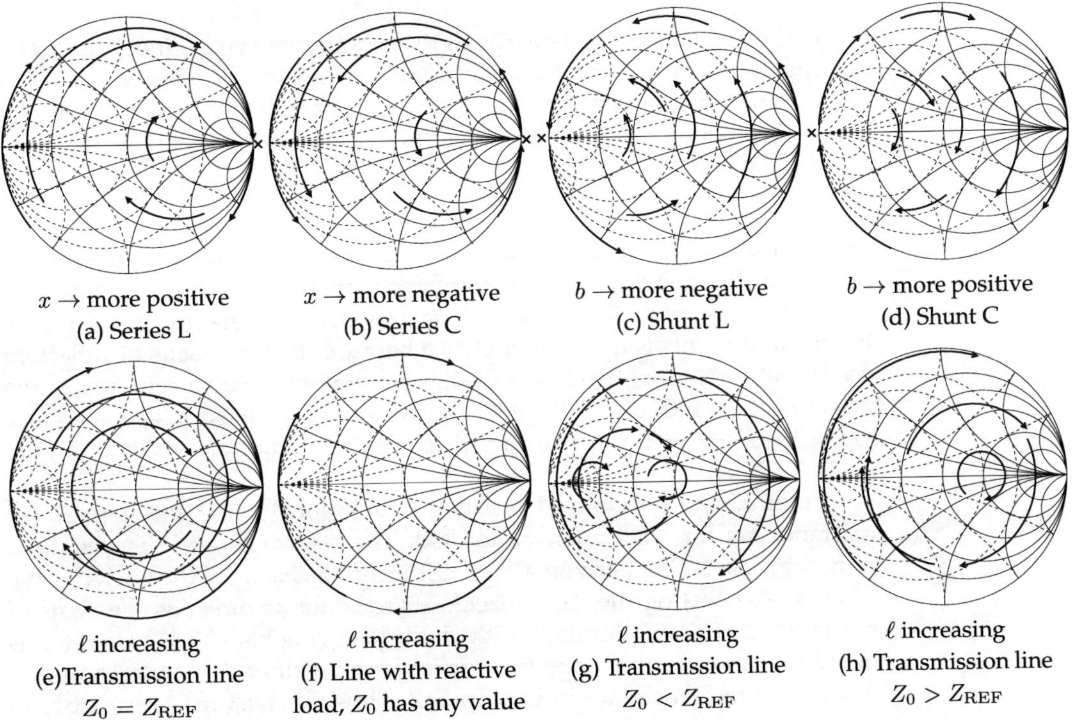

Figure 8-12: Reflection coefficient loci of a load augmented by elements as indicated. The original load is at the start of the arrowed arc. In (a) and (b) each locus is with respect to increasing $|x|$ (normalized reactance magnitude), in (c) and (d) the locus is with respect to increasing $|b|$ (normalized susceptance magnitude), and in (e)–(h) with respect to increasing length ℓ. In (e)–(h) the loci are parts of circles centered on the $x = 0$ line. x indicates that the locus cannot cross infinity (open circuit for (a) and (b), short circuit for (c) and (d)).

8.3.4 *Smith Chart Manipulations*

Microwave design often proceeds by taking a known load and transforming it into another impedance, perhaps for maximum power transfer. Smith charts are used to show these manipulations. In design the manipulations required are not known up front and the Smith chart enables identification of those required. Except for particular situations, lossless elements such as reactances and transmission lines are used. The few situations where resistances are introduced include introducing stability in oscillators and amplifiers, and deliberately reduce signals levels. Most of the time introducing resistances unnecessarily increases noise and reduces signal levels thus reducing critical signal-to-noise ratios.

The circuit in Figure 8-13 will be used to illustrate the manipulations on a Smith chart. The most common view is to consider that the Smith chart is a plot of the reflection coefficient at various stages in the circuit, i.e., Γ_A, Γ_B, Γ_C, Γ_D, and Γ_{IN}. Additionally the load, Z_L, is plotted and reactances, susceptances, or transmission lines transform the reflection coefficient from one stage to the next. Yet another concept is the idea that the effect of an element is regarded as gradually increasing from a negligible value up to the final actual value and in so doing tracing out a locus which ends in an arrow head. This is the approach nearly all RF and microwave engineers use. The manipulations corresponding to the circuit are illustrated in Figure 8-14. The individual steps are identified in Figure 8-15. The first few steps are confined to the top half of the Smith chart which is repeated on a larger scale in Figure 8-16 for the first three steps.

Step 1 L, a, b, c, d, e, f
The first step is to plot the load Z_L on a Smith chart and this is the Point L in Figures 8-14–8-16. The reference impedance Z_{REF} is chosen to be 50 Ω, the same as the characteristic impedance of the transmission line in the circuit. This is a common choice because then the locus of reflection coefficient variation introduced by the line will be a circle centered at the origin of the Smith chart. To plot L derive the normalized load impedance $z_L = Z_L/Z_{\mathrm{REF}} = 0.3 + \jmath 0.6$ (future numerical values are given in Figure 8-13).

To plot z_L the normalized resistance circle for 0.3 and the normalized reactance arc for +0.3 must be located. The resistance circle is identified from the scale on the horizontal axis of the Smith chart, see the circled value labeled 'a'. Locating the +0.3 reactance arc is not as direct as the signs of reactance are missing on the Smith chart. Referring back to Figure 8-6 it is noted that an inductive impedance is in the top half of the Smith chart and so positive reactances are in the top half. Thus the +0.3 reactance arc is in the top half of the Smith chart and the correct arc is identified by 'c'. From the curves identified by 'a' and 'c' the arcs 'b' and 'd' are drawn with the impedance z_L, i.e. point L, at the intersection of the arcs.

It is instructive to determine Γ_L, the reflection coefficient at L. On the Smith chart the reflection coefficient vector Γ_L is drawn from the origin to the point L. Γ_L is evaluated by separately determining its magnitude and angle. To determine the magnitude measure the length of the $\Gamma_L|$ vector either using a ruler, a compass, or marking the edge of a piece of paper. Then this length can be compared against the reflection coefficient scale shown at the bottom of Figure 8-14 yielding $|\Gamma_L| = 0.644$. Alternatively the distance from the origin to the unit circle can be measured using a ruler and the ratio of the $|\Gamma_L|$ measurement and of the unit circle measurement taken as the value of $|\Gamma_L|$. The angle of Γ_L is read by extending the Γ_L vector out to the inner most circular scale on the Smith chart. This extension is labeled as e. The angle is

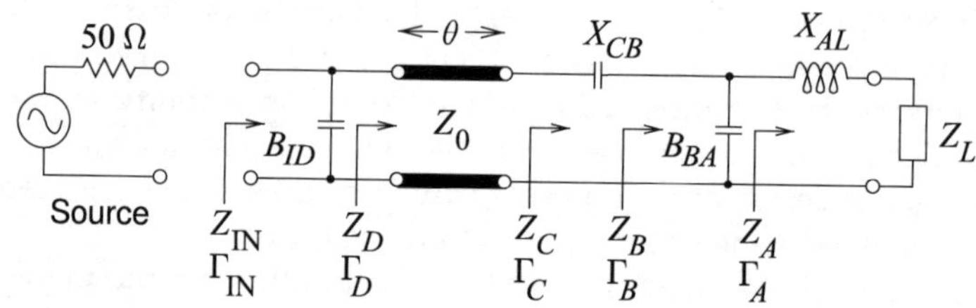

Circuit elements	Normalized	Derived values	
$Z_L = 15 + \jmath 30$ Ω	$z_L = 0.3 + \jmath 0.6$	$\Gamma_L = 0.644\angle 115°$	$z_A = 0.300 + \jmath 1.200$
$X_{AL} = 30$ Ω	$x_{AL} = 0.600$	$\Gamma_A = 0.785\angle 115°$	$y_A = 0.196 - \jmath 0.784$
$B_{BA} = 4.6$ mS	$b_{BA} = 0.230$	$\Gamma_B = 0.741\angle 59°$	$y_B = 0.196 - \jmath 0.554$
$X_{CB} = -180$ Ω	$x_{CB} = -3.600$	$\Gamma_C = 0.805\angle -50°$	$z_B = 0.567 + \jmath 1.603$
$Z_0 = 50$ Ω	$Z_0 = 1$	$\Gamma_D = 0.805\angle 144°$	$z_C = 0.567 - \jmath 1.997$
$\Theta = 83°$	$\Theta = 83°$	$\Gamma_{\mathrm{IN}} = 0$	$y_D = 0.998 - \jmath 2.71$
$B_{ID} = 54.2$ mS	$b_{ID} = 2.71$		$z_{\mathrm{IN}} = 1$, $y_{\mathrm{IN}} = 1$

Figure 8-13: Circuit used in illustrating Smith chart manipulations. $Z_{\mathrm{REF}} = 50$ Ω, $Y_{\mathrm{REF}} = 1/Z_{\mathrm{REF}} = 20$ mS, $z = Z/Z_{\mathrm{REF}}$, $y = Y/Y_{\mathrm{REF}}$, $Z_{\mathrm{IN}} = Z_{\mathrm{REF}} = 50$ Ω. In the circuit B indicates susceptance and X indicates reactance.

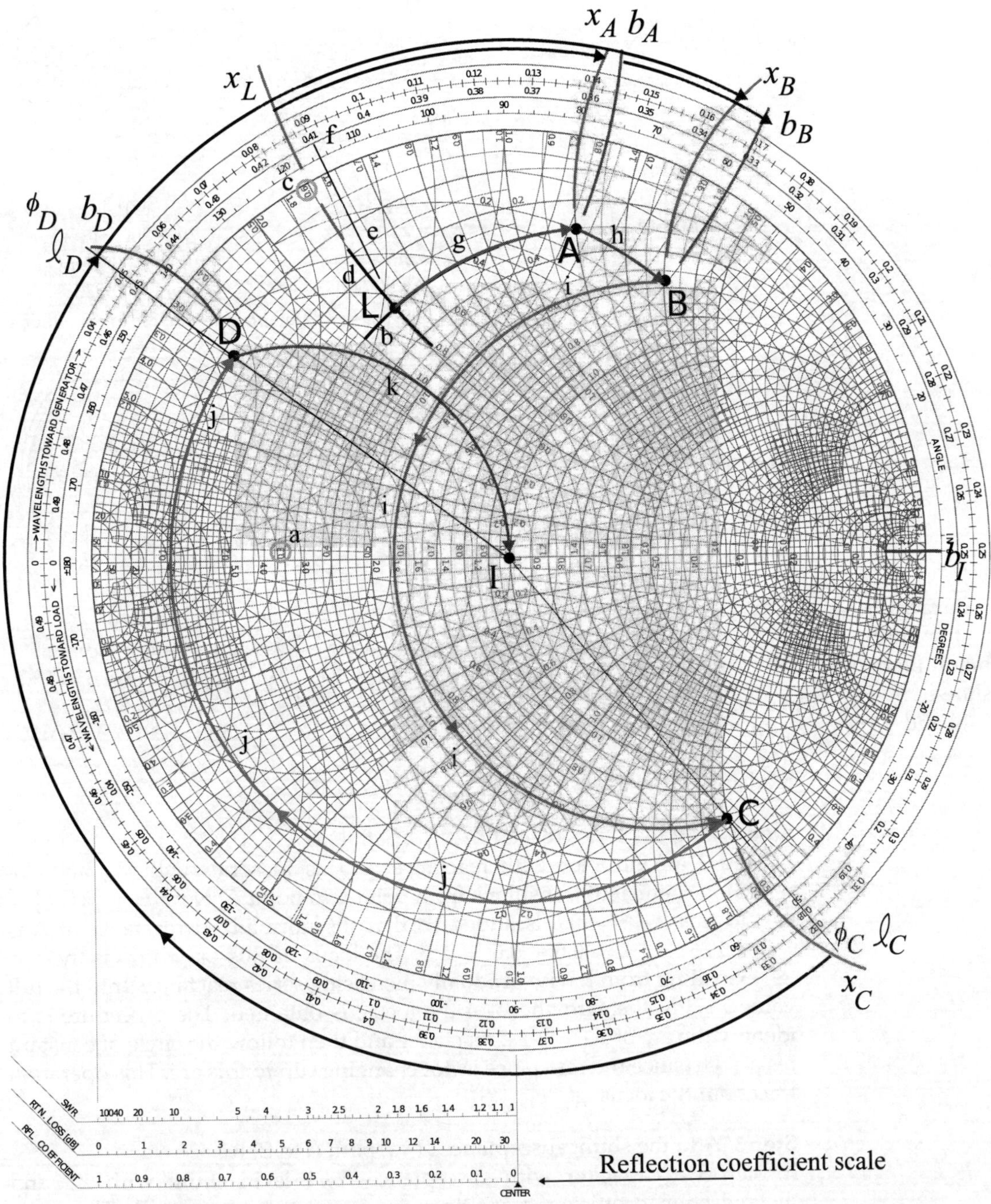

Figure 8-14: Smith chart manipulations corresponding to the circuit in Figure 8-13 with circuit elements added one at a time beginning with the load impedance at Point L.

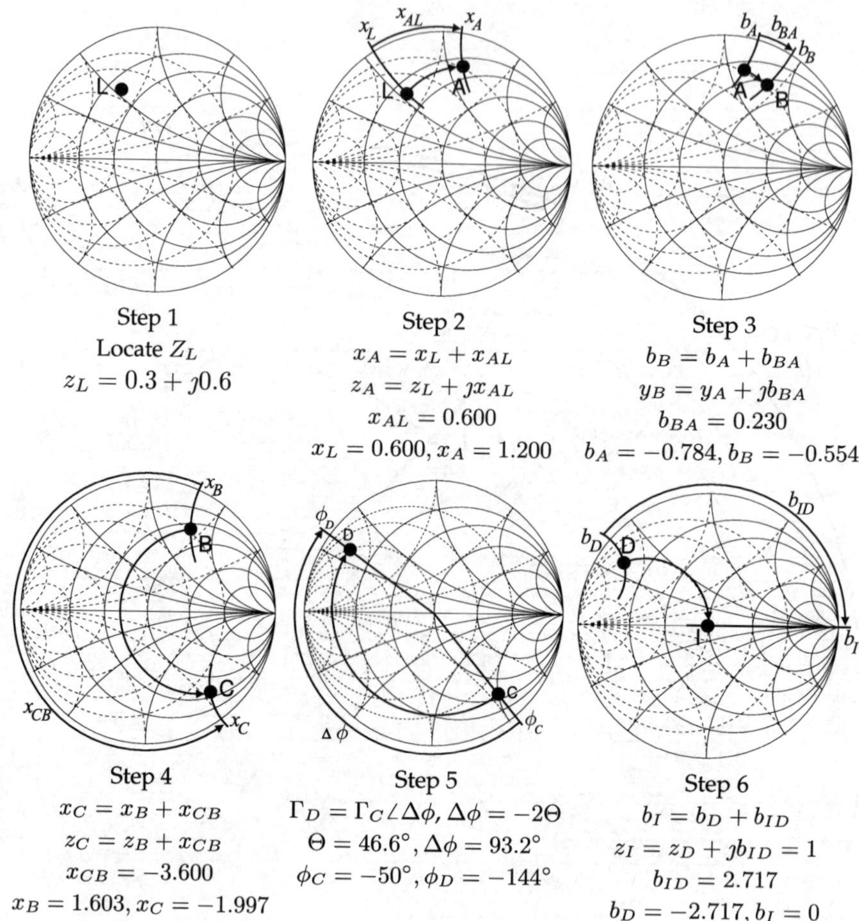

Figure 8-15: Steps in the Smith chart manipulations shown in Figures 8-14 and 8-16.

read at point 'f' as 115°. Thus $\Gamma_L = 0.644\angle 115°$.

Step 2 Add the series reactance X_{AL}, path L to A, g
In step 2 an inductor with reactance X_{AL} is in series with Z_L and the reflection coefficient transitions from point L to point A. Now $x_L = \Im\{z_L\} = 0.6$. To this $x_{AL} = 0.6$ is added so that the normalized reactance at A is $x_A = x_L + x_{AL} = 0.6 + 0.6 = 1.2$. The locus of this operation is the arc, 'g', extending from L with gradually increasing series reactance until the full value x_{AL}, at the arrowhead of the locus, is obtained. The procedure is to identify the arc of $x_A = +1.2$ reactance and then follow the circle of constant resistance (since the resistance is not changing) up to this arc. This operation traces out the locus 'g'.

Step 3 Add the shunt susceptance B_{BA}, path A to B, h
In step 3 a capacitor with susceptance X_{BA} is in shunt with Z_A and the reflection coefficient transitions from point A to point B. The locus of this operation follows a circle of constant conductance with only they susceptance changing. The final value must be calculated as $b_B = b_A + b_{BA}$. The susceptance b_B is read from the graph as -0.784. This value is read by following the arc of constant susceptance out to the unit circle where a value of 0.784 is read. Note that the arc must be interpolated and that the user must realize that the susceptance is negative in the top half of the

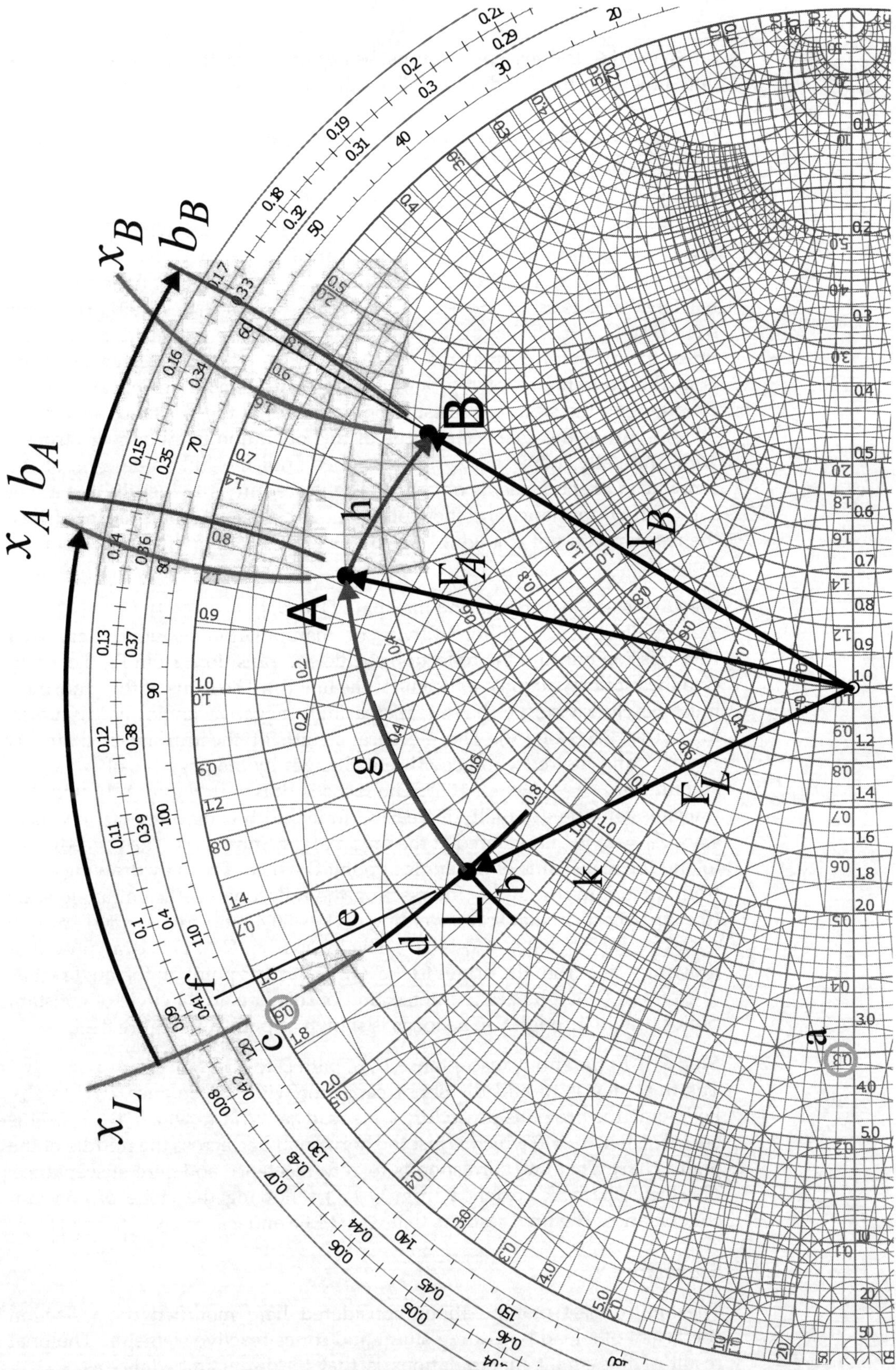

Figure 8-16: The first three steps in the Smith chart manipulations in Figure 8-14.

Smith chart so the susceptance must be negative even though a minus sign is not shown on the Smith chart. Thus the correct reading for x_A is -0.784. Now $b_{BA} = +0.230$ so the locus, 'h', of the transition follows the circle of constant conductance ending at the (interpolated arc) with susceptance $b_B = b_a + b_{BA} = -0.784 + 0.230 = -0.554$.

Step 4 Add the series reactance X_{CB}, path B to C, i
In step 4 a capacitor with reactance X_{CB} is in series with Z_B and the reflection coefficient transitions from point B to point C. Now x_B is read from the graph as $+1.603$. To this add $x_{CB} = -3.600$ so that the normalized reactance at C is $x_C = x_B + x_{CB} = 1.603 + (-3.600) = -1.997$. The locus from B to C, path 'i' begins at B and follows a circle of constant resistance up to the arc with normalized reactance $x_C = -1.997$. This reactance arc is in the bottom half of the Smith chart as reactances are negative there even though the signs are missing on the labels of the reactance arcs in the bottom half of the chart. The locus of this operation is the arc, 'g', from L with gradually increasing series reactance until the full value z_{CB}, at the arrowhead of the locus, is obtained. The procedure is to identify the arc of $x_C = -1.997$ reactance and then follow the circle of constant resistance (since the resistance is not changing) up to this arc. This operation traces out the locus 'i'.

Step 5 Insert the transmission line, path C to D, j
Step 5 illustrates a different type of manipulation as now there is a transmission line and the reflection coefficient transitions from Γ_C, i.e. point C to the input reflection coefficient of the line Γ_D. The locus of this transition must be clockwise, i.e. having increasingly negative angle (as discussed in Section 3.3.3 of [5]). The electrical length of the transmission line is $\Theta = 83°$ and the reflection coefficient changes by the negative of twice this amount, $\phi_{DC} = -2\Theta = -166°$. The locus is drawn in Figure 8-14 with the transmission line gradually increasing in length tracing out a circle which, since $Z_0 = Z_{\text{REF}}$, is centered at the origin of the Smith cart. The procedure is to find the scale value of the angle at point C which is read by drawing a line from the origin through C intersecting the reflection coefficient angle scale (the innermost circular scale) yielding $\phi_C = -50.4°$ and so $\phi_D = \phi_C + \phi_{DC} = -50.4° - 166° + 360°) = +143.6°$. The locus from C to D is drawn by first drawing a line from the origin to the $\phi_D = -143.6°$ point on the angle scale. Then point D will be at the intersection of this line and a circle of constant radius drawn through C. The locus is shown as path 'j' in Figure 8-14.

Step 6 Add the shunt susceptance B_{ID}, path D to I, k
The final step is to add the shunt capacitor with susceptance B_{ID} to Z_D. Following the previous procedure b_D is read as -2.71 to which $b_{ID} = 2.71$ is added so that $b_I = 0$ which is just the horizontal line across the middle of the Smith chart. This line corresponds to zero reactance and zero susceptance. The locus, path 'k', extends from D to I following the circle of constant conductance. The final result is that $z_{\text{IN}} = 1.00$ and $Z_{\text{IN}} = z_{\text{IN}} \cdot Z_0 = 50\ \Omega$.

Summary

The Smith chart manipulations considered here modified the reflection coefficient of a load by adding shunt and series reactive elements. The final result of the circuit manipulations is that the input impedance is $Z_{\text{IN}} =$

50 Ω. If the source has a Thevenin source impedance of 50 Ω then there is maximum power transfer to the circuit. Since all of the elements in the circuit manipulations are lossless this means that there is maximum power transfer from the source to the load Z_L. Of course fewer circuit manipulations could have been used to achieve this result. Note that manipulations resulting from adding series and/or shunt resistances were not considered. It is rare that this would be desired as that simply means that power is absorbed in the resistance and additional noise is added to the circuit.

EXAMPLE 8.4 Reflection Coefficient of a Shorted Microstrip Line on a Smith Chart

AWR Design Environment Project File: RFDesign_Shorted_Microstrip_Line_Smith.emp

A shorted microstrip line on an alumina substrate is shown in Figure 8-17. The line has low loss and so Γ is always close to 1. The microstrip line was designed to be 50 Ω and Γ of the low loss line is plotted on a 50 Ω Smith chart in Figure 8-18(a). Plotting Γ on a 50 Ω Smith chart indicates that the reflection coefficient was calculated with respect to 50 Ω. At a very low frequency, 0.1 GHz is the lowest frequency here, the locus of the reflection coefficient is very close to $\Gamma = -1$, identifying a short circuit. As the electrical length increases, in this case the frequency increases as the physical length of the line is fixed, the locus of the reflection coefficient moves clockwise, hugging the unity reflection coefficient circle. At the highest frequency, 30 GHz, the reflection coefficient is less than 1 and the locus starts moving in from the unity circle. It is interesting to see what happens with a high-loss line, and this is achieved by changing the loss tangent, $\tan\delta$, of the substrate from 0.001 to 0.1. The locus of the reflection coefficient of the high-loss line is shown in Figure 8-18(b). Loss increases as the electrical length of the line increases and the locus of the reflection coefficient traces out a clockwise inward spiral.

8.3.5 An Alternative Admittance Chart

Design often requires switching between admittance and impedance. So it is convenient to use the colored combined Smith chart shown in Figure 8-10. Monochrome charts were once the only ones available and an impedance Smith chart was used for admittance-based calculations provided that reflection coefficients were rotated by 180°. This form of the Smith chart is not used now and is very confusing.

8.3.6 Summary

The Smith chart is the most powerful of tools used in RF and Microwave Design. Design using Smith charts will be considered in other chapters but at this stage the reader should be totally conversant with the techniques described in this section.

8.4 Summary

Graphical representations of power flow enable RF and microwave engineers to quickly ascertain circuit performance and arrive at qualitative design decisions. Humans are very good at processing graphical information and seeing patterns, anomalies, and the path from one point to another.

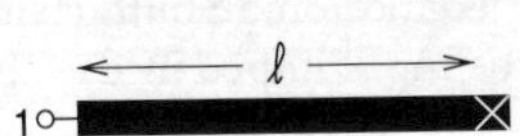

Figure 8-17: A 50 Ω shorted gold microstrip line with width w = 500 μm, length ℓ = 1 cm on a 600 μm thick alumina substrate with relative permittivity $\varepsilon_r = 9.8$ and loss tangent $\tan\delta = 0.001$.

The Smith chart is a richly annotated polar plot for representing reflection and transmission coefficients, and more generally, scattering parameters. The Smith chart representation of scattering parameter data aligns very well with the intuitive understanding of an RF designer. The experienced RF designer is intrinsically familiar with the Smith chart and prefers that circuit performance during design be represented on one. Representing something as simple as an extension of a line length to a two-port is quite complex if described using network parameters other than scattering parameters. However, with scattering parameters this extension results in a change of the angle of a scattering parameter, or on a Smith chart an arc. Scattering parameters relate directly to power flow. So from a Smith chart an experienced designer can ascertain the effect of circuit design on power flow, which then relates to signal-to-noise ratio and power gain.

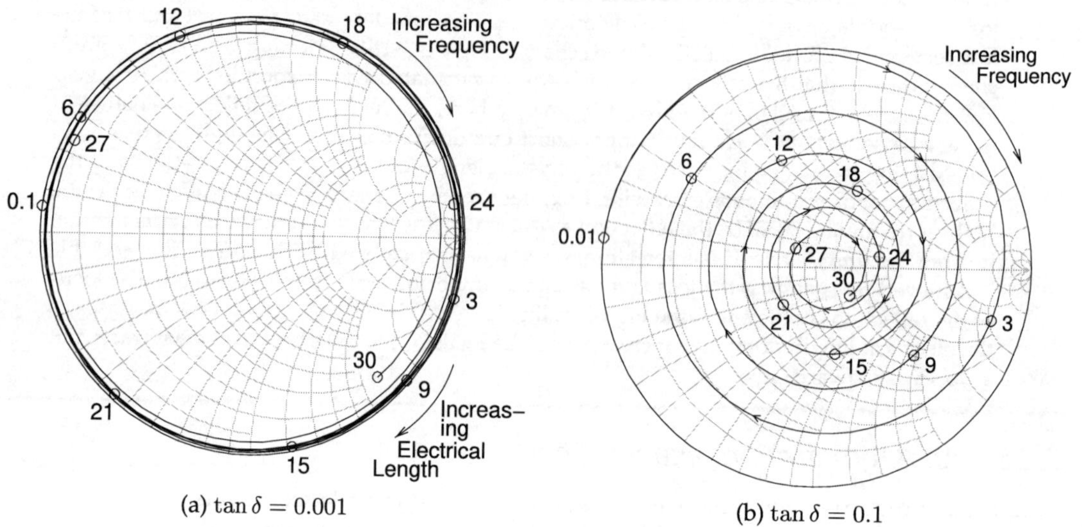

Figure 8-18: Smith chart plot of the reflection coefficient of the shorted 1 cm-long microstrip line in Figure 8-17: (a) a low-loss substrate with $\tan\delta = 0.001$; and (b) a high-loss substrate with $\tan\delta = 0.1$. Frequencies are marked in gigahertz.

8.5 References

[1] P. Smith, *Electronic Applications of the Smith Chart: In Waveguide, Circuit, and Component Analysis*. SciTech Publishing, 2000.

[2] ——, "Transmission line calculator," *Electronics*, 1939.

[3] ——, "An improved transmission-line calculator," *Electronics*, 1944.

[4] ——, "R.F. transmission line nomographs," *Electronics*, vol. 22, pp. 112–117, Feb. 1949.

[5] M. Steer, *Microwave and RF Design, Transmission Lines*, 3rd ed. North Carolina State University, 2019.

8.6 Exercises

1. In the distribution of signals on a cable TV system a 75 Ω coaxial cable is used, with a loss of 0.1 dB/m at 1 GHz. If a subscriber disconnects a television set from the cable so that the load impedance looks like an open circuit, estimate the input impedance of the cable at 1 GHz and 1 km from the subscriber. An answer within 1% is required. Estimate the error of your answer. Indicate the input impedance on a Smith chart, drawing the locus of the input impedance as the

line is increased in length from nothing to 1 km. (Consider that the dielectric filling the line has $\epsilon_r = 1$.)

2. A resistive load has a reflection coefficient with a magnitude of 0.7. If a transmission line is placed in series with the load, on a polar plot sketch the locus of the input reflection coefficient looking into the input of the terminated line as the line increases in electrical length from zero to $\lambda/2$. By reading the Smith chart, determine the normalized input impedance of the line when it has an electrical length of $\pi/2$.

3. A complex load has a reflection coefficient of $0.5+\jmath0.5$. If a transmission line is placed in series with the load, on a polar plot sketch the locus of the input reflection coefficient looking into the input of the terminated line as the line increases in electrical length from zero to $\pi/2$.

4. A resistive load has a reflection coefficient of -0.5. If a transmission line is placed in series with the load, on a polar plot sketch the locus of the input reflection coefficient looking into the input of the terminated line as the line increases in electrical length from 0 to $3\lambda/8$.

5. S_{21} of a two-port is 0.5. If a transmission line is placed in series with Port 1, on a polar plot sketch the locus of S_{21} of the augmented two-port as the electrical length of the line increases from zero to $\lambda/2$.

6. A load has an impedance $Z = 115 - \jmath20\ \Omega$.
   (a) What is the reflection coefficient, Γ_L, of the load in a 50 Ω reference system?
   (b) Plot the reflection coefficient on a polar plot of reflection coefficient.
   (c) If a one-eighth wavelength long lossless 50 Ω transmission line is connected to the load, what is the reflection coefficient, Γ_{in}, looking into the transmission line? (Again, use the 50 Ω reference system.) Plot Γ_{in} on the polar reflection coefficient plot of part (b). Clearly identify Γ_{in} and Γ_L on the plot.
   (d) On the Smith chart, identify the locus of Γ_{in} as the length of the transmission line increases from 0 to $\lambda/8$ long. That is, on the Smith chart, plot Γ_{in} as the length of the transmission line varies.

7. A load has a reflection coefficient with a magnitude of 0.5. If a transmission line is placed in series with the load, on a polar plot sketch the locus of the input reflection coefficient looking into the input of the terminated line as the line increases in electrical length from zero to $\lambda/2$. What is the normalized input impedance of the line when it has an electrical length of $\lambda/2$?

8. A resistive load has a reflection coefficient with a magnitude of 0.7. If a transmission line is placed in series with the load, on a polar plot sketch the locus of the input reflection coefficient looking into the input of the terminated line as the line increases in electrical length from zero to $\lambda/4$. By reading the Smith chart, determine the normalized input impedance of the line when it has an electrical length of $\lambda/4$.

Problems 9–15 refer to the normalized Smith chart in Figure 8-20 with reference impedance $Z_{\text{REF}} = 50\ \Omega$ and reflection coefficient Γ, voltage reflection coefficient ${}^V\Gamma$, current reflection coefficient ${}^I\Gamma$, and normalized impedance $z = r + \jmath x$ and admittance $y = g + \jmath b$. Γ should be given in magnitude-angle format.

9. (a) What is ${}^V\Gamma$ at A?
   (b) What is ${}^I\Gamma$ at A?
   (c) What is r at B?
   (d) What is z at C?
   (e) What is y at D?
   (f) What is $|\Gamma|$ at F?
   (g) What is b at I?
   (h) What is Γ at P?
   (i) What is Γ at D?
   (j) What is Γ at T?

10. (a) What is z at A?
   (b) What is y at I?
   (c) What is z at E?
   (d) What is y at Z?
   (e) What is y at H?
   (f) What is $|\Gamma|$ at W?
   (g) What is b at F?
   (h) What is x at K?
   (i) What is Γ at K?
   (j) What is Γ at R?

11. (a) What is y at A?
   (b) What is y at I?
   (c) What is z at G?
   (d) What is y at O?
   (e) What is y at V?
   (f) What is Γ at B?
   (g) What is x at C?
   (h) What is Γ at G?
   (i) What is z at L, label this z_L?
   (j) Use the Smith chart to find z_{in} of a 50 Ω $\lambda/8$-long line with load z_L?

12. (a) What is ${}^V\Gamma$ at M?
   (b) What is ${}^I\Gamma$ at M?
   (c) What is r at W?
   (d) What is z at Y?
   (e) What is y at V?
   (f) What is $|\Gamma|$ at B?
   (g) What is b at K?
   (h) What is Γ at V?
   (i) What is Γ at P?
   (j) What is Γ at N?

13. (a) What is z at M?
   (b) What is y at K?
   (c) What is z at S?
   (d) What is y at R?
   (e) What is g at B?
   (f) What is $|\Gamma|$ at F?
   (g) What is b at B?
   (h) What is x at I?
   (i) What is Γ at I?
   (j) What is Γ at Q?

14. (a) What is g at M?
   (b) What is r at K?
   (c) What is y at S?
   (d) What is z at R?
   (e) What is y at Z?
   (f) What is Γ at W?
   (g) What is x at Y?
   (h) What is Γ at T?
   (i) What is z at O, label this z_O?
   (j) Using the Smith chart find z_{in} of a $3\lambda/8$-long 50 Ω line with load z_O?

15. (a) What is g at P?
   (b) What is y at J?
   (c) What is Γ at L?
   (d) What is z at N?
   (e) What is y at S?
   (f) What is $|\Gamma|$ at U?
   (g) What is ${}^V\Gamma$ at X?
   (h) What is ${}^I\Gamma$ at X?
   (i) What is g at B?
   (j) What is x at I?

16. Design a short-circuited stub to realize a normalized susceptance of 2.15. Show the locus of the stub as its length increases from zero to its final length. What is the minimum length of the stub in terms of wavelengths?

17. Design a short-circuited stub to realize a normalized susceptance of -0.56. Show the locus of the stub as its length increases from zero to its final length. What is the minimum length of the stub in terms of wavelengths?

18. Design a short-circuited stub to realize a normalized susceptance of -2.2. Show the locus of the stub as its length increases from zero to its final length. What is the minimum length of the stub in terms of wavelengths?

19. A 75 Ω transmission line is terminated in a load with a reflection coefficient, Γ, normalized to 75 Ω, of $0.5\angle 45°$. If Γ at the input of the line is $0.5\angle -135°$, what is the minimum electrical length of the line in degrees.

20. An open-circuited 75 Ω transmission line has an input reflection coefficient with an angle of 40° what is the electrical length of the line in degrees? If there is more than one answer provide at least two correct answers.

21. Repeat Example 8.1 using the full impedance Smith chart of Figure 8-5.

22. Plot the normalized impedances $z_A = 0.5 + \jmath 0.5$, $z_A = 0.5 + \jmath 0.5$, and $z_B = 0.185 - \jmath 1.05$ on the full impedance Smith chart of Figure 8-5. [Parallels Example 8.1]

23. A 50 Ω lossy transmission line is shorted at one end. The line loss is 2 dB per wavelength. Note that since the line is lossy the characteristic impedance will be complex, but close to 50 Ω, since it is only slightly lossy. There is no way to calculate the actual characteristic impedance with the information provided. That is, problems must be solved with small inconsistencies. Microwave engineers do the best they can in design and always rely on measurements to calibrate results.
    (a) What is the reflection coefficient at the load (in this case the short)?
    (b) Consider the input reflection coefficient, Γ_{in}, at a distance ℓ from the load. Determine Γ_{in} for ℓ going from 0.1λ to λ in steps of 0.1λ.
    (c) On a Smith chart plot the locus of Γ_{in} from $\ell = 0$ to λ.
    (d) Calculate the input impedance, Z_{in}, when the line is $3\lambda/8$ long using the telegrapher's equation.
    (e) Repeat part (d) using a Smith chart.

24. In Figure 8-19 the results of several different experiments are plotted on a Smith chart. Each experiment measured the input reflection coefficient from a low frequency (denoted by a circle) to a high frequency (denoted by a square) of a one-port. Determine the load that was measured. The loads that were measured are one of those shown below.

Load	Description
i	An inductor
ii	A capacitor
iii	A reactive load at the end of a transmission line
iv	A resistive load at the end of a transmission line
v	A parallel connection of an inductor, a resistor, and a capacitor going through resonance and with a transmission line offset
vi	A series connection of a resistor, an inductor and a capacitor going through resonance and with a transmission line offset
vii	A series resistor and inductor
viii	An unknown load and not one of the above

For each of the measurements below indicate the load or loads using the load identifier above (e.g., i, ii, etc.).
    (a) What load(s) is indicated by curve A?
    (a) What load(s) is indicated by curve A?
    (b) What load(s) is indicated by curve B?
    (c) What load(s) is indicated by curve C?
    (d) What load(s) is indicated by curve D?
    (e) What load(s) is indicated by curve E?
    (f) What load(s) is indicated by curve F?

25. Design an open-circuited stub with an input impedance of $+\jmath 75$ Ω. Use a transmission line with a characteristic impedance of 75 Ω. [Parallels Example 8.2]

26. Design a short-circuited stub with an input impedance of $-\jmath 50$ Ω. Use a transmission line with a characteristic impedance of 100 Ω. [Parallels Example 8.2]

27. A load has an impedance $Z_L = 25 - \jmath 100$ Ω.
    (a) What is the reflection coefficient, Γ_L, of the load in a 50 Ω reference system?
    (b) If a one-quarter wavelength long 50 Ω transmission line is connected to the load, what is the reflection coefficient, Γ_{in}, looking into the transmission line?
    (c) Describe the locus of Γ_{in}, as the length of the transmission line is varied from zero length

to one-half wavelength long. Use a Smith chart to illustrate your answer

28. A network consists of a source with a Thevenin equivalent impedance of 50 Ω driving first a series reactance of −50 Ω followed by a one-eighth wavelength long transmission line with a characteristic impedance of 40 Ω and an element with a reactive impedance of $\jmath 25\ \Omega$ in shunt with a load having an impedance $Z_L = 25 - \jmath 50\ \Omega$. This problem must be solved graphically and no credit will be given if this is not done.
    (a) Draw the network.
    (b) On a Smith chart, plot the locus of the reflection coefficient first for the load, then with the element in shunt, then looking into the transmission line, and finally the series element. Use letters to identify each point on the Smith chart. Write down the reflection coefficient at each point.
    (c) What is the impedance presented by the network to the source?

8.6.1 Exercises by Section

†challenging

§8.3 1, 2, 3, 4, 5, 6, 7, 8, 9, 10, 11, 12, 13, 14, 15, 16, 17, 18, 19, 20, 21, 22, 23†, 24†, 25†, 26†, 27†, 28†

8.6.2 Answers to Selected Exercises

1 $\Gamma_{IN} = 10^{-10}$

23(c) ii & iii

24(d) $8.37 - \jmath 49.3\ \Omega$

27 $0.825\angle -50.9°$

28 $\approx 250 - \jmath 41\ \Omega$

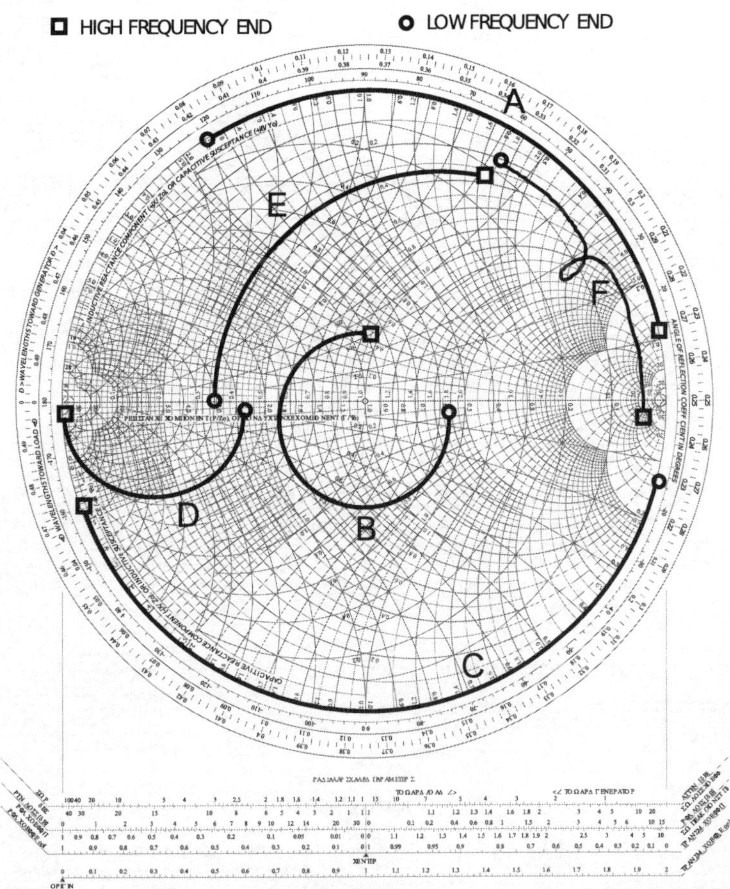

Figure 8-19: The locus of various loads plotted on a Smith chart.

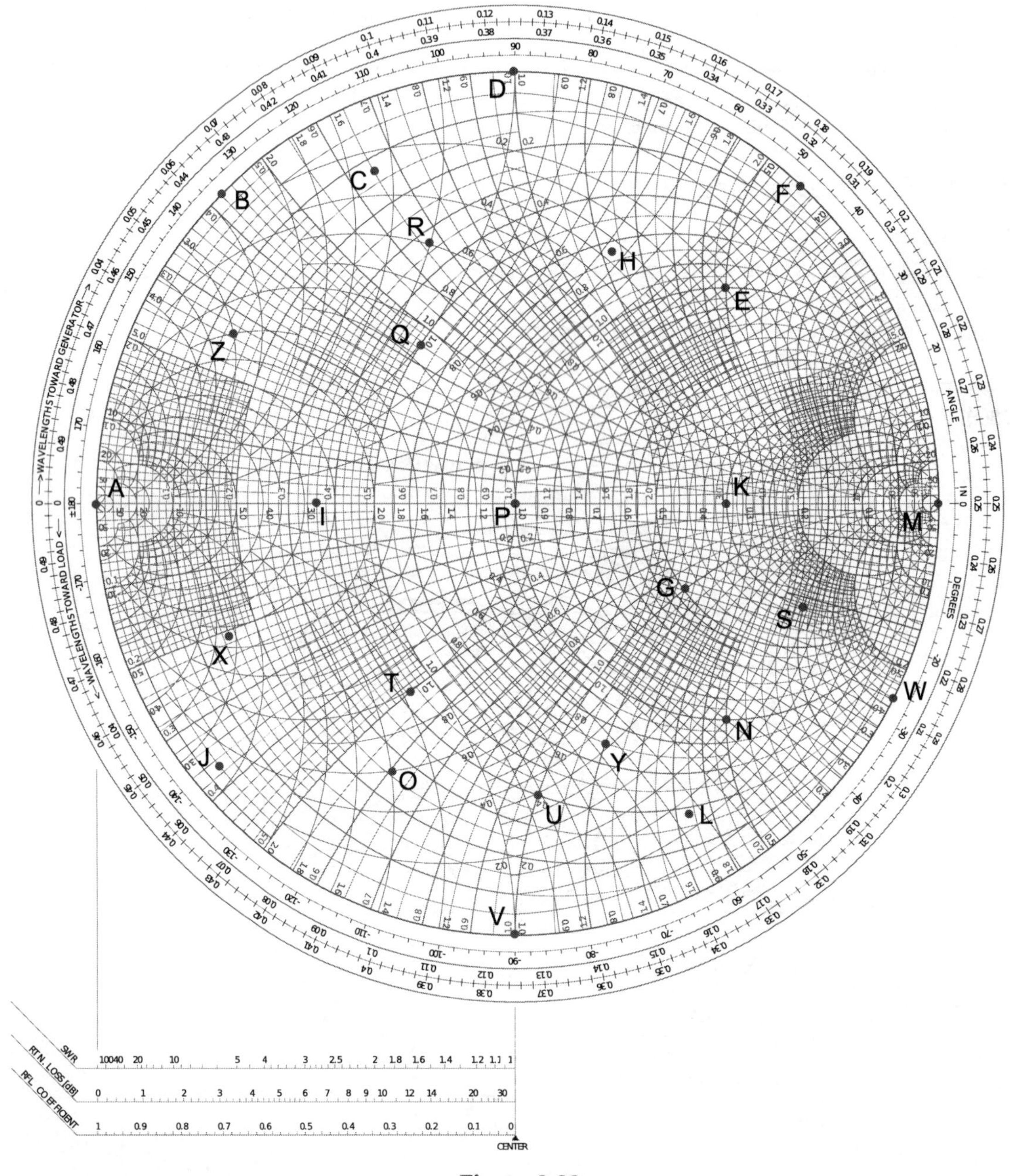
A
B
C
D
E
F
G
H
I
J
K
L
M
N
O
P
Q
R
S
T
U
V
W
X
Y
Z
SWR
RTN. LOSS [dB]
RFL. COEFFICIENT
CENTER

Figure 8-20:

CHAPTER 9

Passive Components

9.1 Introduction

This chapter introduces a wide variety of passive components. It is not possible to be comprehensive, as there is an enormous catalog of microwave elements and scores of variations, and new concepts are introduced every year. At microwave frequencies distributed components can be constructed that have features with particular properties related to coupling, to traveling waves, and to storage of EM energy. Sometimes it is possible to develop lumped-element equivalents of the distributed elements by using the LC ladder model of a transmission line thus realizing lumped-element circuits that would be difficult to imagine otherwise.

9.2 Q Factor

RF inductors and capacitors also have loss and parasitic elements. With inductors there is both series resistance and shunt capacitance mainly from interwinding capacitance, while with capacitors there will be shunt resistance and series inductance. A practical inductor or capacitor is limited to operation below the self-resonant frequency determined by the inductance and capacitance itself resonating with its reactive parasitics. The impact of loss is quantified by the Q factor (the quality factor). Q is loosely related to bandwidth in general and the strict relationship is based on the response of a series or parallel connection of a resistor (R), an inductor (L), and a capacitor (C). The response of an RLC network is described by a second-

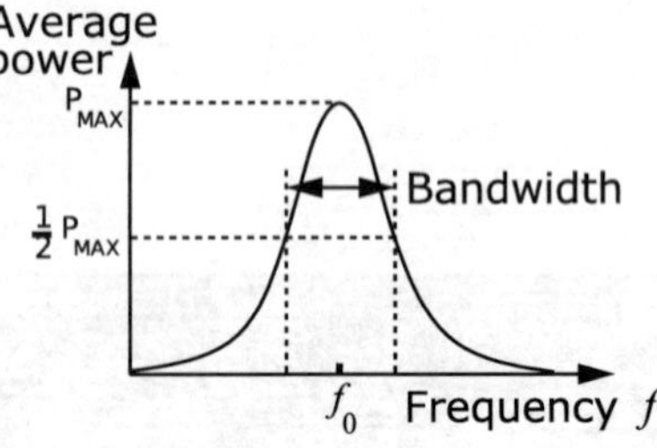

Figure 9-1: Transfer characteristic of a resonant circuit. (The transfer function is V/I for the parallel resonant circuit of Figure 9-2(a) and I/V for the series resonant circuit of Figure 9-2(b).)

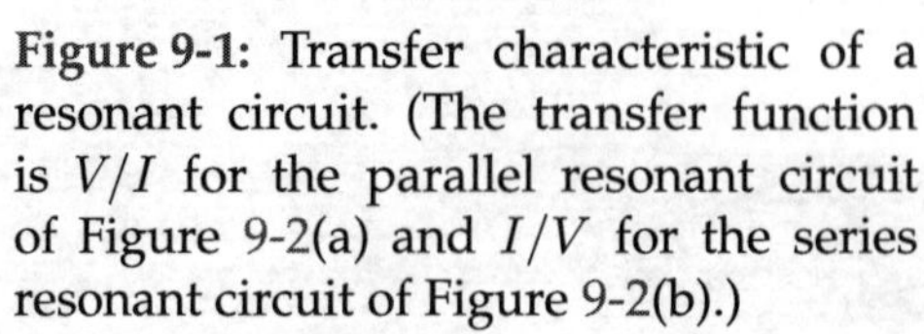

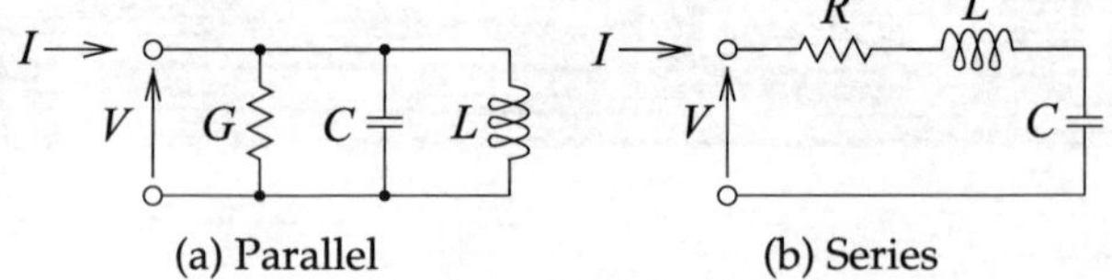

Figure 9-2: Second-order resonant circuits.

order differential equation with the conclusion that the 3 dB fractional bandwidth of the response (i.e., when the power response is at its half-power level below its peak response) is $1/Q$. (The fractional bandwidth is $\Delta f/f_0$ where $f_0 = f_r$ is the resonant frequency at the center of the band and Δf is the 3 dB bandwidth.) This is not true for any network other than a second-order circuit, but as a guiding principle, networks with higher Qs will have narrower bandwidths.

9.2.1 Definition

The Q factor of a component at frequency f is defined as the ratio of $2\pi f$ times the maximum energy stored to the energy lost per cycle. In a lumped-element resonant circuit, stored energy is transferred between an inductor, which stores magnetic energy, and a capacitor, which stores electric energy, and back again every period. Distributed resonators function the same way, exchanging energy stored in electric and magnetic forms, but with the energy stored spatially. The quality factor is

$$Q = 2\pi f_0 \left(\frac{\text{average energy stored in the resonator at } f_0}{\text{power lost in the resonator}} \right), \tag{9.1}$$

where f_0 is the resonant frequency.

A simple response is shown in Figure 9-1 for a parallel resonant circuit with elements L, C, and $G = 1/R$ (see Figure 9-2(a)),

$$Q = \omega_r C/G = 1/(\omega_r LG), \tag{9.2}$$

where $f_r = \omega_r/(2\pi)$ is the resonant frequency and is the frequency at which the maximum amount of energy is stored in a resonator. The conductance, G, describes the energy lost in a cycle. For a series resonant circuit (Figure 9-2(b)) with L, C, and R elements,

$$Q = \omega_r L/R = 1/(\omega_r CR). \tag{9.3}$$

These second-order resonant circuits have a bandpass transfer characteristic (see Figure 9-1) with Q being the inverse of the fractional bandwidth of the resonator. The fractional bandwidth, $\Delta f/f$, is measured at the half-power points as shown in Figure 9-1. (Δf is also referred to as the two-sided -3 dB

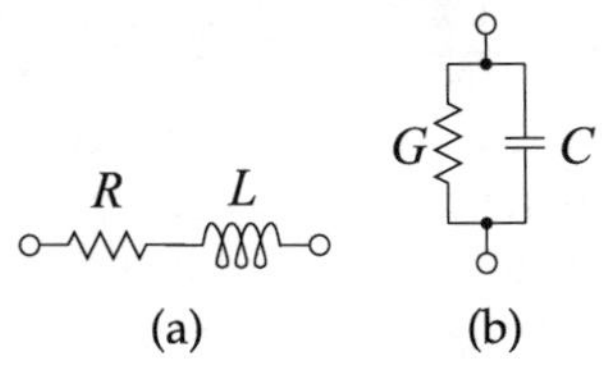

Figure 9-3: Loss elements of practical inductors and capacitors: (a) an inductor has a series resistance R; and (b) for a capacitor, the dominant loss mechanism is a shunt conductance $G = 1/R$.

bandwidth.) Then

$$Q = f_r/\Delta f. \tag{9.4}$$

Thus the Q is a measure of the sharpness of the bandpass frequency response. The determination of Q using the measurement of bandwidth together with Equation (9.4) is often not very precise, so another definition that uses the much more sensitive phase change at resonance is preferred when measurements are being used. With ϕ being the phase (in radians) of the transfer characteristic, the definition of Q is now

$$Q = \frac{\omega_r}{2}\left|\frac{d\phi}{d\omega}\right|. \tag{9.5}$$

Equation (9.5) is another equivalent definition of Q for parallel RLC or series RLC resonant circuits. It is meaningful to talk about the Q of circuits other than three-element RLC circuits, and then its meaning is always a ratio of the energy stored to the energy dissipated per cycle. The Q of these structures can no longer be determined by bandwidth or by the rate of phase change.

9.2.2 Q of Lumped Elements

Q is also used to characterize the loss of lumped inductors and capacitors. Inductors have a series resistance R, and the main loss mechanism of a capacitor is a shunt conductance G (see Figure 9-3).
The Q of an inductor at frequency $f = \omega/(2\pi)$ with a series resistance R and inductance L is

$$Q_{\text{INDUCTOR}} = \frac{\omega L}{R}. \tag{9.6}$$

Since R is approximately constant with respect to frequency for an inductor, the Q will vary with frequency.
The Q of a capacitor with a shunt conductance G and capacitance C is

$$Q_{\text{CAPACITOR}} = \frac{\omega C}{G}. \tag{9.7}$$

G is due mainly to relaxation loss mechanisms of the dielectric of a capacitor and so varies linearly with frequency but also it is usually very small. Thus the Q of a capacitor is almost constant with respect to frequency. For microwave components $Q_{\text{CAPACITOR}} \gg Q_{\text{INDUCTOR}}$, and both are smaller than the Q of most transmission line networks. Thus, if the length of a transmission line is not too long for an application, transmission line networks are preferred. If lumped elements must be used, the use of inductors should be minimized.

9.2.3 *Loaded Q Factor*

The Q of a component as defined in the previous section is called the unloaded Q, Q_U. However, if a component is to be measured or used in any way, it is necessary to couple energy in and out of it. The Q is reduced and thus the resonator bandwidth is increased by the power lost to the external circuit so that the loaded Q, the Q that is measured, is

$$Q_L = 2\pi f_0 \left(\frac{\text{average energy stored in the resonator at} f_0}{\text{power lost in the resonator and to the external circuit}} \right)$$

$$= \frac{1}{1/Q + 1/Q_X}, \tag{9.8}$$

and

$$Q_X = \left(\frac{1}{Q_L} - \frac{1}{Q_U} \right)^{-1}. \tag{9.9}$$

where Q_X is called the external Q. Q_L accounts for the power extracted from the resonant circuit. If the loading is kept very small, $Q_L \approx Q_U$.

9.2.4 *Summary of the Properties of Q*

In summary:

(a) Q is properly defined and related to the energy stored in a resonator for a second-order network, one with two reactive elements of opposite types.

(b) Q is not well defined for networks with three or more reactive elements.

(e) It is only used (as defined or some approximation of it) for guiding the design.

9.3 Surface-Mount Components

The majority of the RF and microwave design effort goes into developing modules and interconnecting modules on circuit boards. With these the most common type of component to use is a surface-mount component. Figure 9-4(a) shows a two-terminal element, such as a resistor or capacitor, in the form of a surface-mount component. Figure 9-4(b and c) show the use of surface-mount components on a microwave circuit board.

A two-terminal surface-mount resistor or capacitor is commonly called a chip resistor or chip capacitor. These can be very small, and the smaller the component often the higher the operating frequency due to reduced parasitic capacitance or inductance. Common sizes of two-terminal chip components are listed in Table 9-1. The resonance when a chip capacitor (inductor) resonates with the parasitic inductance (capacitance) is called self-resonance and the resonant frequency is called the self-resonance frequency. Figure 9-5(a) shows an inductor in a surface-mount package and details are shown in Figure 9-5(b) and the inductor is useable as an inductor at a frequency backed-off from the SRF, see Table 9-2.

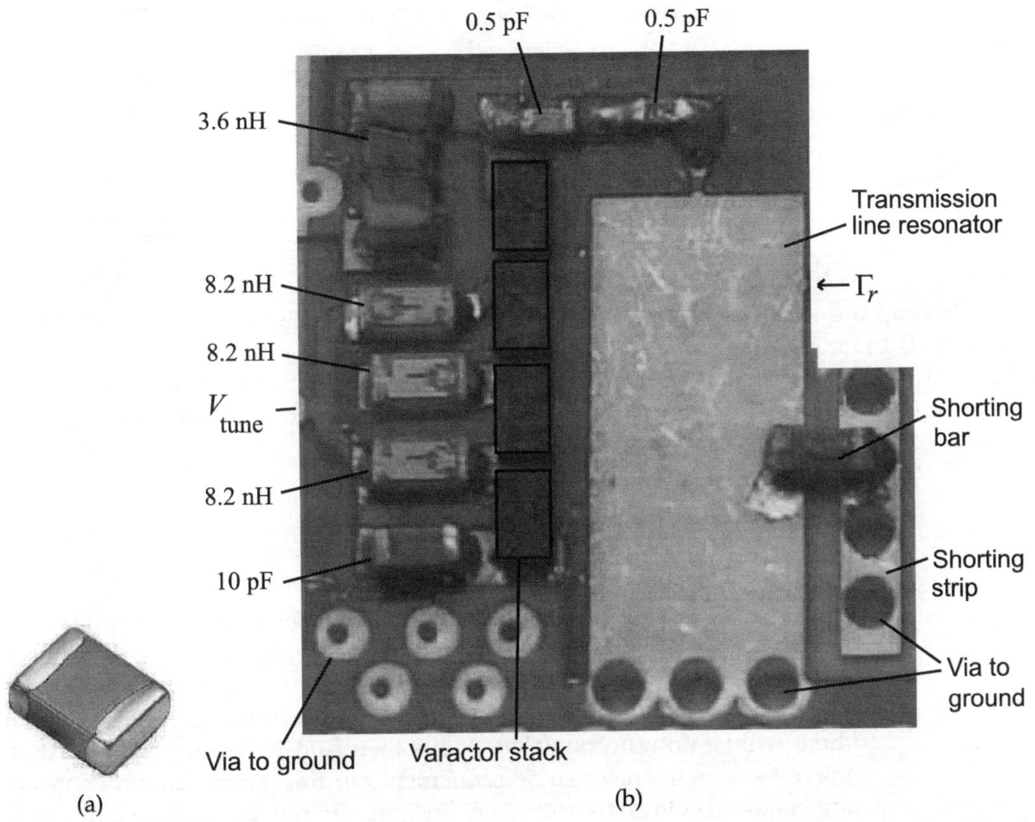

Figure 9-4: Circuit board showing the use of surface-mount components: (a) chip resistor or capacitor with metal terminals at the two ends; and (b) a populated RF microstrip circuit board of a 5 GHz tunable resonator [1]. The larger components have dimensions 1.6 mm × 0.8mm. The shorting bar is a 0 Ω chip resistor.

Designation	Size (inch × inch)	Metric designation	Size (mm × mm)
01005	0.016 × 0.0079	0402	0.4 × 0.2
0201	0.024 × 0.012	0603	0.6 × 0.3
0402	0.039 × 0.020	1005	1.0 × 0.5
0603	0.063 × 0.031	1608	1.6 × 0.8
0805	0.079 × 0.049	2012	2.0 × 1.25
1008	0.098 × 0.079	2520	2.5 × 2.0
1206	0.13 × 0.063	3216	3.2 × 1.6
1210	0.13 × 0.098	3225	3.2 × 2.5
1806	0.18 × 0.063	4516	4.5 × 1.6
1812	0.18 × 0.13	4532	4.5 × 3.2
2010	0.20 × 0.098	5025	5.0 × 2.5
2512	0.25 × 0.13	6432	6.4 × 3.2
2920	0.29 × 0.20	–	7.4 × 5.1

Table 9-1: Sizes and designation of two-terminal surface mount components. Note the designation of a surface-mount component refers (approximately) to its dimensions in hundredths of an inch.

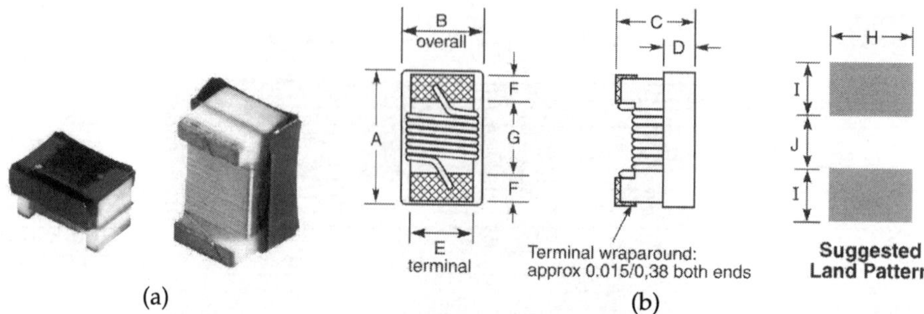

Figure 9-5: Chip inductor: (a) in an 0603 surface-mount package; and (b) schematic showing sizes and pads to be provided on a circuit board (A=64 mils (1.63 mm), B=33 mils (0.84 mm), C=24 mils (0.61 mm), D=13 mils (0.33 mm), E=30 mils (0.76 mm), F=25 mils (0.64 mm), G=25 mils (0.64 mm), and H=40 mils (1.02 mm)). Copyright Coilcraft, Inc., used with permission [2].

9.4 Terminations and Attenuators

9.4.1 Terminations

Terminations are used to completely absorb a forward-traveling wave and the reflection coefficient of a termination is ideally zero. If a transmission line has a resistive characteristic impedance $R_0 = Z_0$, then terminating the line in a resistance R_0 will fully absorb the forward-traveling wave and there will be no reflection. The line is then said to be matched. At RF and microwave frequencies some refinements to this simple circuit connection are required. On a transmission line the energy is contained in the EM fields. For the coaxial line, a simple resistive connection between the inner and outer conductors would not terminate the fields and there would be some reflection. Instead, coaxial line terminations generally comprise a disk of resistive material (see Figure 9-6(a)). The total resistance of the disk from the inner to the outer conductor is the characteristic resistance of the line, however, the resistive material is distributed and so creates a good termination of the fields.

Terminations are a problem with microstrip, as the characteristic impedance varies with frequency, is in general complex, and the vias that would be required if a lumped resistor was used has appreciable inductance at frequencies above a few gigahertz. A high-quality termination is realized using a section of lossy line as shown in Figure 9-6(b). Here lossy material is deposited on top of an open-circuited microstrip line. This increases the loss of the line appreciably without significantly affecting the characteristic

Table 9-2: Parameters of the inductors in Figure 9-5. L_{nom} is the nominal inductance, SRF is the self-resonance frequency, R_{DC} is the inductor's series resistance, and I_{max} is the maximum RMS current supported.

L_{nom}	900 MHz		1.7 GHz		SRF	R_{DC}	I_{max}
(nH)	L (nH)	Q	L (nH)	Q	(GHz)	(Ω)	(mA)
1.0	0.98	39	0.99	58	16.0	0.045	1600
2.0	1.98	46	1.98	70	12.0	0.034	1900
5.1	5.12	68	5.18	93	5.50	0.050	1400
10	10.0	67	10.4	85	3.95	0.092	1100
20	20.2	67	21.6	80	2.90	0.175	760
56	59.4	54	75.4	48	1.75	0.700	420

Component	Symbol	Alternate
Attenuator, fixed		
Attenuator, balanced		
Attenuator, unbalanced		
Attenuator, variable		
Attenuator, continuously variable		
Attenuator, stepped variable		

Table 9-3: IEEE standard symbols for attenuators [3].

impedance of the line. If the length of the lossy line is sufficiently long, say one wavelength, the forward-traveling wave will be totally absorbed and there will be no reflection. Tapering the lossy material, as shown in Figure 9-6(b), reduces the discontinuity between the lossless microstrip line and the lossy line by ensuring that some of the power in the forward-traveling wave is dissipated before the maximum impact of the lossy material occurs. Thus a matched termination is achieved without the use of a via.

9.4.2 Attenuators

An attenuator is a two-port network that reduces the amplitude of a signal and it does this by absorbing power and without distorting the signal. The input and output of the attenuator are both matched, so there are no reflections. An attenuator may be fixed, continuously variable, or discretely variable. The IEEE standard symbols for attenuators are shown in Table 9-3. When the attenuation is fixed, an attenuator is commonly called a **pad**.
Balanced and unbalanced resistive pads are shown in Figures 9-7 and 9-8 together with their design equations. The attenuators in Figure 9-7 are T or Tee attenuators, where Z_{01} is the system impedance to the left of the pad and Z_{02} is the system impedance to the right of the pad. The defining characteristic is that the reflection coefficient looking into the pad from the

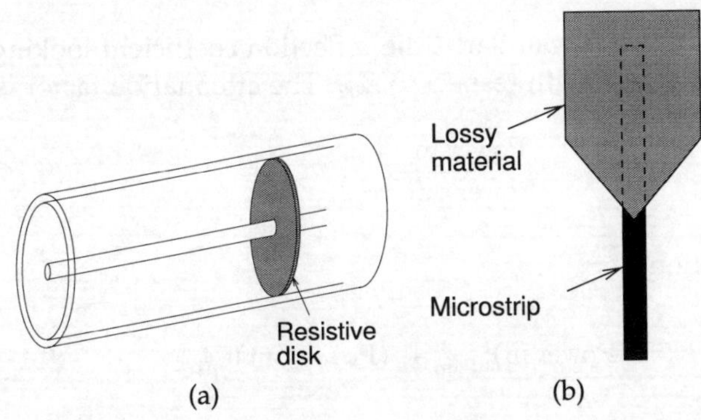

Figure 9-6: Terminations: (a) coaxial line resistive termination; (b) microstrip matched load.

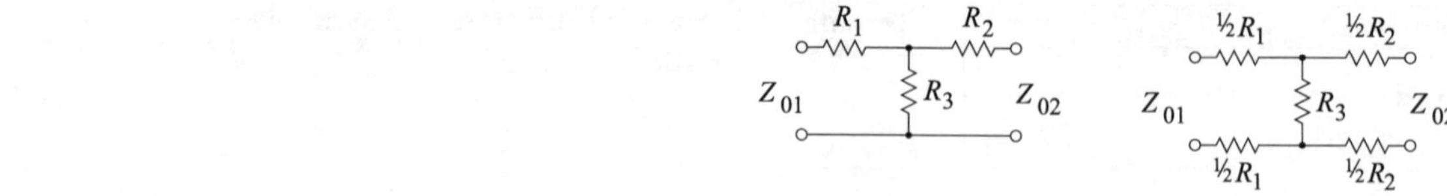

(a) Unbalanced (b) Balanced

$$R_1 = \frac{Z_{01}(K+1) - 2\sqrt{KZ_{01}Z_{02}}}{K-1} \qquad R_2 = \frac{Z_{02}(K+1) - 2\sqrt{KZ_{01}Z_{02}}}{K-1}$$

$$R_3 = \frac{2\sqrt{KZ_{01}Z_{02}}}{K-1} \tag{9.10}$$

If $Z_{01} = Z_{02} = Z_0$, then

$$R_1 = R_2 = Z_0\left(\frac{\sqrt{K}-1}{\sqrt{K}+1}\right) \quad \text{and} \quad R_3 = \frac{2Z_0\sqrt{K}}{K-1} \tag{9.11}$$

Figure 9-7: T (Tee) attenuator. K is the (power) attenuation factor, e.g. a 3 dB attenuator has $K = 10^{3/10} = 1.995$.

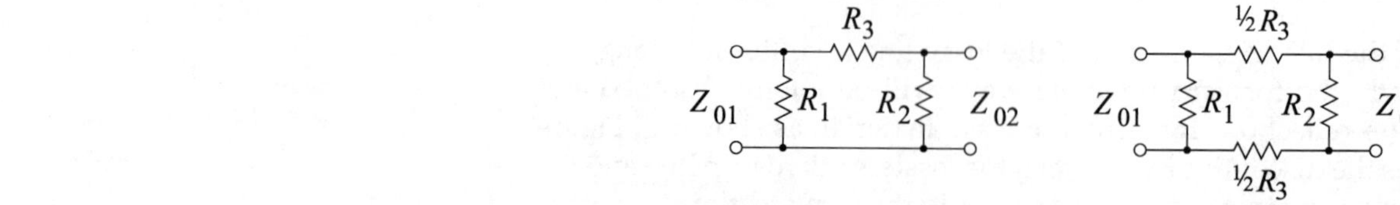

(a) Unbalanced (b) Balanced

$$R_1 = \frac{Z_{01}(K-1)\sqrt{Z_{02}}}{(K+1)\sqrt{Z_{02}} - 2\sqrt{KZ_{01}}} \qquad R_2 = \frac{Z_{02}(K-1)\sqrt{Z_{01}}}{(K+1)\sqrt{Z_{01}} - 2\sqrt{KZ_{02}}} \tag{9.12}$$

$$R_3 = \frac{(K-1)}{2}\sqrt{\frac{Z_{01}Z_{02}}{K}} \tag{9.13}$$

If $Z_{01} = Z_{02} = Z_0$, then

$$R_1 = R_2 = Z_0\left(\frac{\sqrt{K}+1}{\sqrt{K}-1}\right) \quad \text{and} \quad R_3 = \frac{Z_0(K-1)}{2\sqrt{K}}. \tag{9.14}$$

If $R_1 = R_2$, then $Z_{01} = Z_{02} = Z_0$

$$Z_0 = \sqrt{\frac{R_1^2 R_3}{2R_1 + R_3}}, \quad \text{and} \quad K = \left(\frac{R_1 + Z_0}{R_1 - Z_0}\right)^2. \tag{9.15}$$

Figure 9-8: Pi (Π) attenuator. K is the (power) attenuation factor, e.g. a 20 dB attenuator has $K = 10^{20/10} = 100$.

left is zero when referred to Z_{01}. Similarly, the reflection coefficient looking into the right of the pad is zero with respect to Z_{02}. The attenuation factor is

$$K = \frac{\text{Power in}}{\text{Power out}}. \tag{9.16}$$

In decibels, the attenuation is

$$K|_{\text{dB}} = 10\log_{10} K = (\text{Power in})|_{\text{dBm}} - (\text{Power out})|_{\text{dBm}}. \tag{9.17}$$

If the left and right system impedances are different, then there is a minimum

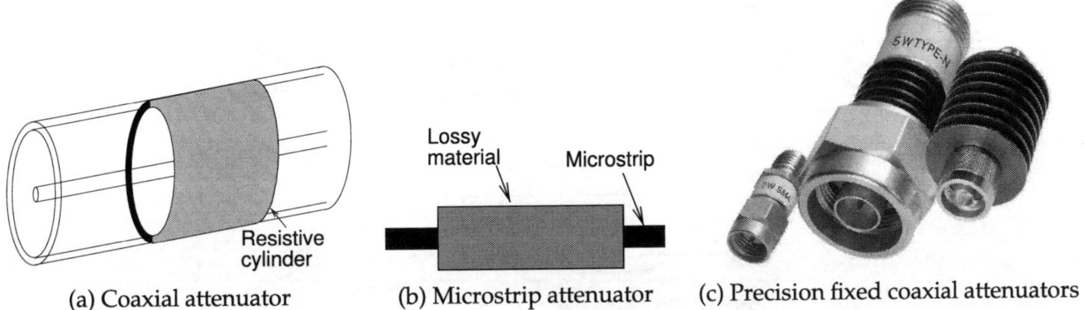

(a) Coaxial attenuator (b) Microstrip attenuator (c) Precision fixed coaxial attenuators

Figure 9-9: Distributed attenuators. The attenuators in (c) have power handling ratings of 2W, 5W and 20 W (left to right). Copyright 2012 Scientific Components Corporation d/b/a Mini-Circuits, used with permission [4].

attenuation factor that can be achieved:

$$K_{\text{MIN}} = \frac{2Z_{01}}{Z_{02}} - 1 + 2\sqrt{\frac{Z_{01}}{Z_{02}}\left(\frac{Z_{01}}{Z_{02} - 1}\right)}. \tag{9.18}$$

This limitation comes from the simultaneous requirement that the pad be matched. If there is a single system impedance, $Z_0 = Z_{01} = Z_{02}$, then $K_{\text{MIN}} = 1$, and so any value of attenuation can be obtained.

Lumped attenuators are useful up to 10 GHz above which the size of resistive elements becomes large compared to a wavelength. Also, for planar circuits, vias are required, and these are undesirable from a manufacturing standpoint, and electrically they have a small inductance. Fortunately attenuators can be realized using a lossy section of transmission line, as shown in Figure 9-9. Here, lossy material results in a section of line with a high-attenuation constant. Generally the lossy material has little effect on the characteristic impedance of the line, so there is little reflection at the input and output of the attenuator. Distributed attenuators can be used at frequencies higher than lumped-element attenuators can, and they can be realized with any transmission line structure.

EXAMPLE 9.1 Pad Design

Design an unbalanced 20 dB pad in a 75 Ω system.

Solution:

There are two possible designs using resistive pads. These are the unbalanced Tee and Pi pads shown in Figures 9-7 and 9-8. The Tee design will be chosen. The K factor is

$$K = 10^{(K|_{\text{dB}}/10)} = 10^{(20/10)} = 100. \tag{9.19}$$

Since $Z_{01} = Z_{02} = 75\ \Omega$, Equation (9.11) yields

$$R_1 = R_2 = 75\left(\frac{\sqrt{100} - 1}{\sqrt{100} + 1}\right) = 75\left(\frac{9}{11}\right) = 61.4\ \Omega \tag{9.20}$$

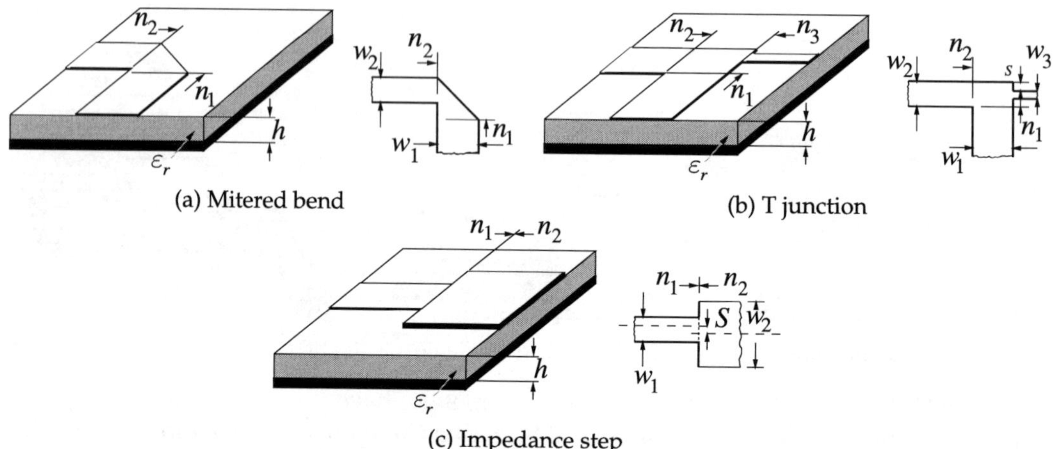

Figure 9-10: Microstrip discontinuities.

$$R_3 = \frac{2 \cdot 75\sqrt{100}}{100 - 1} = 150\left(\frac{10}{99}\right) = 15.2\ \Omega. \tag{9.21}$$

The final design is 61.4 Ω 61.4 Ω 15.2 Ω

9.5 Transmission Line Stubs and Discontinuities

Interruptions of the magnetic or electric field create regions where additional magnetic energy or electric energy is stored. If the additional energy stored is predominantly magnetic, the discontinuity will introduce an inductance. If the additional energy stored is predominantly electric, the discontinuity will introduce a capacitance. Such discontinuities occur with all transmission lines. In some cases transmission line discontinuities introduce undesired parasitics, but they also provide an opportunity to effectively introduce lumped-element components.

The simplest microwave circuit element is a uniform section of transmission line that can be used to introduce a time delay or a frequency-dependent phase shift. More commonly it is used to interconnect other components. Other line segments including bends and junctions are shown in Figure 9-10.

9.5.1 Open

Many transmission line discontinuities arise from fringing fields. One element is the microstrip open, shown in Figure 9-11. The fringing fields at the end of the transmission line in Figure 9-11(a) store energy in the electric field, and this can be modeled by the fringing capacitance, C_F, shown in Figure 9-11(b). This effect can also be modeled by an extended transmission line, as shown in Figure 9-11(c). For a typical microstrip line with ε_r = 9.6, h = 600 μm, and w/h = 1, C_F is approximately 36 fF. However, C_F varies

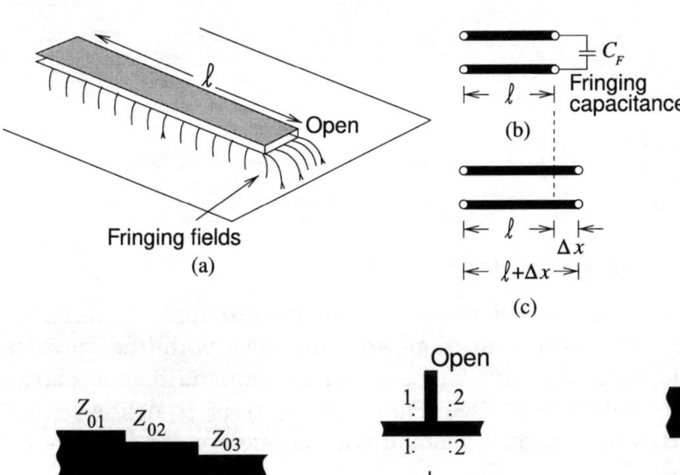

Figure 9-11: An open on a microstrip transmission line: (a) microstrip line showing fringing fields at the open; (b) fringing capacitance model of the open; and (c) an extended line model of the open with Δx being the extra transmission line length that captures the open.

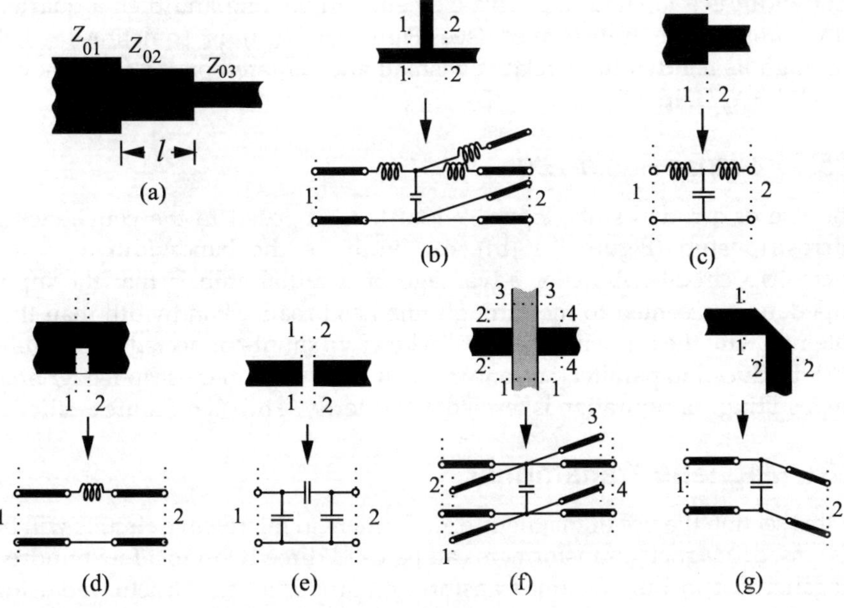

Figure 9-12: Microstrip discontinuities: (a) quarter-wave impedance transformer; (b) open microstrip stub; (c) step; (d) notch; (e) gap; (f) crossover; and (g) bend.

with frequency, and the extended length is a much better approximation to the effect of fringing [5]. For the same dimensions, the length section is approximately $0.35h$ and almost independent of frequency [6]. As with many fringing effects, a capacitance or inductance can be used to model the effect of fringing, but generally a distributed model is better.

9.5.2 Discontinuities

Several microstrip discontinuities and their equivalent circuits are shown in Figure 9-12. Discontinuities (Figure 9-12(b–g)) are modeled by capacitive elements if the E field is affected and by inductive elements if the H field (or current) is disturbed. The stub shown in Figure 9-12(b), for example, is best modeled using lumped elements describing the junction as well as the transmission line of the stub itself. Current bunches at the right angle bends from the through line to the stub. The current bunching leads to excess energy being stored in the magnetic field, and hence an inductive effect. There is also excess charge storage in the parallel plate region bounded by the left- and right-hand through lines and the stub. This is modeled by a capacitance.

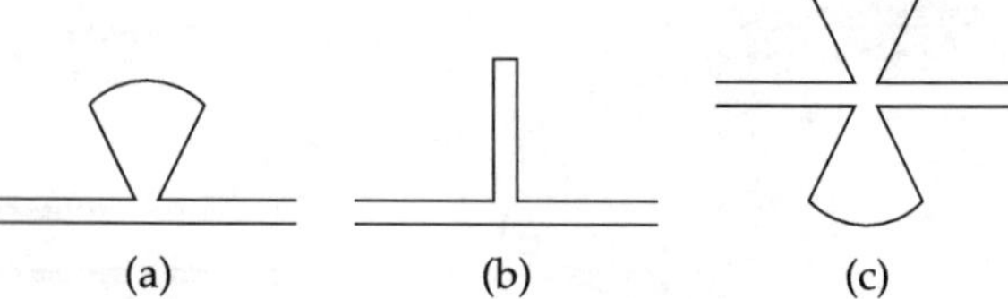

Figure 9-13: Microstrip stubs: (a) radial shunt-connected stub; (b) conventional shunt stub; and (c) butterfly radial stub.

9.5.3 Impedance Transformer

Impedance transformers interface two lines of different characteristic impedance. The smoothest transition and the one with the broadest bandwidth is a tapered line. This element can be long and then a quarter-wave impedance transformer (see Figure 9-12(a)) is sometimes used, although its bandwidth is relatively small and centered on the frequency at which $l = \lambda_g/4$. Ideally $Z_{0,2} = \sqrt{Z_{0,1}Z_{0,3}}$.

9.5.4 Planar Radial Stub

The use of a radial stub (Figure 9-13(a)), as opposed to the conventional microstrip stub (Figure 9-13(b)), can improve the bandwidth of many microstrip circuits. A major advantage of a radial stub is that the input impedance presented to the through line has broader bandwidth than that obtained with the conventional stub. When two shunt-connected radial stubs are introduced in parallel (i.e., one on each side of the microstrip feeder line) the resulting configuration is termed a "butterfly" stub (see Figure 9-13(c)).

9.6 Magnetic Transformer

In this section the use of magnetic transformers in microwave circuits will be discussed. Magnetic transformers can be used directly up to a few hundred megahertz or so, but the same transforming properties can be achieved using coupled transmission lines.

9.6.1 Properties of a Magnetic Transformer

A magnetic transformer (see Figure 9-14) magnetically couples the current in one wire to current in another. The effect is amplified using coils of wires and using a core of magnetic material (material with high permeability) to create greater magnetic flux density. When coils are used, the symbol shown in Figure 9-14(a) is used, with one of the windings called the primary winding and the other called the secondary winding. If there is a **magnetic core** around which the coils are wound, then the symbol shown in Figure 9-14(b) is used, with the vertical lines indicating the core. However, even if there is a core, the simpler transformer symbol in Figure 9-14(a) is more commonly used. Magnetic cores are useful up to several hundred megahertz and rely on the alignment of magnetic dipoles in the core material. Above a few hundred megahertz the magnetic dipoles cannot react quickly enough and so the core looks like an open circuit to magnetic flux. Thus the core is not useful for magnetically coupling signals above a few hundred megahertz. As mentioned, the dots above the coils in Figure 9-14(a) indicate the polarity of the magnetic flux with respect to the currents in the coils so that, as shown, V_1 and V_2 will have the same sign. Even if the magnetic polarity is not specifically shown, it is implied (see Figure 9-14(c)). There are two ways of showing inversion of the magnetic polarity, as shown in Figure 9-14(d), where a negative number of windings indicates opposite magnetic polarity. The interest in using magnetic transformers in high-frequency circuit design is that configurations of magnetic transformers can be realized using coupled transmission lines to extend operation to hundreds of gigahertz. The transformer is easy to conceptualize, so it is convenient to first develop circuits using the transformer and then translate it to transmission line form. That is, in "back-of-the-envelope" microwave design, transformers can be used to indicate coupling, with the details of the coupling left until later when the electrical design is translated into a physical design.

The following notation is used with a magnetic transformer:

L_1, L_2: the self-inductances of the two coils
M: the mutual inductance
k: the coupling factor,

$$k = \frac{M}{\sqrt{L_1 L_2}}. \tag{9.22}$$

Referring to Figure 9-14(e), the voltage transformer ratio is

$$V_2 = nV_1, \tag{9.23}$$

where n is the ratio of the number of secondary to primary windings. An ideal transformer has "perfect coupling," indicated by $k = 1$, and the self-inductances are proportional to the square of the number of windings, so

$$\frac{V_2}{V_1} = \sqrt{\frac{L_2}{L_1}}. \tag{9.24}$$

The general equation relating the currents of the circuit in Figure 9-14(e) is

$$RI_2 + \jmath\omega L_2 I_2 + \jmath\omega M I_1 = 0, \tag{9.25}$$

and so

$$\frac{I_1}{I_2} = -\frac{R + \jmath\omega L_2}{\jmath\omega M}. \tag{9.26}$$

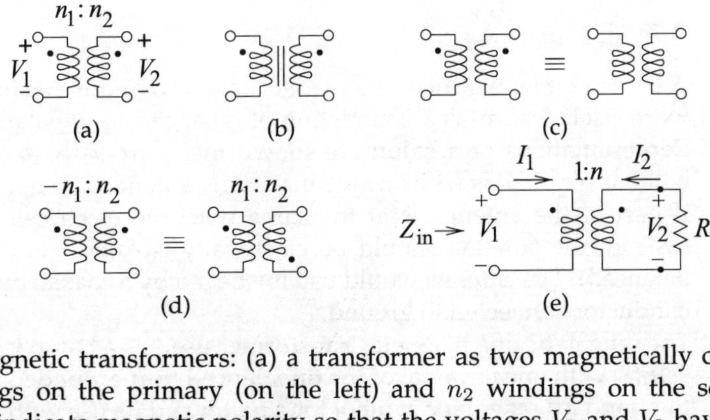

$n_1 : n_2$ + V_1 − + V_2 −

(a)

(b)

≡

(c)

$-n_1 : n_2$ ≡ $n_1 : n_2$

(d)

I_1 1:n I_2 + Z_{in} → V_1 − + V_2 − R

(e)

Figure 9-14: Magnetic transformers: (a) a transformer as two magnetically coupled windings with n_1 windings on the primary (on the left) and n_2 windings on the secondary (on the right) (the dots indicate magnetic polarity so that the voltages V_1 and V_2 have the same sign); (b) a magnetic transformer with a magnetic core; (c) identical representations of a magnetic transformer with the magnetic polarity implied for the transformer on the right; (d) two equivalent representations of a transformer having opposite magnetic polarities (an **inverting transformer**); and (e) a magnetic transformer circuit.

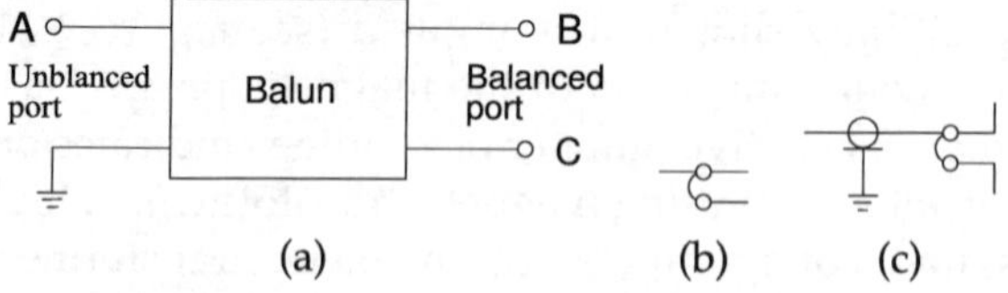

Figure 9-15: A balun: (a) as a two-port with four terminals; (b) IEEE standard schematic symbol for a balun [3]; and (c) an (unbalanced) coaxial cable driving a dipole antenna through a balun.

If $R \ll \omega L_2$, then the current transformer ratio is

$$\frac{I_1}{I_2} \approx -\frac{L_2}{M} = -\sqrt{\frac{L_2}{L_1}}. \tag{9.27}$$

Notice that combining Equations (9.24) and (9.27) leads to calculation of the transforming effect on impedance. On the Coil 1 side, the input impedance is

$$Z_{\text{in}} = \frac{V_1}{I_1} = -\frac{V_2}{I_2}\left(\frac{L_1}{L_2}\right) = R\frac{L_1}{L_2}. \tag{9.28}$$

In practice, however, there is always some magnetic field leakage—not all of the magnetic field created by the current in Coil 1 goes through (or links) Coil 2—and so $k < 1$. Then from Equations (9.24)–(9.28),

$$V_1 = \jmath\omega L_1 I_1 + \jmath\omega M I_2 \tag{9.29}$$

$$0 = R I_2 + \jmath\omega L_2 I_2 + \jmath\omega M I_1. \tag{9.30}$$

Again, assuming that $R \ll \omega L_2$, a modified expression for the input impedance is obtained that accounts for nonideal coupling:

$$Z_{\text{in}} = R\frac{L_1}{L_2} + \jmath\omega L_1(1 - k^2). \tag{9.31}$$

Imperfect coupling, $k < 1$, causes the input impedance to be reactive and this limits the bandwidth of the transformer. Stray capacitance is another factor that impacts the bandwidth of the transformer.

9.7 Baluns

A balun [7, 8] is a structure that joins balanced and unbalanced circuits. The word itself (balun) is a contraction of <u>bal</u>anced-to-<u>un</u>balanced transformer. Representations of a balun are shown in Figure 9-15. A situation when a balun is required is with an antenna. Many antennas do not operate correctly if part of the antenna is at the same potential electrically as the ground. Instead, the antenna should be electrically isolated from the ground (i.e., balanced). The antenna would usually be fed by a coaxial cable with its outer conductor connected to ground.

The schematic of a magnetic transformer used as a balun is shown in Figure 9-16(a) with one terminal of the unbalanced port grounded. Figure 9-15(b) is the standard schematic symbol for a balun and its use with a dipole antenna is shown in Figure 9-15(c). The second port is floating and is not referenced to ground. An example of the use of a balun is shown in Figure 9-16(b), where the output of a single-ended amplifier is unbalanced, being referred to ground. A balun transitions from the unbalanced transistor output to a balanced output.

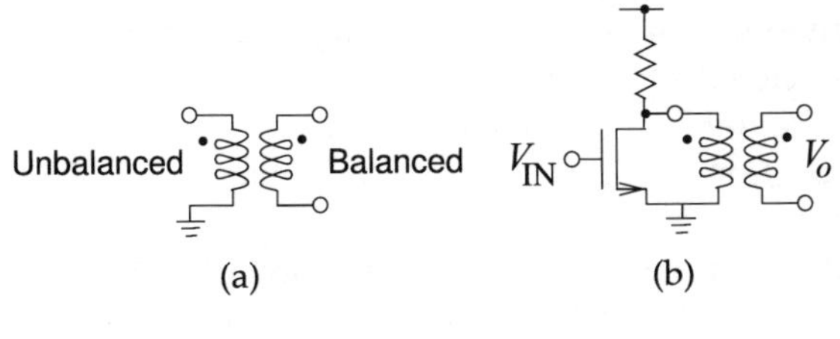

Figure 9-16: Balun: (a) schematic representation as a transformer showing unbalanced and balanced ports; and (b) connected to a single-ended unbalanced amplifier yielding a balanced output.

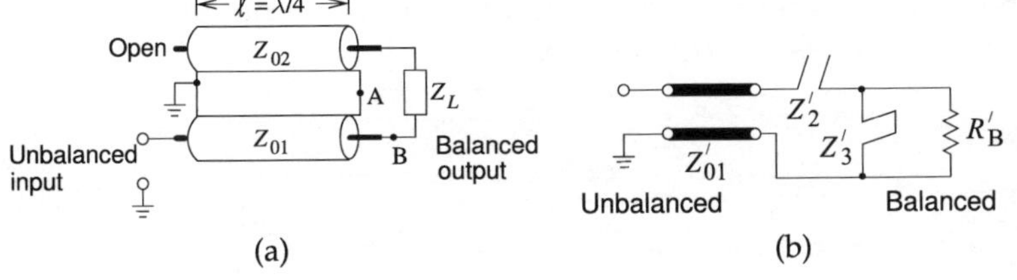

Figure 9-17: Marchand balun: (a) coaxial form of the Marchand balun; and (b) its equivalent circuit.

9.7.1 *Marchand Balun*

The most common form of microwave balun is the Marchand balun [7, 9–12]. An implementation of the Marchand balun using coaxial transmission lines is shown in Figure 9-17(a) [13]. The Z_2 line acts as both a series stub and a shunt stub. Thus the model of the Marchand balun is as shown in Figure 9-17(b).

9.8 Wilkinson Combiner and Divider

A combiner is used to combine power from two or more sources. A typical use is to combine power from two high power amplifiers to obtain a higher power than would be otherwise be available. Dividers divide power so that the power from an amplifier can be routed to different parts of a circuit.

The Wilkinson divider can be used as a combiner or divider that divides input power among the output ports [14]. Figure 9-18(a) is a two-way divider that splits the power at Port 1 equally between the two output ports at Ports 2 and 3. A particular insight that Wilkinson brought was the introduction of the resistor between the output ports and this acts to suppress any imbalance between the output signals due to nonidealities. If the division is exact, no current will flow in the resistor. The circuit works less well as a general purpose combiner. Ideally power entering Ports 2 and 3 would combine losslessly and appear at Port 1. A typical application is to combine the power at the output of two matched transistors where the amplitude and the phase of the signals can be expected to be closely matched. However, if the signals are not identical, the portion that is mismatched will be absorbed in the resistor. The bandwidth of the Wilkinson combiner/divider is limited by the one-quarter wavelength long lines.

The S parameters of the two-way Wilkinson power divider with an equal split of the output power are therefore

$$\mathbf{S} = \begin{bmatrix} 0 & -\jmath/\sqrt{2} & -\jmath/\sqrt{2} \\ -\jmath/\sqrt{2} & 0 & 0 \\ -\jmath/\sqrt{2} & 0 & 0 \end{bmatrix}. \tag{9.32}$$

Figure 9-18(b) is a compact representation of the two-way Wilkinson divider, and a three-way Wilkinson divider is shown in Figure 9-18(d). This pattern can be repeated to produce N-way power dividing. The lumped-element version of the Wilkinson divider shown in Figure 9-18(c) is based on the LC model of a one-quarter wavelength long transmission line segment. With a 50 Ω system impedance and center frequency of 400 MHz, the elements of the lumped element are (from Section 22.7.2 of [15]) $L = 28.13$ nH, $C_1 = 11.25$ pF, $C_2 = 5.627$ pF, and $R = 100\ \Omega$.

Figure 9-19(a) is the layout of a direct microstrip realization of a Wilkinson divider. A high-performance microstrip layout is shown in Figure 9-19(b), where the transmission lines are curved to bring the output ports near each other so that a chip resistor can be used.

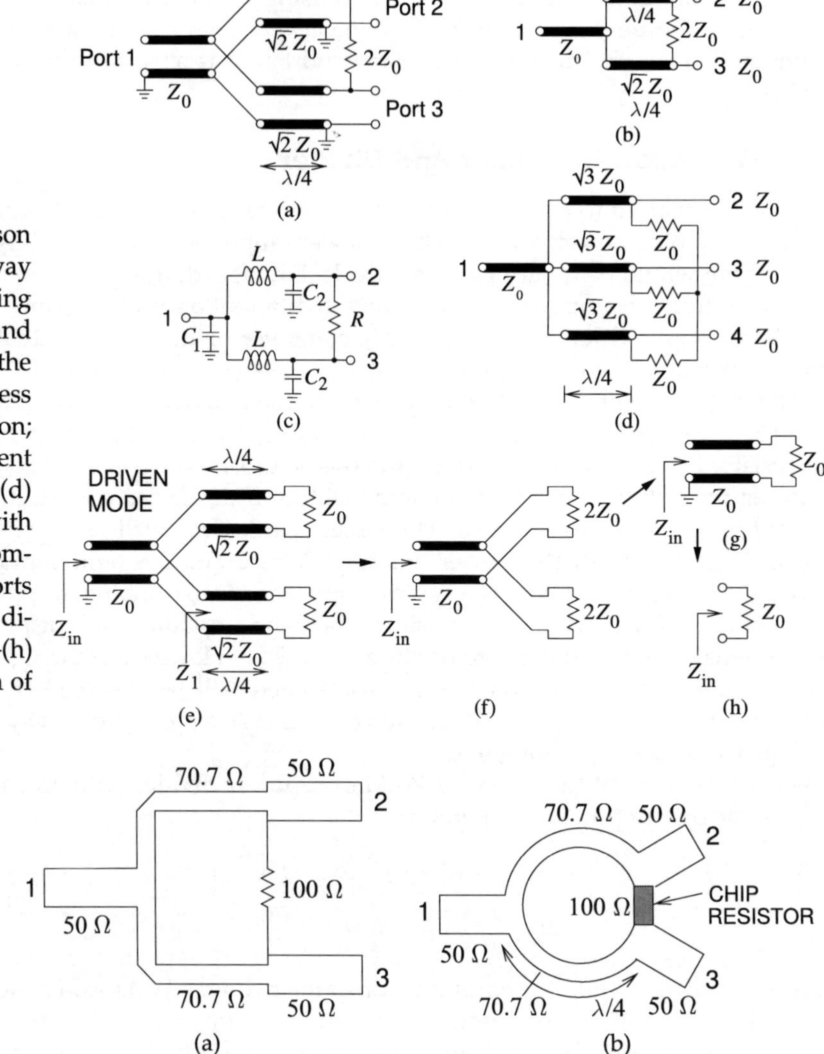

Figure 9-18: Wilkinson divider: (a) two-way divider with Port 1 being the combined signal and Ports 2 and 3 being the divided signals; (b) less cluttered representation; (c) lumped-element implementation; (d) three-way divider with Port 1 being the combined signal and Ports 2, 3, and 4 being the divided signals; and (e)–(h) steps in the derivation of the input impedance.

Figure 9-19: Wilkinson combiner and divider: (a) microstrip realization; and (b) higher performance microstrip implementation.

EXAMPLE 9.2 Lumped-Element Wilkinson Divider

Design a lumped-element 2-way Wilkinson divider in a 60 Ω system. The center frequency of the design should be 10 GHz.

Solution:

The design begins with the transmission line form of the Wilkinson divider which will be converted to a lumped-element form latter. The design parameters are $Z_0 = 60\ \Omega$, $f = 10$ GHz, $\omega = 2\pi 10^{10} = 2.283 \cdot 10^{10}$ and so

$$Z_{01} = \sqrt{2}Z_0 = 84.85\ \Omega, \quad R = 2Z_0 = 120\ \Omega.$$

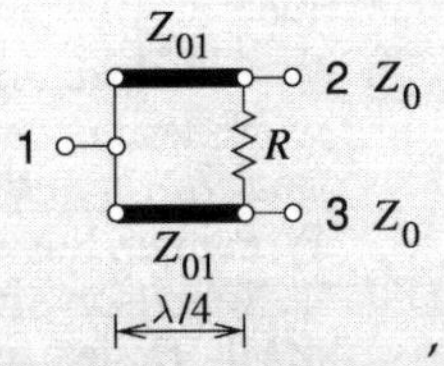

The next stage is to convert the transmission lines to lumped elements. A broadband design of a quarter-wavelength transmission line is presented in Section 22.7.2 of [15]. That is, each of the quarter-wave lines has the model

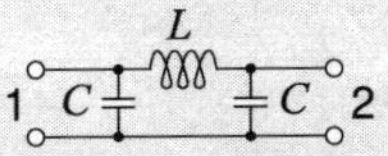

with $L = Z_{01}/\omega = 84.85/\omega = 954.9$ pH, $C = 1/(Z_{01}\omega) = 1/(84.85\omega) = 265.3$ fF.

So the final lumped element design is

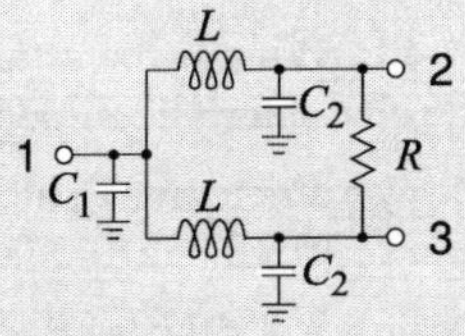

with
$C_1 = 2C = 530.6$ fF,
$C_2 = C = 265.3$ fF,
$L = 954.9$ pH,
$R = 120\ \Omega$.

9.9 Summary

This chapter introduced the richness of microwave components available to the RF designer. Microwave lumped-element R, L, and C components are carefully constructed so that they function as intended up to 10 or so. They are usually in surface-mount form so that they can be integrated in design while minimizing the parasitic effects introduced by leads. To a limited extent, transmission line discontinuities can be used as lumped elements. Even if the transmission line discontinuities are not specifically introduced for this purpose, their lumped-element equivalent circuits must be included in circuit analysis. Transmission line stubs are widely used to introduce capacitance and inductance in circuits. In most transmission line technologies only shunt stubs are available, and thus there is a strong preference for shunt elements in circuit designs.

9.10 References

[1] A. Victor and M. Steer, "Reflection coefficient shaping of a 5-GHz voltage-tuned oscillator for improved tuning," *IEEE Trans. on Microwave Theory and Techniques*, vol. 55, no. 12, pp. 2488–2494, Dec. 2007.

[2] http://www.coilcraft.com.

[3] IEEE Standard 315-1975, Graphic Symbols for Electrical and Electronics Diagrams (Including Reference Designation Letters), Adopted Sept. 1975, Reaffirmed Dec. 1993. Approved by American National Standards Institute, Jan. 1989. Approved adopted for mandatory use, Department of Defense, United States of America, Oct. 1975. Approved by Canadian Standards Institute, Oct. 1975.

[4] http://www.minicircuits.com.

[5] P. Katehi and N. Alexopoulos, "Frequency-dependent characteristics of microstrip discontinuities in millimeter-wave integrated circuits," *IEEE Trans. on Microwave Theory and Techniques*, vol. 33, no. 10, pp. 1029–1035, Oct.

1985.

[6] T. Edwards and M. Steer, *Foundations for Microstrip Circuit Design*. John Wiley & Sons, 2016.

[7] R. Mongia, I. Bahl, and P. Bhartia, *RF and Microwave Coupled-line Circuits*. Artech House, 1999.

[8] W. Fathelbab and M. Steer, "Broadband network design," in *Multifunctional Adaptive Microwave Circuits and Systems*, M. Steer and W. Palmer, Eds., 2008, ch. 8.

[9] I. Robertson and S. Lucyszyn, *RFIC and MMIC Design and Technology*. IEE, 2001, iEE Circuits, Devices and Systems Series.

[10] C. Goldsmith, A. Kikel, and N. Wilkens, "Synthesis of marchand baluns using multilayer microstrip structures," *Int. J. of Microwave & Millimeter-Wave Computer-Aided Eng*, pp. 179–188, 1992.

[11] N. Marchand, "Transmission line conversion transformers," *Electronics*, pp. 142–145, 1944.

[12] W. Fathelbab and M. Steer, "New classes of miniaturized planar marchand baluns," *IEEE Trans. on Microwave Theory and Techniques*, vol. 53, no. 4, pp. 1211–1220, Apr. 2005.

[13] J. J. Cloete, "Graphs of circuit elements for the marchand balun," *Microwave Journal*, vol. 24, no. 5, pp. 125–128, May 1981.

[14] E. Wilkinson, "An n-way hybrid power divider," *IRE Trans. on Microwave Theory and Techniques*, vol. 8, no. 1, pp. 116–118, Jan. 1960.

[15] M. Steer, *Microwave and RF Design, Transmission Lines*, 3rd ed. North Carolina State University, 2019.

9.11 Exercises

1. A spiral inductor is modeled as an ideal inductor of 10 nH in series with a 5 Ω resistor. What is the Q of the spiral inductor at 1 GHz?
2. Consider the design of a 50 dB resistive T attenuator in a 75 Ω system. [Parallels Example 9.1]
   (a) Draw the topology of the attenuator.
   (b) Write down the design equations.
   (c) Complete the design of the attenuator.
3. Consider the design of a 50 dB resistive Pi attenuator in a 75 Ω system. [Parallels Example 9.1]
   (a) Draw the topology of the attenuator.
   (b) Write down the design equations.
   (c) Complete the design of the attenuator.
4. A 20 dB attenuator in a 17 Ω system is ideally matched at both the input and output. Thus there are no reflections and the power delivered to the load is reduced by 20 dB from the applied power. If a 5 W signal is applied to the attenuator, how much power is dissipated in the attenuator?
5. A resistive Pi attenuator has shunt resistors of $R_1 = R_2 = 294\ \Omega$ and a series resistor $R_3 = 17.4\ \Omega$. What is the attenuation (in decibels) and the characteristic impedance of the attenuator?
6. Design a resistive Pi attenuator with an attenuation of 10 dB in a 100 Ω system.
7. Design a 3 dB resistive Pi attenuator in a 50 Ω system.
8. A resistive Pi attenuator has shunt resistors $R_1 = R_2 = 86.4\ \Omega$ and a series resistor $R_3 = 350\ \Omega$. What is the attenuation (in decibels) and the system impedance of the attenuator?
9. A balun can be realized using a wire-wound transformer, and by changing the number of windings on the transformer it is possible to achieve impedance transformation as well as balanced-to-unbalanced functionality. A 500 MHz balun based on a magnetic transformer is required to achieve impedance transformation from an unbalanced impedance of 50 Ω to a balanced impedance of 200 Ω. If there are 20 windings on the balanced port of the balun transformer, how many windings are there on the unbalanced port of the balun?

9.11.1 Exercises by Section

†challenging

§9.2 1, 2, 3, 4 §9.4 5†, 6, 7, 8† §9.7 9†

9.11.2 Answers to Selected Exercises

1 12.57
2 R_1=R_2=74.5 Ω
3 75.48 Ω
4 4.95 W

CHAPTER 10

Impedance Matching

10.1 Introduction

The maximum transfer of signal power is one of the prime objectives in RF and microwave circuit design. Power traverses a network from a source to a load generally through a sequence of two-port networks. Maximum power transfer requires that the Thevenin equivalent impedance of a source be matched to the impedance seen from the source. That is, the source should be presented with the complex conjugate of the source impedance. This is achieved by designing what is called a matching network inserted between the source and load. Design of the matching network is called impedance matching.

Section 10.2 describes two common matching objectives. Then design approaches for impedance matching are presented first with an algorithmic approach in Sections 10.3–10.6 and then a graphical approach based on using a Smith chart in Sections 10.7 and 10.9.

10.2 Matching Networks

Matching networks are constructed using lossless elements such as lumped capacitors, lumped inductors and transmission lines and so have, ideally, no loss and introduce no additional noise. This section discusses matching objectives and the types of matching networks.

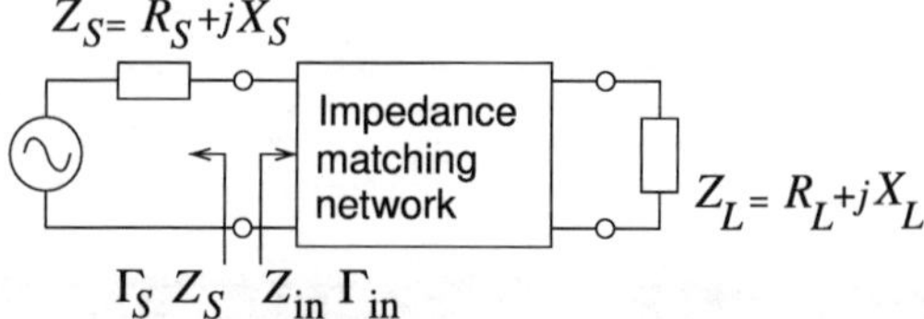

Figure 10-1: A source with Thevenin equivalent impedance Z_S and load with impedance Z_L interfaced by a matching network presenting an impedance Z_{in} to the source.

10.2.1 Matching for Zero Reflection or for Maximum Power Transfer

With RF circuits the aim of matching is to achieve maximum power transfer. With reference to Figure 10-1 the condition for maximum power transfer is $Z_{\text{in}} = Z_S^*$. which is equivalent to $\Gamma_{\text{in}} = \Gamma_S^*$. The proof is as follows:

$$\Gamma_{\text{in}} = \left(\frac{Z_{\text{in}} - Z_{\text{REF}}}{Z_{\text{in}} + Z_{\text{REF}}}\right), \tag{10.1}$$

and for maximum power transfer $Z_{\text{in}} = Z_S^*$, so

$$\begin{aligned}\Gamma_{\text{in}}^* &= \frac{Z_{\text{in}} - Z_{\text{REF}}}{Z_{\text{in}} + Z_{\text{REF}}} = \left(\frac{Z_S^* - Z_{\text{REF}}}{Z_S^* + Z_{\text{REF}}}\right)^* = \frac{(Z_S^* - Z_0)^*}{(Z_S^* + Z_0)^*} \\ &= \frac{(Z_S^*)^* - Z_{\text{REF}}^*}{(Z_S^*)^* + Z_{\text{REF}}^*} = \frac{Z_S - Z_{\text{REF}}^*}{Z_S + Z_{\text{REF}}^*} = \Gamma_S.\end{aligned} \tag{10.2}$$

If Z_{REF} is real, $Z_{\text{REF}}^* = Z_{\text{REF}}$ and then the condition for maximum power transfer is

$$\Gamma_{\text{in}}^* = \frac{Z_S - Z_{\text{REF}}}{Z_S + Z_{\text{REF}}} = \Gamma_S. \tag{10.3}$$

Thus, provided that Z_{REF} is real, the condition for maximum power transfer in terms of reflection coefficients is $\Gamma_{\text{in}}^* = \Gamma_S$ or $\Gamma_{\text{in}} = \Gamma_S^*$.

10.2.2 Types of Matching Networks

Up to a few gigahertz, lumped inductors and capacitors can be used in matching networks. Above a few gigahertz parasitics result in self-resonance. Lumped elements are also lossy. Segments of transmission lines are also used in matching networks as the loss of a transmission line component is always much less than the loss of a lumped inductor. However the length of a transmission line segement is up to $\lambda/4$ which is far too large to fit in consumer wireless products operating below a few gigahertz.

An impedance matching network may consist of

(a) Lumped elements only. These are the smallest networks, but have the most stringent limit on the maximum frequency of operation. The relatively high resistive loss of an inductor is the main limiting factor limiting performance. The self resonant frequency of an inductor limits operation to low microwave frequencies.

(b) Distributed elements (microstrip or other transmission line circuits) only. These have excellent performance, but their size restricts their use in systems to above a few gigahertz.

(c) A hybrid design combining lumped and distributed elements, primarily small sections of lines with capacitors. These lines are shorter than in a design with distributed elements only, but the hybrid design has higher performance than a lumped-element-only design.

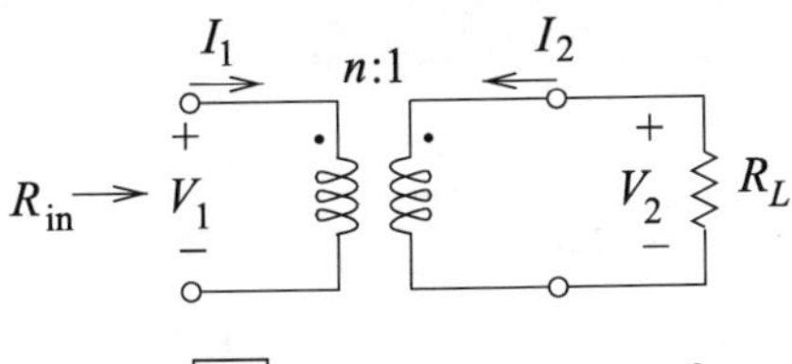

Figure 10-2: A transformer as a matching network. Port 1 is on the left or primary side and Port 2 is on the right or secondary side.

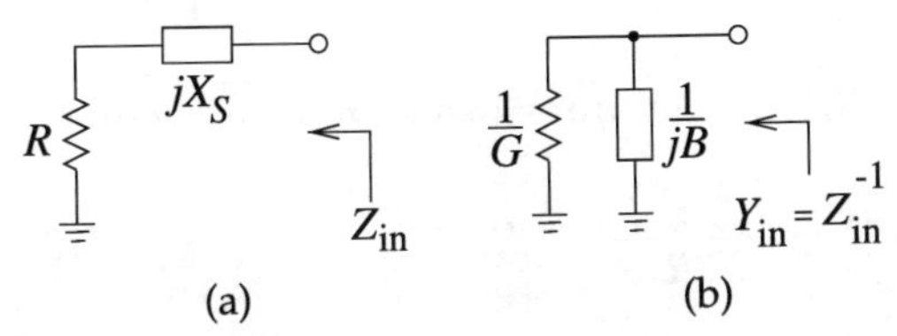

Figure 10-3: Matching using a series reactance: (a) the series reactive element; and (b) its equivalent shunt circuit.

10.3 Impedance Transforming Networks

Transformers and reactive elements considered in this section can be used to losslessly transform impedance levels. his is a basic aspect of network design.

10.3.1 The Ideal Transformer

The ideal transformer shown in Figure 10-2 can be used to match a load to a source if the source and load impedances are resistances. This will be shown by starting with the constitutive relations of the transformer:

$$V_1 = nV_2 \quad \text{and} \quad I_1 = -I_2/n. \tag{10.4}$$

Here n is the transformer ratio. For a wire-wound transformer, n is the ratio of the number of windings on the primary side, Port 1, to the number of windings on the secondary side, Port 2. Thus the input resistance, R_{in}, is related to the load resistance, R_L, by

$$R_{\text{in}} = \frac{V_1}{I_1} = -n^2\frac{V_2}{I_2} = n^2 R_{\text{L}}. \tag{10.5}$$

The matching problem with purely resistive load and source impedances is solved by choosing the appropriate winding ratio, n. However, resistive-only problems are rare at RF, and so other matching circuits must be used.

10.3.2 A Series Reactive Element

Matching using lumped elements is based on the impedance and admittance transforming properties of series and shunt reactive elements. Even a single reactive element can achieve limited impedance matching. Consider the series reactive element shown in Figure 10-3(a). Here the reactive element, X_S, is in series with a resistance R. The shunt equivalent of this network is shown in Figure 10-3(b) with a shunt susceptance of B. In this transformation the resistance R has been converted to a resistance $R_P = 1/G$. The mathematics describing this transformation is as follows. The input admittance of the series connection (Figure 10-3(a)) is

$$Y_{\text{in}}(\omega) = \frac{1}{Z_{\text{in}}(\omega)} = \frac{1}{R + \jmath X_S} = \frac{R}{R^2 + X_S^2} - \jmath\frac{X_S}{R^2 + X_S^2}. \tag{10.6}$$

Thus the elements of the equivalent shunt network, Figure 10-3(b), are

$$G = \frac{R}{R^2 + X_S^2} \quad \text{and} \quad B = -\frac{X_S}{R^2 + X_S^2}. \tag{10.7}$$

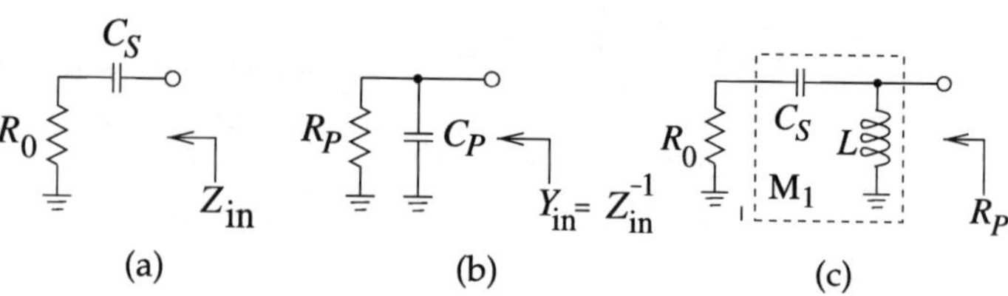

Figure 10-4: Impedance transformation by a series reactive element: (a) a resistor with a series capacitor; (b) its equivalent shunt circuit; and (c) an LC network.

The "resistance" of the network, R, has been transformed to a new value,

$$R_P = G^{-1} = \frac{R^2 + X_S^2}{R} > R. \tag{10.8}$$

This is an important start to matching, as X_S can be chosen to convert R (a load, for example) to any desired resistance value (such as the resistance of a source). However there is still a residual reactance that must be removed to complete the matching network design. Before moving on to the solution of this problem consider the following example.

EXAMPLE 10.1 Capacitive Impedance Transformation

Consider the impedance transforming properties of the capacitive series element in Figure 10-4(a). Show that the capacitor can be adjusted to obtain any positive shunt resistance.

Solution:

The concept here is that the series resistor and capacitor network has an equivalent shunt circuit that includes a capacitor and a resistor. By adjusting C_S any value can be obtained for R_P. From Equation (10.8),

$$R_P = \frac{R_0^2 + (1/\omega^2 C_S^2)}{R_0} = \frac{1 + \omega^2 C_S^2 R_0^2}{\omega^2 C_S^2 R_0} \tag{10.9}$$

and the susceptance is
$$B = \frac{(1/\omega C_S)}{R_0^2 + 1/\omega^2 C_S^2} = \omega \frac{C_S}{1 + \omega^2 C_S^2 R_0^2}. \tag{10.10}$$

Thus
$$C_P = \frac{B}{\omega} = \frac{C_S}{1 + \omega^2 C_S^2 R_0^2}. \tag{10.11}$$

To match R_0 to a resistive load R_P ($> R_0$) at a radian frequency ω_d, then, from Equation (10.9), the series capacitance required, i.e. the design equation for C_S, comes from

$$\omega_d C_S = 1/\sqrt{R_0 R_P - R_0^2}, \tag{10.12}$$

To complete the matching design, use a shunt inductor L, as shown in Figure 10-4(c), where $\omega_d C_P = 1/(\omega_d L)$. The equivalent impedance in Figure 10-4(c) is a resistor of value R_P, with a value that can be adjusted by choosing C_S which then requires L to be adjusted.

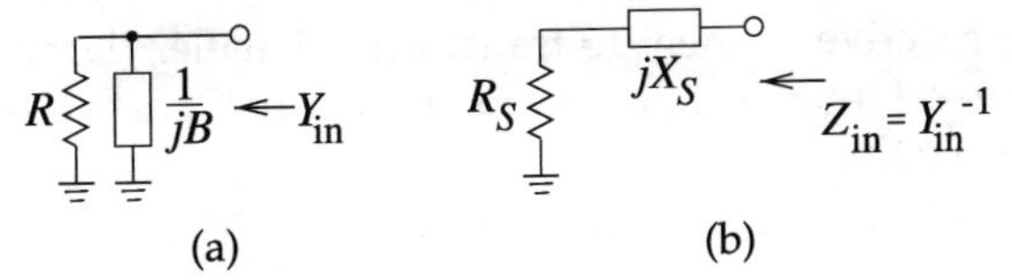

Figure 10-5: A resistor with (a) a shunt parallel reactive element where B is a susceptance, and (b) its equivalent series circuit.

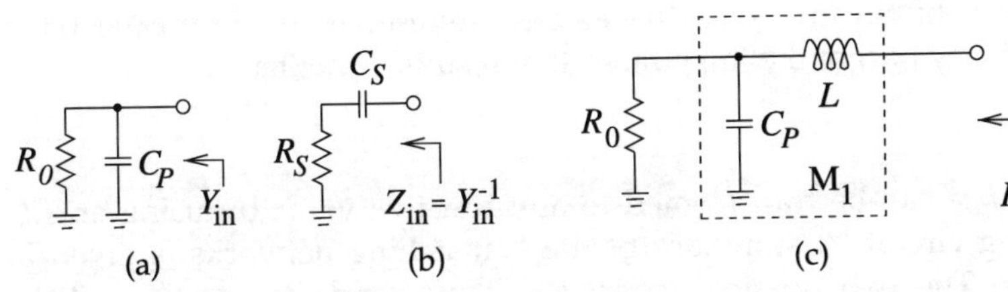

Figure 10-6: Parallel-to-series transformation: (a) resistor with shunt capacitor; (b) its equivalent series circuit; and (c) the transforming circuit with added series inductor.

10.3.3 A Parallel Reactive Element

The dual of the series matching procedure is the use of a parallel reactive element, as shown in Figure 10-5(a). The input admittance of the shunt circuit

$$Y_{\text{in}} = \frac{1}{R} + \jmath B. \tag{10.13}$$

This can be converted to a series circuit by calculating $Z_{\text{in}} = 1/Y_{\text{in}}$:

$$Z_{\text{in}} = \frac{R}{1 + \jmath BR} = \frac{R}{1 + B^2R^2} - \jmath\frac{BR^2}{1 + B^2R^2}. \tag{10.14}$$

$$\text{So} \quad R_S = \frac{R}{1 + B^2R^2} \quad \text{and} \quad X_S = \frac{-BR^2}{1 + B^2R^2}. \tag{10.15}$$

Notice that $R_S < R$.

EXAMPLE 10.2 Parallel Tuning

As an example of the use of a parallel reactive element to tune a resistance value, consider the circuit in Figure 10-6(a) where a capacitor tunes the effective resistance value so that the series equivalent circuit (Figure 10-6(b)) has elements

$$R_S = \frac{R_0}{1 + \omega^2 C_P^2 R_0^2} \quad \text{and} \quad X_S = -\frac{\omega C_P R_0^2}{1 + \omega^2 C_P^2 R_0^2} = -\frac{1}{\omega C_S}. \tag{10.16}$$

$$\text{So} \quad C_S = \frac{1 + \omega^2 C_P^2 R_0^2}{\omega^2 C_P R_0^2}. \tag{10.17}$$

Now consider matching R_0 to a resistive load R_S, which is less than R_0 at a given frequency ω_d. This requires that

$$\omega_d C_P = \sqrt{1/(R_S R_0) - 1/R_0^2}.$$

To complete the design, use a series inductor to remove the reactive effect of the capacitor, as shown in Figure 10-6(C). The value of the inductor required is found from

$$\omega_d L = \frac{1}{\omega_d C_S}, \quad \text{that is,} \quad L = \frac{1}{\omega_d^2 C_S}. \tag{10.18}$$

10.4 The L Matching Network

The examples in the previous two sections suggest the basic concept behind lossless matching of two different resistance levels using an L network:

Step 1: Use a series (shunt) reactive element to transform a smaller (larger) resistance up (down) to a larger (smaller) value with a real part equal to the desired resistance value.

Step 2: Use a shunt (series) reactive element to resonate with (or cancel) the imaginary part of the impedance that results from Step 1.

So a resistance can be transformed to any resistive value by using an LC transforming circuit. A summary of the L matching networks is given in Figure 10-7. The two possible cases, $R_S < R_L$ and $R_L < R_S$, will be considered in the following subsections.

10.4.1 Design Equations for $R_S < R_L$

Consider the matching network topology of Figure 10-8. Here

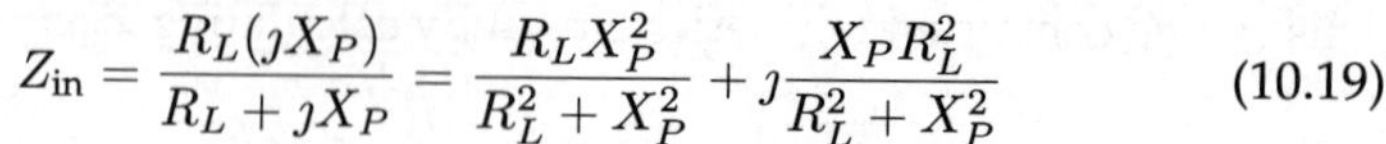

$$Z_{\text{in}} = \frac{R_L(\jmath X_P)}{R_L + \jmath X_P} = \frac{R_L X_P^2}{R_L^2 + X_P^2} + \jmath \frac{X_P R_L^2}{R_L^2 + X_P^2} \tag{10.19}$$

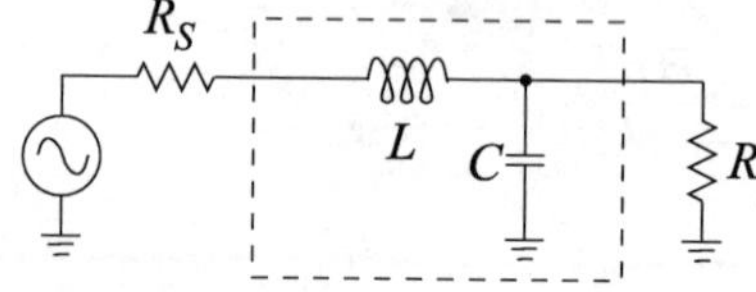

(a) Lowpass $R_S < R_L$.
$Q = \sqrt{R_L/R_S - 1}$
$Q = |X_L|/R_S = R_L/|X_C|$

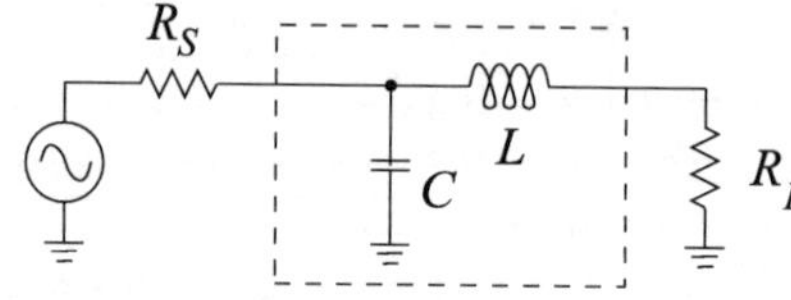

(b) Lowpass $R_S > R_L$.
$Q = \sqrt{R_S/R_L - 1}$
$Q = |X_L|/R_L = R_S/|X_C|$

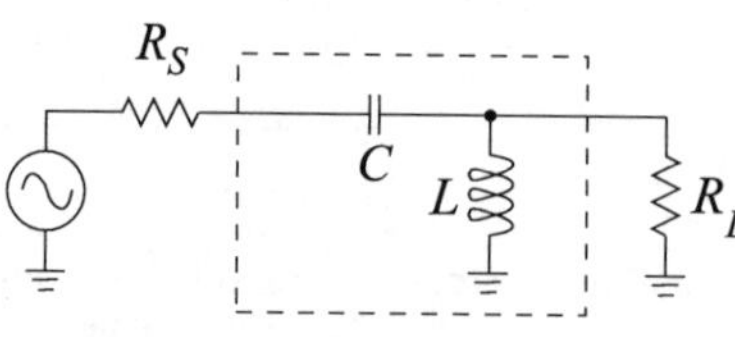

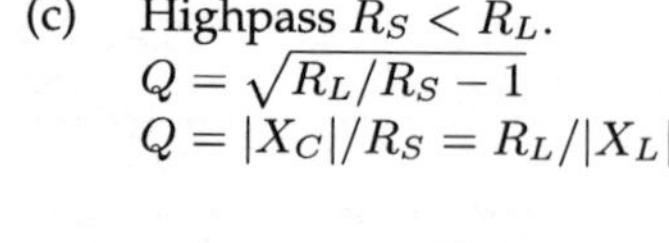

(c) Highpass $R_S < R_L$.
$Q = \sqrt{R_L/R_S - 1}$
$Q = |X_C|/R_S = R_L/|X_L|$

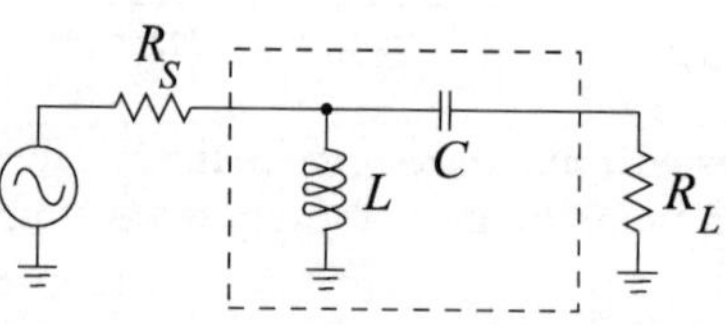

(d) Highpass $R_S > R_L$.
$Q = \sqrt{R_S/R_L - 1}$
$Q = |X_C|/R_L = R_S/|X_L|$

Figure 10-7: L matching networks consisting of one shunt reactive element and one series reactive element. (R_S is matched to R_L.) X_C is the reactance of the capacitor C, and X_L is the reactance of the inductor L. Note that with a two-element matching network the Q and thus bandwidth of the match is fixed.

R_S jX_S jX_P Z_{in} R_L

Figure 10-8: Two-element matching network topology for $R_S < R_L$. X_S is the series reactance and X_P is the parallel reactance.

and the matching objective is $Z_{\text{in}} = R_S - \jmath X_S$ so that

$$R_S = \frac{R_L X_P^2}{R_L^2 + X_P^2} \quad \text{and} \quad X_S = \frac{-X_P R_L^2}{X_P^2 + R_L^2}. \tag{10.20}$$

From these $$\frac{R_S}{R_L} = \frac{1}{(R_L/X_P)^2 + 1} \quad \text{and} \quad -\frac{X_S}{R_S} = \frac{R_L}{X_P}. \tag{10.21}$$

Introducing the quantities

$$Q_S = \text{the } Q \text{ of the series leg} = |X_S/R_S| \tag{10.22}$$
$$Q_P = \text{the } Q \text{ of the shunt leg} = |R_L/X_P| \tag{10.23}$$

leads to the final design equations for $R_S < R_L$:

$$|Q_S| = |Q_P| = \sqrt{\frac{R_L}{R_S} - 1}. \tag{10.24}$$

The L matching network principle is that X_P and X_S will be either capacitive or inductive and they will have the opposite sign (i.e., the L matching network comprises one inductor and one capacitor). Also, once R_S and R_L are given, the Q of the network and thus bandwidth is defined; with the L network, the designer does not have a choice of circuit Q.

EXAMPLE 10.3 Matching Network Design

Design a circuit to match a 100 Ω source to a 1700 Ω load at 900 MHz. Assume that a DC voltage must also be transferred from the source to the load.

Solution:

Here $R_S < R_L$, and so the topology of Figure 10-9(a) can be used and there are two versions, one with a series inductor and one with a series capacitor. The series inductor version (see Figure 10-9(b)) is chosen as this enables DC bias to be applied. From Equations (10.22)–(10.24) the design equations are

$$|Q_S| = |Q_P| = \sqrt{\frac{1700}{100}} - 1 = \sqrt{16} = 4, \quad \frac{X_S}{R_S} = 4, \quad \text{and} \quad X_S = 4 \cdot 100 = 400. \tag{10.25}$$

This indicates that $\omega L = 400\ \Omega$, and so the series element is

$$L = \frac{400}{2\pi \cdot 9 \cdot 10^8} = 70.7 \text{ nH}. \tag{10.26}$$

For the shunt element next to the load, $|R_L/X_C| = 4$, and so

$$|X_C| = \frac{R_L}{4} = \frac{1700}{4} = 425. \tag{10.27}$$

Thus $1/\omega C = 425$ and $$C = \frac{1}{2\pi \cdot 9 \cdot 10^8 \cdot 425} = 0.416 \text{ pF}. \tag{10.28}$$

The final matching network design is shown in Figure 10-9(c).

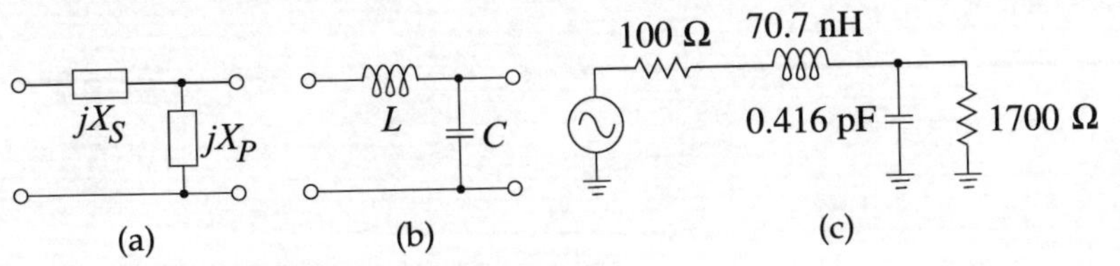

Figure 10-9: Matching network development for Example 10.3.

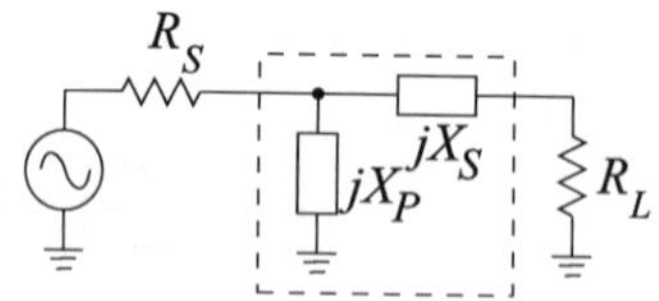

Figure 10-10: Two-element matching network topology for $R_S > R_L$.

10.4.2 L Network Design for $R_S > R_L$

For $R_S > R_L$, the topology shown in Figure 10-10 is used. The design equations for the L network for $R_S > R_L$ are similarly derived and are

$$|Q_S| = |Q_P| = \sqrt{\frac{R_S}{R_L} - 1} \tag{10.29}$$

$$-Q_S = Q_P, \quad Q_S = \frac{X_S}{R_L}, \quad \text{and} \quad Q_P = \frac{R_S}{X_P}. \tag{10.30}$$

EXAMPLE 10.4 Two-Element Matching Network

Design a passive two-element matching network that will achieve maximum power transfer from a source with an impedance of 50 Ω to a load with an impedance of 75 Ω. Choose a matching network that will not allow DC to pass.

Solution:

$R_L > R_S$, so, from Figure 10-7, the appropriate matching network topology is

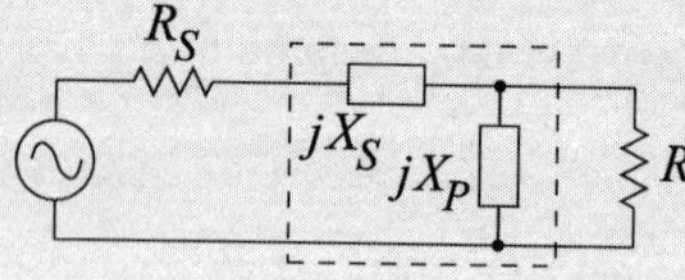

This topology can be either high pass or low pass depending on the choice of X_S and X_P. Design proceeds by finding the magnitudes of X_S and X_P. In two-element matching the circuit Q is fixed. With $R_L = 75\ \Omega$ and $R_S = 50\ \Omega$.

The Q of the matching network is the same for the series and parallel elements:

$$|Q_S| = \frac{|X_S|}{R_S} = \sqrt{\frac{R_L}{R_S} - 1} = 0.7071 \quad \text{and} \quad |Q_P| = \frac{R_L}{|X_P|} = |Q_S| = 0.7071,$$

therefore $|X_S| = R_S \cdot |Q_S| = 50 \cdot 0.7071 = 35.35\ \Omega$. Also

$$|X_P| = R_L/|Q_P| = 75/0.7071 = 106.1\ \Omega.$$

Specific element types can now be assigned to X_S and X_P, and note that they must be of opposite type.

The lowpass matching network is

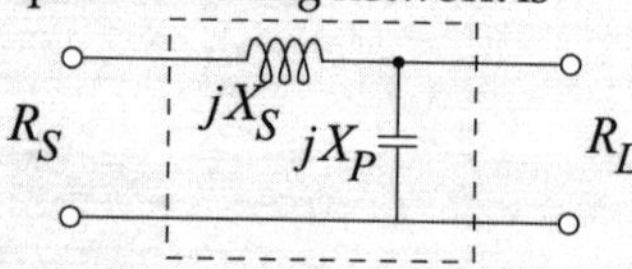

$X_S = +35.35\ \Omega, \quad X_P = -106.1\ \Omega.$

The highpass matching network is

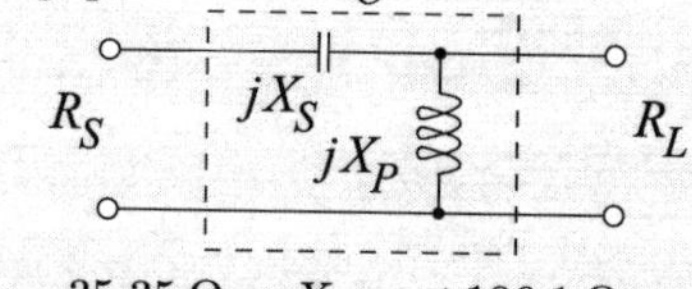

$X_S = -35.35\ \Omega, \quad X_P = +106.1\ \Omega.$

This highpass design satisfies the design criterion that DC is not passed, as DC is blocked by the series capacitor.

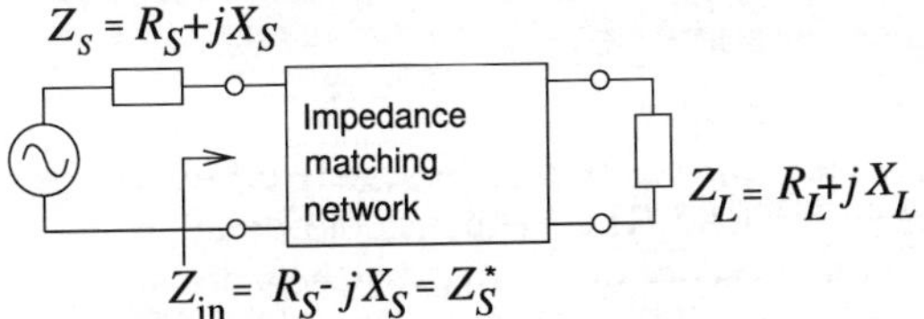

Figure 10-11: A matching network matching a complex load to a source with a complex Thevenin impedance.

10.5 Dealing with Complex Sources and Loads

This section presents strategies for dealing with complex loads. In the algorithmic matching approach design proceeds first by ignoring the complex load and source and then accounting for them either through topology choice or canceling their effect through resonance.

10.5.1 Matching

Input and output impedances of transistors, mixers, antennas, etc., contain both resistive and reactive components. Thus a realistic impedance matching problem looks like that shown in Figure 10-11. The matching approaches that were presented in the previous sections can be directly applied if X_S and X_L are treated as stray reactances that need to be either canceled or, ideally, used as part of the matching network. There are two basic approaches to handling complex impedances:

(a) Absorption: Absorb source and load reactances into the impedance matching network itself. This is done through careful placement of each matching element such that capacitors are placed in parallel with source and load capacitances, and inductors in series with source and load inductances. The stray values are then subtracted from the L and C values for the matching network calculated on the basis of the resistive parts of Z_S and Z_L only. The new (smaller) values, L' and C', constitute the elements of the matching network. Sometimes it is necessary to perform a series-to-parallel, or parallel-to-series, conversion of the source or load impedances so that the reactive elements are in the correct series or shunt arrangement for absorption.

(b) Resonance: Resonate source and load reactances with an equal and opposite reactance at the frequency of interest.

The presence of reactance in a load indicates energy storage, and therefore bandwidth limiting. In the above approaches to handling a reactive load, the resonance approach could easily result in a narrowband matching solution. The major problem in matching is often to obtain sufficient bandwidth. What is sufficient will vary depending on the application. To maximize bandwidth the general goal is to minimize the total energy storage. Roughly the total energy stored will be proportional to the sum of the magnitudes of the reactances in the circuit. Of course, the actual energy storage depends on the voltage and current levels, which will themselves vary in the circuit. A good approach leading to large bandwidths is to incorporate the load reactance into the matching network. Thus the choice of appropriate matching network topology is critical. However, if the source and load reactance value is larger than the calculated matching network element value, then absorption on its own cannot be used. In this situation resonance must be combined with absorption. The majority of impedance matching designs are based on a combination of resonance and absorption.

EXAMPLE 10.5 Matching Network Design Using Resonance

For the configuration shown in Figure 10-12, design an impedance matching network that will block the flow of DC current from the source to the load. The frequency of operation is 1 GHz. Design the matching network, neglecting the presence of the 10 pF capacitance at the load. Since $R_S = 50\ \Omega < R_L = 500\ \Omega$, and from Figure 10-7, consider the topologies of Figures 10-13(a) and 10-13(b). The design criterion of blocking flow of DC from the source to the load narrows the choice to the topology of Figure 10-13(b).

Solution:

Step 1: $$|Q_S| = \left|\frac{X_S}{R_S}\right| = |Q_P| = \left|\frac{R_L}{X_P}\right| = \sqrt{\frac{R_L}{R_S} - 1} = 3 \quad \text{and} \quad Q_P = \frac{R_L}{X_P}. \tag{10.31}$$

So $X_P = \omega L = R_L/Q_P = 500/3$. Reducing this gives

$$\omega L = \frac{500}{3}, \quad \text{and so} \quad L = \frac{500}{3 \times 2\pi \times 10^9} = 26.5\ \text{nH}. \tag{10.32}$$

Similarly $-X_S/R_S = 3$ and so

$$\frac{1/(\omega\, C)}{R_S} = 3 \quad \text{or} \quad C = \frac{1}{3\omega R_S} = \frac{1}{3 \times 2\pi \times 10^9 \times 50} = 1.06\ \text{pF}. \tag{10.33}$$

Step 2:

Resonate the 10 pF capacitor using an inductor in parallel:

$$(\omega L')^{-1} = \omega \times 10 \times 10^{-12} \tag{10.34}$$

$$L' = \frac{1}{\omega^2 10^{-11}} = \frac{1}{(2\pi)^2\, 10^{18} \times 10^{-11}} = 2.533\ \text{nH}. \tag{10.35}$$

Thus Figure 10-13(c) is the required matching network. Two inductors are in parallel and the circuit can be simplified to that shown in Figure 10-14, where

$$L_X = (26.5\ \text{nH} \parallel 2.533\ \text{nH}) = \frac{2.533 \times 26.5}{2.533 + 26.5}\ \text{nH} = 2.312\ \text{nH}. \tag{10.36}$$

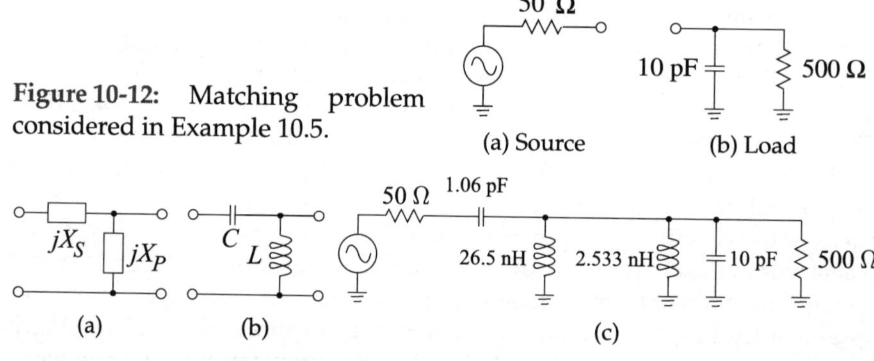

Figure 10-12: Matching problem considered in Example 10.5.

Figure 10-13: Matching network topology used in Example 10.5: a) and (b) topology; and (c) intermediate matching network.

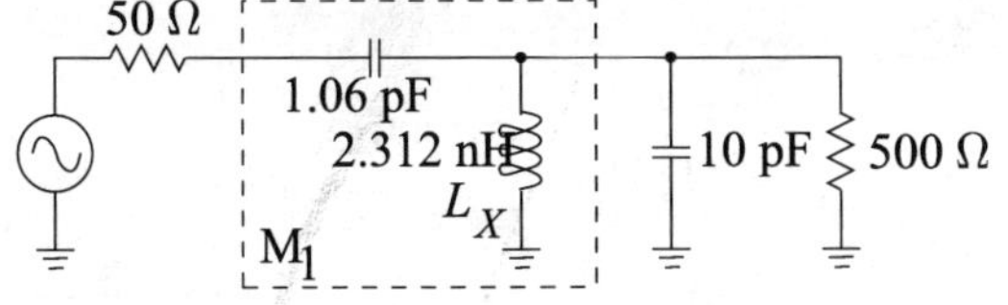

Figure 10-14: Final matching network in Example 10.5.

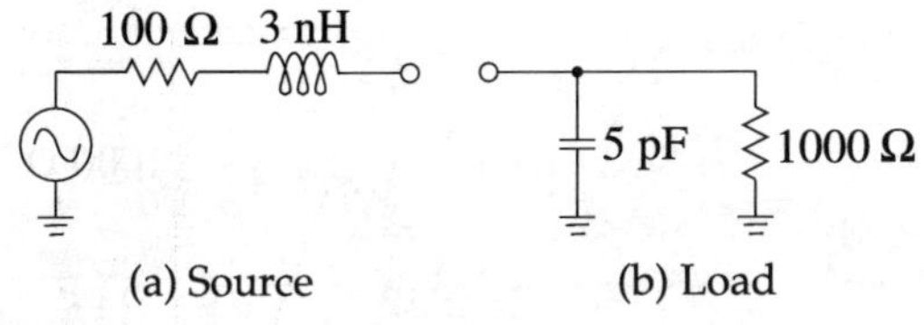

Figure 10-15: Matching problem in Example 10.6.

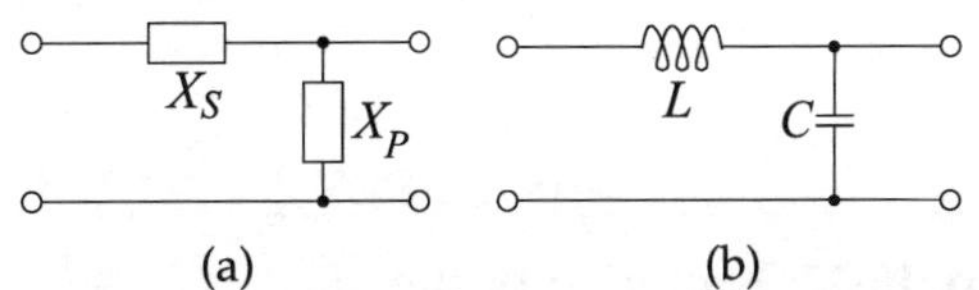

Figure 10-16: Topologies referred to in Example 10.6.

EXAMPLE 10.6 Matching Network Design Using Resonance and Absorption

For the source and load configurations shown in Figure 10-15, design a lowpass impedance matching network at $f = 1$ GHz.

Solution:

Since $R_S < R_L$, use the topology shown in Figure 10-16(a). For a lowpass response, the topology is that of Figure 10-16(b). Notice that absorption is the natural way of handling the 3 nH at the source and the 5 pF at the load. The design process is as follows:

Step 1:

Design the matching network, neglecting the reactive elements at the source and load:

$$|Q_S| = |Q_P| = \sqrt{\frac{R_L}{R_S} - 1} = \sqrt{10 - 1} = 3 \tag{10.37}$$

$$\frac{X_S}{R_S} = 3, \quad X_S = 3 \times 100, \quad \omega L = 300 \quad \text{and} \quad L = \frac{300}{2\pi \times 10^9} = 47.75\text{ nH} \tag{10.38}$$

$$\frac{R_P}{X_P} = -3 \quad \text{and} \quad \frac{1000}{-(1/\omega C)} = -3 \quad \text{and} \quad C = \frac{3}{1000 \times 2\pi \times 10^9} = 0.477\text{ pF}. \tag{10.39}$$

This design is shown in Figure 10-17(a). This is the matching network that matches the 100 Ω source resistance to the 1000 Ω load with the source and load reactances ignored.

Step 2:

Figure 10-17(b) is the interim matching solution. The source inductance is absorbed into the matching network, reducing the required series inductance of the matching network. The capacitance of the load cannot be fully absorbed. The design for the resistance-only case requires a shunt capacitance of 0.477 pF, but 5 pF is available from the load. Thus there is an excess capacitance of 4.523 pF that must be resonated out by the inductance L'':

$$\frac{1}{\omega L''} = \omega 4.523 \times 10^{-12}. \quad \text{So} \quad L'' = \frac{1}{(2\pi)^2 \times 10^{18} \times 4.523 \times 10^{-12}} = 5.600\text{ nH}. \tag{10.40}$$

The final matching network design (Figure 10-17(c)) fully absorbs the source inductance into the matching network, but only partly absorbs the load capacitance.

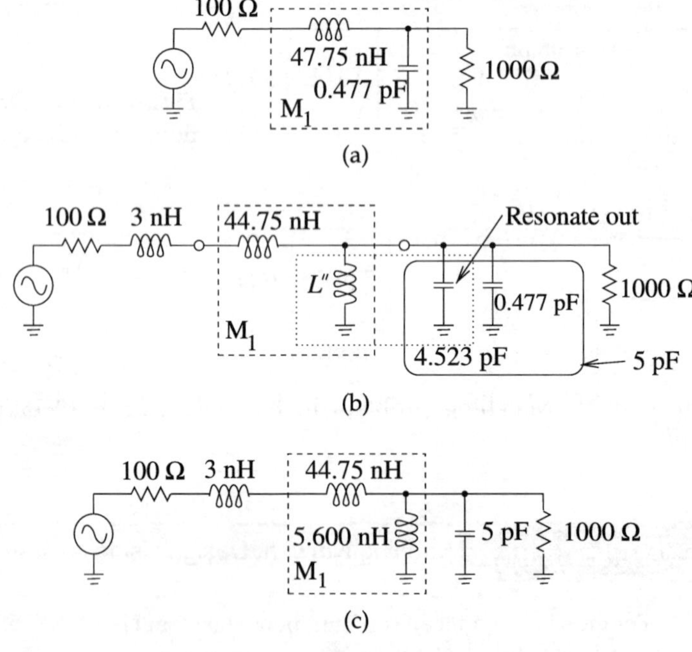

Figure 10-17: Evolution of the matching network in Example 10.6: (a) matching network design considering only the source and load resistors; (b) matching network with the reactive parts of the source and load impedances included; and (c) final design.

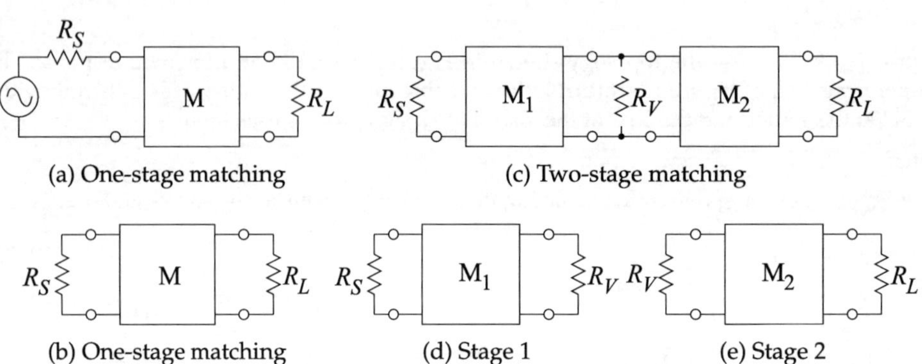

Figure 10-18: Matching in stages: (a) matching network M matching R_S to R_L; (b) without an explicit source; (c) two-stage matching with a virtual resistor R_V; (d) matching R_S to R_V; and (e) matching R_V to R_L.

10.6 Multielement Matching

The bandwidth of a matching network can be controlled by using multiple matching stages either making the matching bandwidth wider or narrower. This concept is elaborated on in this and several design approaches presented.

10.6.1 Design Concept for Manipulating Bandwidth

The concept for manipulating matching network bandwidth is to do the matching in stages as shown in Figure 10-18. Figure 10-18(a) shows the one-stage matching problem using the common identification of the matching network as 'M'. The one-stage matching problem is repeated in Figure 10-

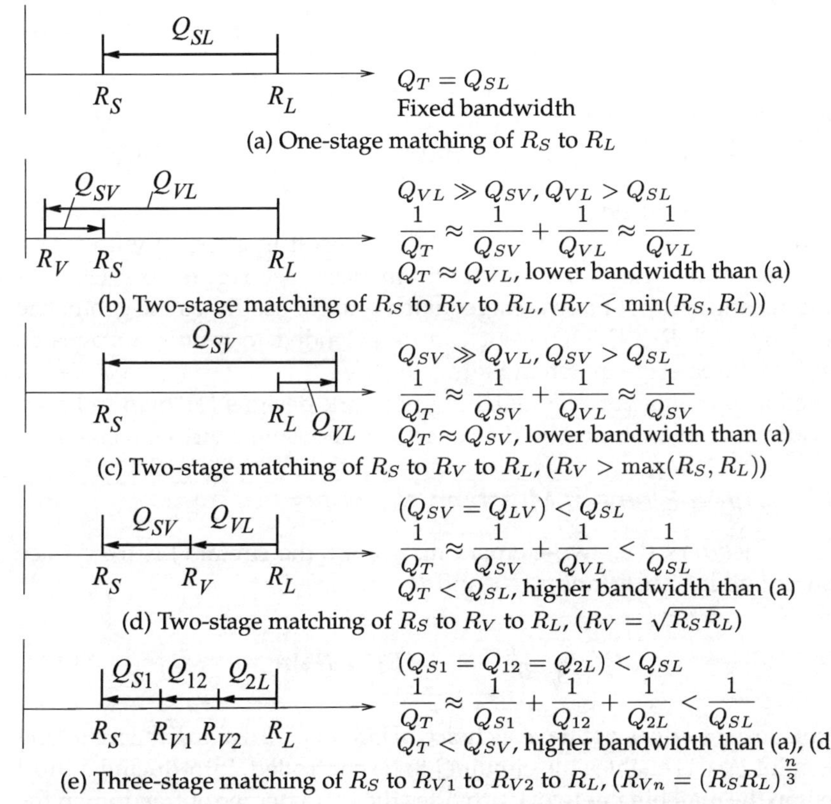

Figure 10-19: Effect of multi-stage matching on total circuit Q, Q_T, and matching bandwidth (which is approximately inversely proportional to Q_T.)

18(b) without explicitly showing the source generator. A two-stage matching problem is shown in Figure 10-18(c) with the introduction of a virtual resistor R_V between the first, M_1, and second, M_2, stage matching networks. R_V is shown as a virtual connection as it is not actually inserted in the circuit. Instead this is a short-hand way of indicating the matching problem to be done in two stages as shown in Figure 10-18(d and e) with the first stage matching the source resistor R_S to R_V and the second stage matching R_V to the load resistor R_L. After M_1 has been designed the resistance looking into the right-hand port of M_1, see Figure 10-18(d), will be R_V so R_V is the Thevenin equivalent source resistance to M_2. Similarly the input impedance looking into the left-hand port of M_2 is R_V so R_V is the effective load resistor of M_1. Of course these are the impedances at the center frequency and away from the center frequency of the match the input impedances will be complex.

The concept behind multi-stage matching network design is shown in Figure 10-19 where the standard one-stage match is shown in Figure 10-19(a). While this is shown for $R_L > R_S$ the concept holds for $R_L < R_S$. The arrows follow the that design begins with the load and ends at the source. With the one stage match the circuit Q is fixed and designated here as the total circuit Q, Q_T being the same as the Q of the one-stage R_L to R_S matching network, Q_{SL}. The two-stage match that reduces bandwidth (compared to the one-stage match) is shown in Figure 10-19(b). The total Q, Q_T, of the second stage is higher than for the one-stage design because the ratio of R_L to R_V is greater than the ratio of R_V to R_S. Bandwidth can also be reduced relative to the one-stage match by assigning R_V to be greater than both R_L

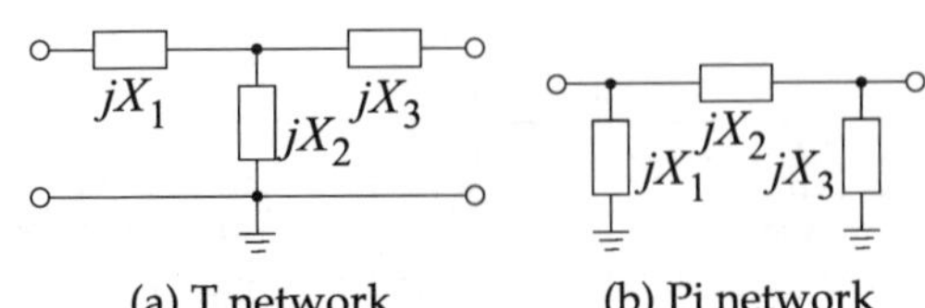

Figure 10-20: Two three-element matching networks.

and R_S, see Figure 10-19(c).

Choosing R_V to be between R_S and R_L will result in a circuit with lower Q_T and the bandwidth of the match will increase, see Figure 10-19(d). The maximum bandwidth for a two-stage match is to choose R_V as the geometric mean of R_S and R_L. This concept can be extended to multiple stages as shown for a three-stage match in Figure 10-19(e).

This section presents various matching network designs for manipulating bandwidth and all are based on the concept of choosing a virtual resistor.

10.6.2 Three-Element Matching Networks

With the L network (i.e., two-element matching), the circuit Q is fixed once the source and load resistances, R_S and R_L, are fixed:

$$Q = \sqrt{\frac{R_L}{R_S} - 1}, \quad (R_L > R_S). \tag{10.41}$$

Thus the designer does not have a choice of circuit Q. Breaking the matching problem into parts enables the circuit Q to be controlled. Introducing a third element in the matching network provides the extra degree of freedom in the design for adjusting Q, and hence bandwidth.

Two three-element matching networks, the T network and the Pi network, are shown in Figure 10-20. Which network is used depends on

(a) the realization constraints associated with the specific design, and

(b) the nature of the reactive parts of the source and load impedances and whether they can be used as part of the matching network.

The three-element matching network comprises 2 two-element (or L) matching networks and is used to increase the overall Q and thus narrow bandwidth. Given R_S and R_L, the circuit Q established by an L matching network is the minimum circuit Q available in the three-element matching arrangement. With three-element matching, the Q can only increase, so three-element matching is used for narrowband (high-Q) applications. However, lower Q can be obtained with more than three elements. The next subsections consider matching with more than three elements.

10.6.3 The Pi Network

The Pi network may be thought of as two back-to-back L networks that are used to match the load and the source to a virtual resistance, R_V, placed at the junction between the two networks, as shown in Figure 10-21(b). The design of each section of the Pi network is as for the L network matching. R_S is matched to R_V and R_V is matched to R_L.

R_V must be selected smaller than R_S and R_L since it is connected to the series arm of each L section. Furthermore, R_V can be any value that is smaller than the smaller of R_S, R_L. However, it is customarily used as the design parameter for specifying the desired Q.

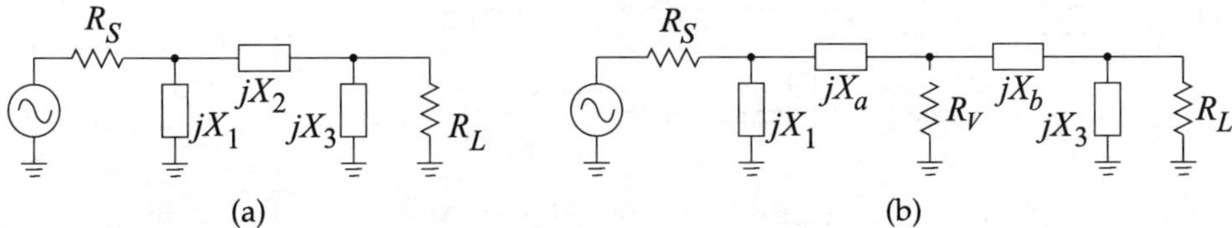

Figure 10-21: Pi matching networks: (a) view of a Pi network; and (b) as two back-to-back L networks with a virtual resistance, R_V, between the networks.

As a useful design approximation, the loaded Q of the Pi network can be taken as the Q of the L section with the highest Q:

$$Q = \sqrt{\frac{\max(R_S, R_L)}{R_V} - 1}. \tag{10.42}$$

Given R_S, R_L, and Q, the above equation yields the value of R_V.

EXAMPLE 10.7 Three-Element Matching Network Design

Design a Pi network to match a 50 Ω source to a 500 Ω load. The desired Q is 10. A suitable matching network topology is shown in Figure 10-22 together with the virtual resistance, R_V, to be used in design.

Solution: $R_S = 50\ \Omega$ and $R_L = 500\ \Omega$ so $\max(R_S, R_L) = 500\ \Omega$ and so the virtual resistor is

$$R_V = \frac{\max(R_S, R_L)}{Q^2 + 1} = \frac{500}{101} = 4.95\ \Omega. \tag{10.43}$$

Design proceeds by separately designing the L networks to the left and right of R_V. For the L network on the left,

$$Q_{\text{left}} = \sqrt{\frac{50}{4.95} - 1} = 3.017. \quad \text{so} \quad Q_{\text{left}} = \frac{|X_a|}{R_V} = \frac{R_S}{|X_1|} = 3.017, \tag{10.44}$$

Note that X_1 and X_a must be of opposite types (one is capacitive and the other is inductive). The left L network has elements

$$|X_a| = 14.933\ \Omega \quad \text{and} \quad |X_1| = 16.6\ \Omega. \tag{10.45}$$

For the L network on the right of R_V,

$$Q_{\text{right}} = Q = 10, \quad \text{thus} \quad \frac{|X_b|}{R_V} = \frac{R_L}{X_3} = 10. \tag{10.46}$$

X_b, X_3 are of opposite types, and

$$|X_b| = 49.5\ \Omega \quad \text{and} \quad |X_3| = 50\ \Omega. \tag{10.47}$$

The resulting Pi network is shown in Figure 10-23 with the values

$$|X_1| = 16.6\ \Omega, \quad |X_3| = 50\ \Omega, \quad |X_a| = 14.933\ \Omega, \quad \text{and} \quad |X_b| = 49.5\ \Omega. \tag{10.48}$$

Note that the pair X_a, X_1 are of opposite types and similarly X_b, X_3 are of opposite types. So there are four possible realizations, as shown in Figure 10-24.

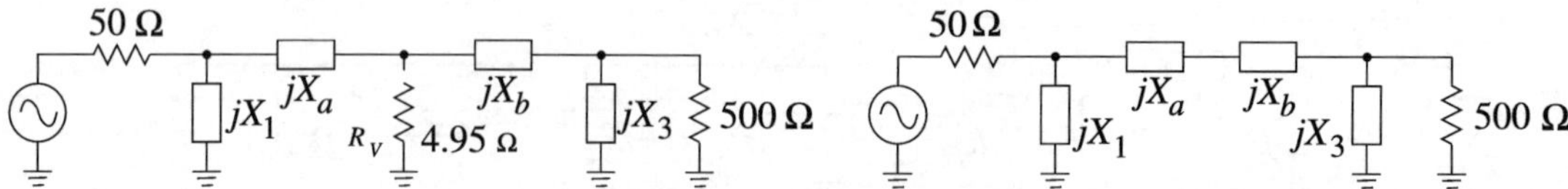

Figure 10-22: Matching network problem of Example 10.7.

Figure 10-23: Final matching network in Example 10.7.

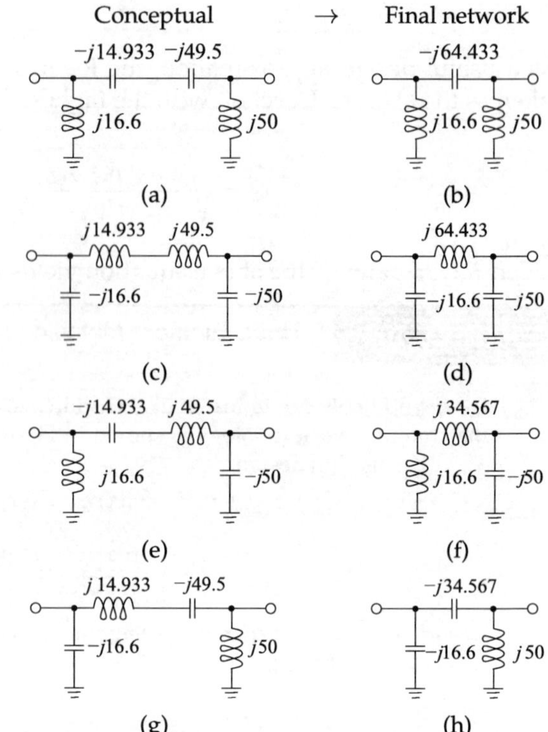

Figure 10-24: Four possible Pi matching networks: (a), (c), (e), and (g) conceptual circuits; and (b), (d), (f), and (h), respectively, their final reduced Pi networks.

In the previous example there were four possible realizations of the three-element matching network, and this is true in general. The specific choice of one of the four possible realizations will depend on specific application-related factors such as

(a) elimination of stray reactances,
(b) the need to pass or block DC current, and
(c) the need for harmonic filtering.

It is fortunate that it may be possible to achieve multiple functions with the same network.

EXAMPLE 10.8 Three-Element Matching with Reactive Source and Load

Design a Pi network to match the source to the load shown. The design frequency is 900 MHz and the desired Q is 10.

50 Ω, 2 pF, Source; 1 pF, 500 Ω, Load

Solution:

The design objective is to arrive at an overall network which has a Q of 10. To achieve this it is necessary to absorb the source and load reactances into the matching network. If they were resonated instead, the overall Q of the network can be expected to higher than the Q of the L matching network on its own.

Design begins by considering the matching of $R_S = 50\ \Omega$ to $R_L = 500\ \Omega$. Since the Q is specified, three (or more) matching elements must be used. The design starting point is shown on the right:

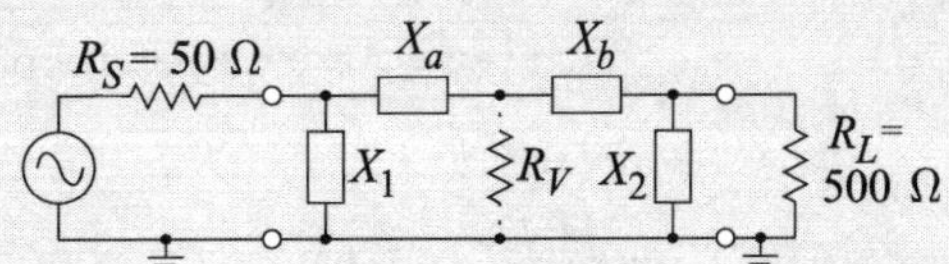

The virtual resistor $R_V = \max(R_s, R_L)/(1+Q^2) = (500\ \Omega)/(1+100) = 4.95\ \Omega$. The left subnetwork with X_1 and X_a has $Q_{\text{LEFT}} = \sqrt{R_S/R_V - 1} = \sqrt{50/4.95 - 1} = 3.017$. The right subnetwork with X_2 and X_b has $Q_{\text{RIGHT}} = \sqrt{R_L/R_V - 1} = \sqrt{500/4.95 - 1} = 10.001$.

Note that Q_{RIGHT} is almost exactly the desired Q of the network and Q_{LEFT} will have little effect on the Q of the overall circuit. Now $Q_{\text{LEFT}} = |X_a|/R_V = R_S/|X_1|$, so $|X_a| = 14.9\ \Omega$ and $|X_1| = 16.57\ \Omega$. $Q_{\text{RIGHT}} = |X_b|/R_V = R_S/|X_2|$, so $|X_b| = 49.5\ \Omega$ and $|X_2| = 50.0\ \Omega$.

X_1 must be chosen to be a capacitor $C_1 = 10.67$ pF so that the 2 pF source capacitance can be absorbed. Similarly X_2 is a capacitor $C_2 = 3.53$ pF. X_a and X_b are both inductors that combine in series for a total inductance $L_3 = 11.38$ nH. This leads to the final design shown on the right where the matching network is in the dashed box.

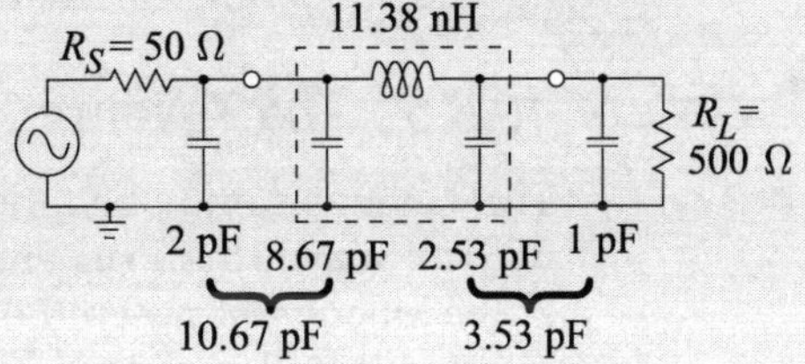

10.6.4 Matching Network Q Revisited

To demonstrate that the circuit Q established by an L matching network is the minimum circuit Q for a network having at most three elements, consider the design equations for $R_S > R_L$. Referring to Figure 10-25,

$$X_1 = \frac{R_S}{Q},\ X_3 = R_L\left(\frac{R_S/R_L}{Q^2+1-R_S/R_L}\right)^{\frac{1}{2}},\ X_2 = \frac{QR_S + R_SR_L/X_3}{Q^2+1}. \tag{10.49}$$

Notice that the denominator of X_3 can be written as

$$Q^2 + 1 - \frac{R_S}{R_L} = \left(Q + \sqrt{\frac{R_S}{R_L} - 1}\right)\left(Q - \sqrt{\frac{R_S}{R_L} - 1}\right). \tag{10.50}$$

$R_S \leftarrow$ jX_1 jX_2 jX_3 $\rightarrow R_L$

Figure 10-25: A three-element matching network.

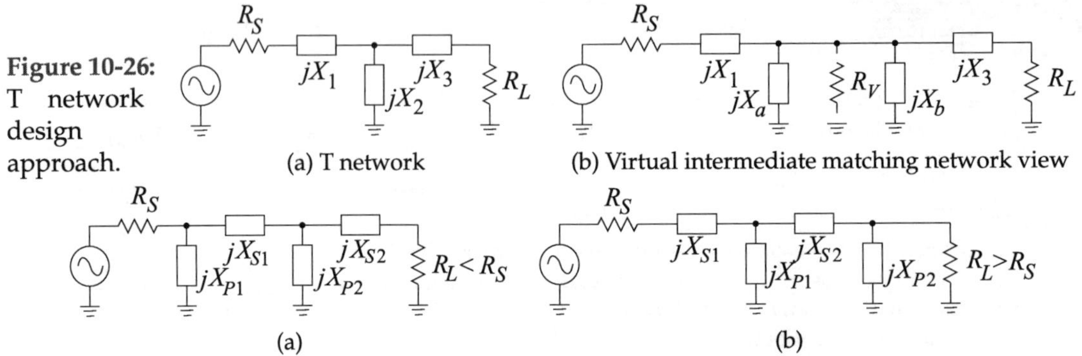

Figure 10-26: T network design approach.

Figure 10-27: Broadband matching networks.

Then for a real solution we must have

$$Q \geq \sqrt{\frac{R_S}{R_L} - 1}, \tag{10.51}$$

and so $Q \geq Q_{\text{L network}}$. For $Q = Q_{\text{L network}}$, $X_3 \to \infty$ and the Pi network reduces to an L network that has two elements. Thus it is not possible to have a lower Q with a three-element matching network than the Q of a two-element matching network. Thus a three-element matching network must have lower bandwidth than that of a two-element matching network.

10.6.5 The T Network

The T network may be thought of as two back-to-back L networks that are used to match the load and the source to a virtual resistance, R_V, placed at the junction between the two L networks (see Figure 10-26). R_V must be selected to be larger than both R_S and R_L since it is connected to the shunt leg of each L section. R_V is chosen according to the equation

$$Q = \sqrt{\frac{R_V}{\min(R_S, R_L)} - 1}, \tag{10.52}$$

where Q is the desired loaded Q of the network. Each L network is calculated in exactly the same manner as was done for the Pi network matching. That is, R_S is matched to R_V and R_V is matched to R_L. Once again there will be four possible designs for the T network, given R_S, R_L, and Q.

10.6.6 Broadband (Low Q) Matching

L network matching does not allow the circuit Q, and hence bandwidth, to be selected. However, Pi network and T network matching allows the circuit Q to be selected independent of the source and load impedances, provided that the chosen Q is larger than that which can be obtained with an L network. Thus the Pi and T networks result in narrower bandwidth designs.

One design solution for broadband matching is to use two (or more) series-connected L sections (see Figure 10-27). Design is still based on the concept of a virtual resistor, R_V, placed at the junction of the two L networks (as in Figure 10-28), but now R_V is chosen to be between R_S and R_L:

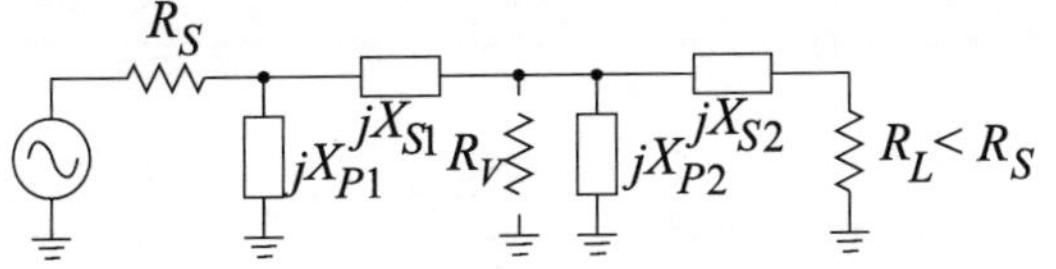

Figure 10-28: Matching network with two L networks.

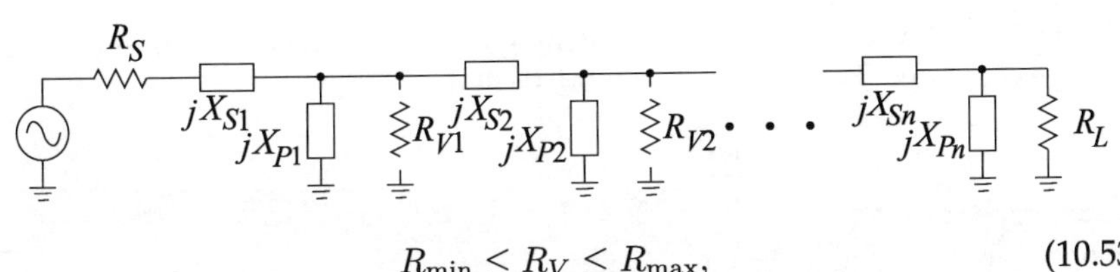

Figure 10-29: Cascaded L networks for broadband matching.

$$R_{\min} \leq R_V \leq R_{\max}, \tag{10.53}$$

where $R_{\min} = \min(R_L, R_S)$ and $R_{\max} = \max(R_L, R_S)$. Then one of the two networks will have

$$Q_1 = \sqrt{\frac{R_V}{R_{\min}} - 1} \quad \text{and the other} \quad Q_2 = \sqrt{\frac{R_{\max}}{R_V} - 1}. \tag{10.54}$$

The maximum bandwidth (minimum Q) available is obtained when

$$Q_1 = Q_2 = \sqrt{\frac{R_V}{R_{\min}} - 1} = \sqrt{\frac{R_{\max}}{R_V} - 1}\,. \tag{10.55}$$

That is, the maximum matching bandwidth is obtained when R_V is the geometric mean of R_S and R_L:

$$R_V = \sqrt{R_L R_S}. \tag{10.56}$$

Even wider bandwidths can be obtained by cascading more than two L networks, as shown in Figure 10-29. In this circuit

$$R_S < R_{V1} < R_{V2} \ldots < R_{Vn-1} < R_L. \tag{10.57}$$

For optimum bandwidth the ratios should be equal,

$$\frac{R_{V_1}}{R_S} = \frac{R_{V_2}}{R_{V_1}} = \frac{R_{V_3}}{R_{V_2}} = \cdots = \frac{R_L}{R_{V_{n-1}}}, \tag{10.58}$$

and the Q is given by

$$Q = \sqrt{\frac{R_{V_1}}{R_S} - 1} = \sqrt{\frac{R_{V_2}}{R_1} - 1} = \cdots = \sqrt{\frac{R_L}{R_{V_{n-1}}} - 1}. \tag{10.59}$$

If there are N L networks used in the match, the maximum bandwidth will be obtained if the ith virtual resistor is

$$R_{Vi} = (R_S R_L)^{i/N}, \quad i = 1, \ldots, (N-1). \tag{10.60}$$

EXAMPLE 10.9 Two-Section Matching Network Design

Consider matching a 10 Ω source to a 1000 Ω load using two L matching networks and designing for a Q of 3. How many matching sections are required?

Solution:

Here the approximate Qs achieved with a single L matching network and with an optimum two-section design are compared. For a single L network design

$$Q = \sqrt{\frac{R_L}{R_S} - 1} = 9.95. \tag{10.61}$$

Now consider an optimum two-section design:

$$R_V = \sqrt{R_S R_L};\ Q_2 = \sqrt{\frac{R_L}{R_V} - 1} = \sqrt{\sqrt{\frac{R_L}{R_S}} - 1} = 3. \tag{10.62}$$

Thus the Q is 3 compared to the Q of an L section of 9.95. If the fractional bandwidth is inversely proportional to Q, then the bandwidth of the two-section design is 9.95/3 = 3.32 times more than that of the L section.

Now consider how many sections are required to obtain a Q of 2:

$$(1 + Q^2) = \frac{R_{V_1}}{R_S} = \frac{R_{V_2}}{R_{V_1}} = \ldots = \frac{R_L}{R_{V_{n-1}}} \Rightarrow \tag{10.63}$$

$$(1 + Q^2)^n = \frac{R_L}{R_S} \Rightarrow n \ln\left(1 + Q^2\right) = \ln \frac{R_L}{R_S} \Rightarrow n = \frac{\ln(R_L/R_S)}{\ln\left(1 + Q^2\right)}. \tag{10.64}$$

For $Q = 2$ and $R_L/R_S = 100$, $n = 2.86$, which rounds to $n = 3$, and three sections are required.

10.7 Impedance Matching Using Smith Charts

The lumped-element matching networks presented up to now can also be developed using Smith charts which provide a fairly intuitive approach to network design. With experience it will be found that this is the preferred approach to developing designs, as trade-offs can be captured graphically. Smith chart-based design will be presented using examples.

10.7.1 Two-Element Matching

The examples here build on the preceding lumped-element matching network design and now use the Smith. Capacitive and inductive regions on the Smith chart are shown in Figure 10-30. In the design examples presented here, circles of constant resistance or constant conductance are followed and these correspond to varying reactance or susceptance, respectively.

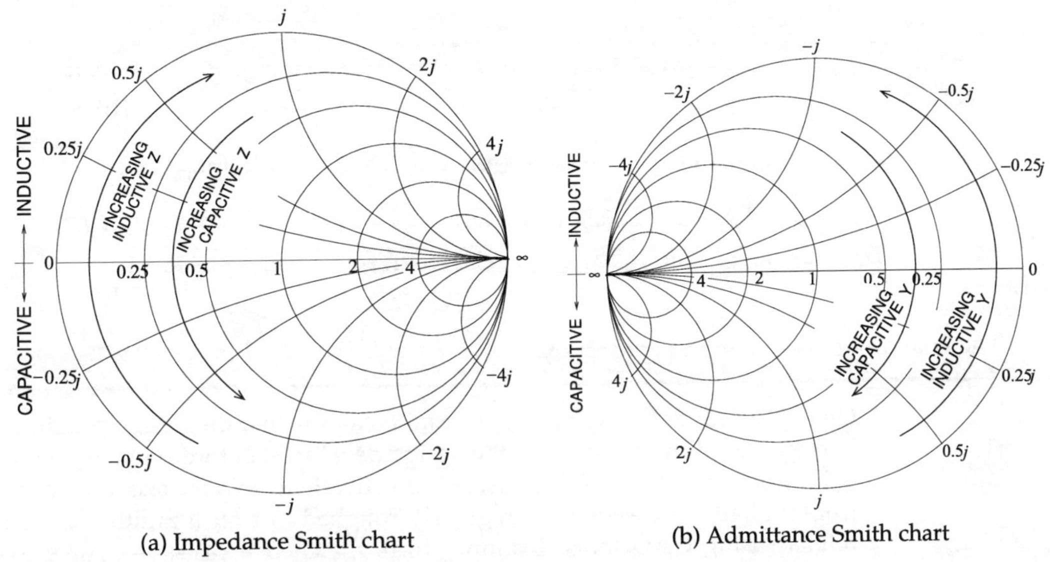

(a) Impedance Smith chart (b) Admittance Smith chart

Figure 10-30: Inductive and capacitive regions on Smith charts. Increasing capacitive impedance (Z) indicates smaller capacitance; increasing inductive admittance (Y) indicates smaller inductance.

EXAMPLE 10.10 Two-Element Matching Network Design Using a Smith Chart

Develop a two-element matching network to match a source with an impedance of $R_S =$ 25 Ω to a load $R_L = 200$ Ω (see Figure 10-31).

Solution:

The design objective is to present conjugate matched impedances to the source and load. However, since here the source and load impedances are real, the design objective is $Z_1 = R_S$ and $Z_2 = R_L$. The load and source resistances are plotted on the Smith chart in Figure 10-33(a) after choosing a normalization impedance of $Z_0 = 50$ Ω (and so $r_S = R_S/Z_0 = 0.5$ and $r_L = R_L/Z_0 = 4$). The normalized source impedance, r_S, is Point A, and the normalized load impedance, r_L, is Point C. The matching network must be lossless, which means that the design must follow lines of constant resistance (on the impedance part of the Smith chart) or constant conductance (on the admittance part of the Smith chart). So Points A and C must be on the above circles and the circles must intersect if a design is possible. The design can be viewed as moving back from the source toward the load or moving back from the load toward the source. (The views result in identical designs.) Here the view taken is moving back from the source toward the load.

One possible design is shown in Figure 10-33(a). From Point A, the line of constant resistance is followed to Point B (there is increasing series reactance along this path). From Point B, the locus follows a line of constant conductance to the final point, Point C. There is also an alternative design that follows the path shown in Figure 10-33(b). There are only two designs that have a path from A to B following just two arcs. At this point two designs have been outlined. The next step is assigning element values.

The design shown in Figure 10-33(a) begins with r_S followed by a series reactance, x_S, taking the locus from A to B. Then a shunt capacitive susceptance, b_P, takes the locus from B to C and r_L. At Point A the reactance $x_A = 0$, at Point B the reactance $x_B = 1.323$. This value is read off the Smith chart, requiring that an arc as shown be interpolated between the arcs provided. It should be noted that not all versions of Smith charts include negative signs, as the chart becomes too complicated. Thus the user needs to be aware and add signs where appropriate. The normalized series reactance is

$$x_S = x_B - x_A = 1.323 - 0 = 1.323, \tag{10.65}$$

that is,

$$X_S = x_s Z_0 = 1.323 \times 50 = 66.1\,\Omega. \tag{10.66}$$

A shunt capacitive element takes the locus from Point B to Point C and

$$b_P = b_C - b_B = 0 - (-0.661) = 0.661, \tag{10.67}$$

so

$$B_P = b_P/Z_0 = 0.661/50 = 13.22\text{ mS} \quad \text{or} \quad X_P = -1/B_P = -75.6\,\Omega. \tag{10.68}$$

The final design is shown in Figure 10-32.

One of the advantages of using the Smith chart is that the design progresses in stages, with the structure of the design developed before actual numerical values are calculated. Of course, it is difficult to extract accurate values from a chart, so designs are regularly roughed out on a Smith chart and refined using CAD tools. Example 10.10 matched a resistive source to a resistive load. The next example considers the matching of complex load

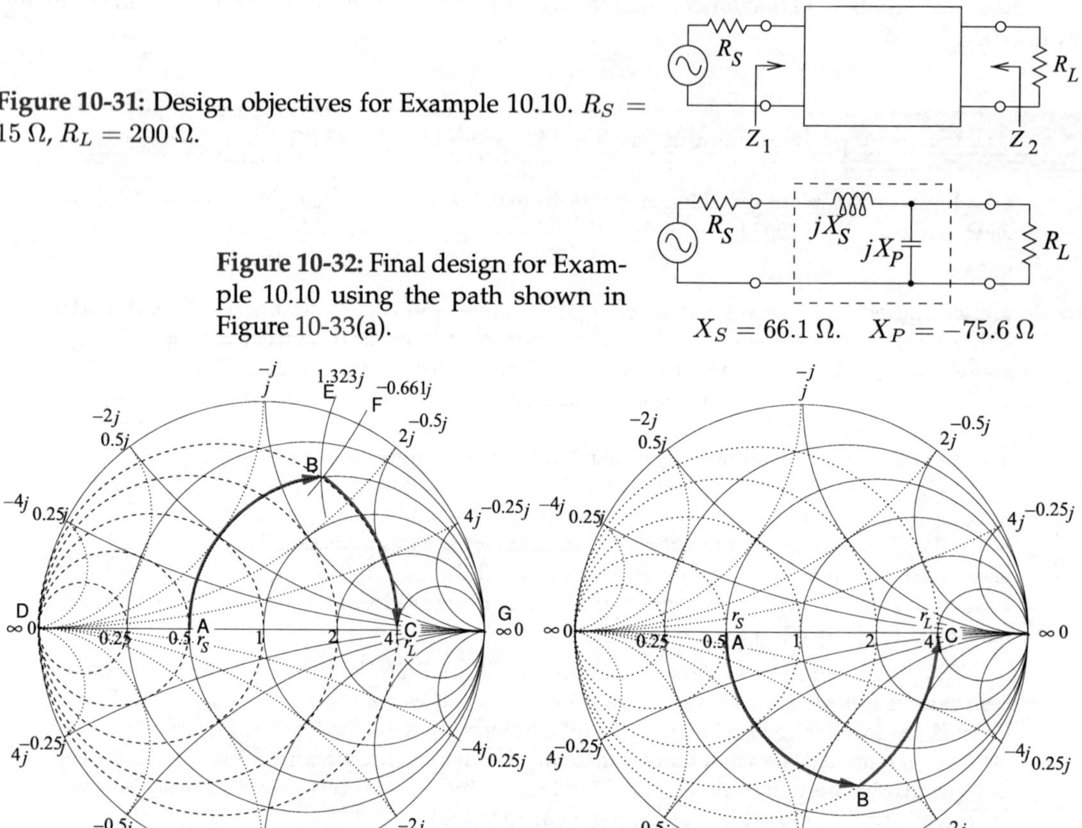

Figure 10-31: Design objectives for Example 10.10. $R_S = 15\,\Omega$, $R_L = 200\,\Omega$.

Figure 10-32: Final design for Example 10.10 using the path shown in Figure 10-33(a).

Figure 10-33: Alternative designs for Example 10.10. The normalization impedance is 50 Ω.

and source impedances. In the earlier algorithmic approach to matching network design absorption and resonance were introduced as strategies for dealing with complex terminations. Design was not always straightforward. This complication disappears with a Smith chart-based design, as it is conceptually not much different from the resistive problem of Example 10.10.

EXAMPLE 10.11 Matching Network Design With Complex Impedances

Develop a two-element matching network to match a source with an impedance of $Z_S = 12.5 + 12.5\jmath\ \Omega$ to a load $Z_L = 50 - 50\jmath\ \Omega$, as shown in Figure 10-34.

Solution:

The design objective is to present conjugate matched impedances to the source and load; that is, $Z_1 = Z_S^*$ and $Z_2 = Z_L^*$. The choice here is to design for Z_1; that is, elements will be inserted in front of Z_L to produce the impedance Z_1. The normalized source and load impedances are plotted in Figure 10-35(a) using a normalization impedance of $Z_0 = 50\ \Omega$, so $z_S = Z_S/Z_0 = 0.25 + 0.25\jmath$ (Point S) and $z_L = Z_L/Z_0 = 1 - \jmath$ (Point C).

The impedance to be synthesized is $z_1 = Z_1/Z_0 = z_S^* = 0.25 - 0.25\jmath$ (Point A). The matching network must be lossless, which means that the lumped-element design must follow lines of constant resistance (on the impedance part of the Smith chart) or constant conductance (on the admittance part of the Smith chart). Points A and C must be on the above circles and the circles must intersect if a design is possible.

The design can be viewed as moving back from the load impedance toward the conjugate of the source impedance. The direction of the impedance locus is important. One possible design is shown in Figure 10-35(a). From Point C the line of constant conductance is followed to Point B (there is increasing positive [i.e., capacitive] shunt susceptance along this path). From Point B the locus follows a line of constant resistance to the final point, Point A.

The design shown in Figure 10-35(a) begins with a shunt susceptance, b_P, taking the locus from Point C to Point B and then a series inductive reactance, x_S, taking the locus to Point A. At Point C the susceptance $b_C = 0.5$, at Point B the susceptance $b_B = 1.323$. This value is read off the Smith chart, requiring that an arc of constant susceptance, as shown, be interpolated between the constant susceptance arcs provided. The normalized shunt susceptance is

$$b_P = b_B - b_C = 1.323 - 0.5 = 0.823, \tag{10.69}$$

that is,
$$B_P = b_P/Z_0 = 0.823/(50\ \Omega) = 16.5\ \text{mS or } X_P = -1/B_P = -60.8\ \Omega. \tag{10.70}$$

A series reactive element takes the locus from Point B to Point A, so

$$x_S = x_A - x_B = -0.25 - (-0.661) = 0.411, \tag{10.71}$$

so
$$X_S = x_S Z_0 = 0.411 \times 50\ \Omega = 20.6\ \Omega. \tag{10.72}$$

The final design is shown in Figure 10-36.

There are only two designs that have a path from Point C to Point A following just two arcs. In Design 1, shown in Figure 10-35(a), Path CBA is much shorter than Path CHA for Design 2 shown in Figure 10-35(b). The path length is an approximate indication of the total reactance required, and the higher the reactance, the greater the energy storage and hence the narrower the bandwidth of the design. (The actual relative bandwidth depends on the voltage and current levels in the network; the path length criteria, however, is an important rule of thumb.) Thus Design 1 can be expected to have a much higher bandwidth than Design 2. Since designing broader bandwidth is usually an objective, a design requiring a shorter path on a Smith chart is usually preferable.

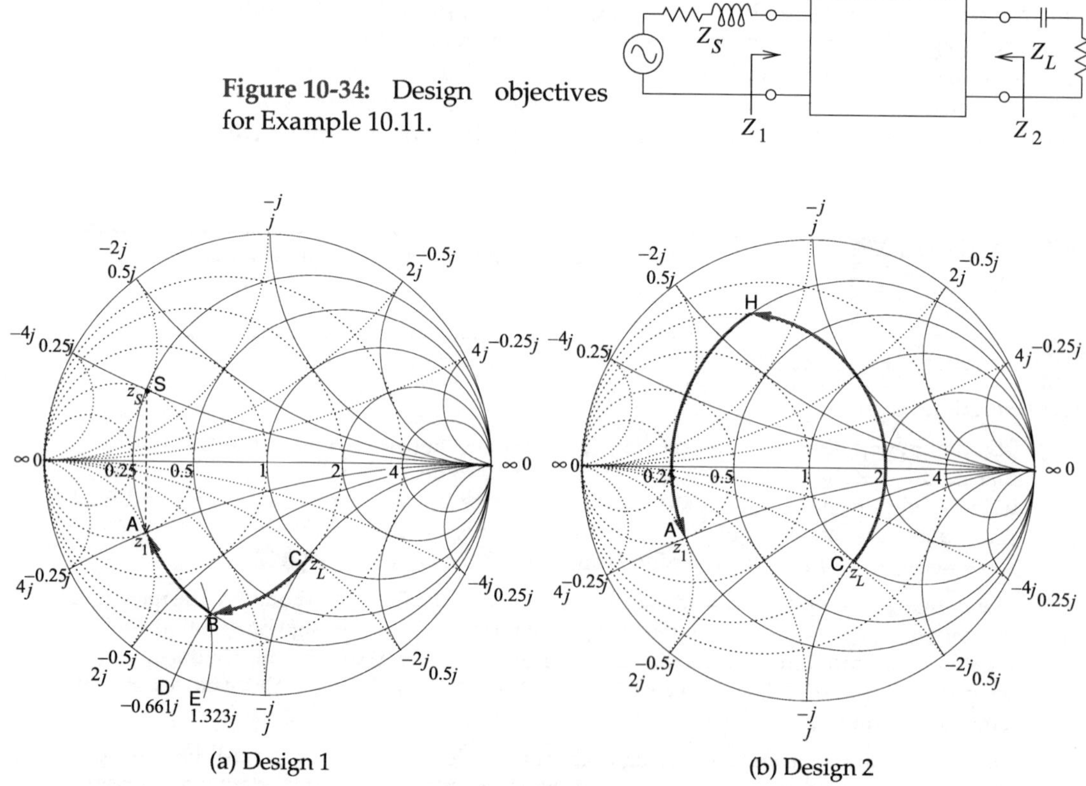

Figure 10-34: Design objectives for Example 10.11.

Figure 10-35: Smith chart-based designs used in Example 10.11. (50 Ω normalization used.)

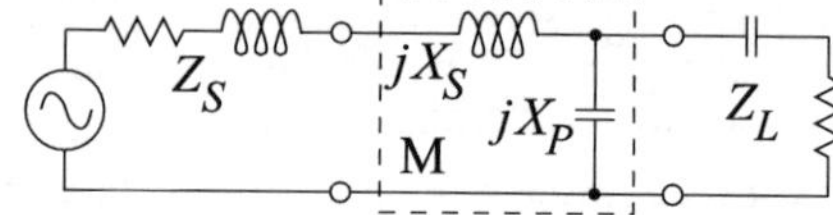

Figure 10-36: Final circuit for Design 1 of Example 10.11. $X_S = 20.6\ \Omega$, $X_P = -60.8\ \Omega$.

10.8 Distributed Matching

Matching using lumped elements leads to series and shunt lumped elements. The shunt elements can be implemented using shunt transmission lines, as a short length (less than one-quarter wavelength long) of short-circuited transmission line looks like an inductor and a short section of open-circuited transmission line looks like a capacitor. However, in microstrip it is not possible to realize the series elements as lengths of transmission lines. The solution is to use lengths of transmission line together with shunt elements. If space is not at a premium, this is an optimum solution, as transmission lines have much lower loss than a lumped inductor. The series transmission lines rotate the reflection coefficient on the Smith chart.

As with all matching design, using transmission lines begins with a topology in mind. Several topologies are shown in Figure 10-37. Figure 10-37(a) is the top view of a microstrip matching network with a series transmission line and stub realized as an open-circuited transmission line. Figure 10-37(b) is a shorthand schematic for this circuit. Matching network design then becomes a problem of choosing the lengths and characteristic impedances of the lines.

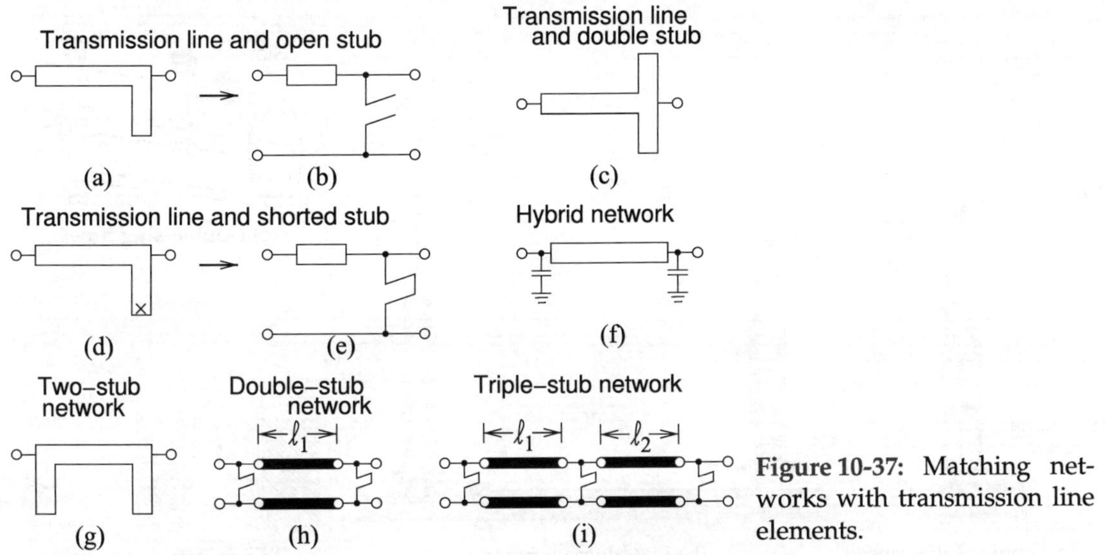

Figure 10-37: Matching networks with transmission line elements.

The stub here is used to realize a capacitive shunt element. This network corresponds to two-element matching with a shunt capacitor. The value of the shunt capacitance can be increased using a dual stub, as shown in Figure 10-37(c), where the capacitive input impedances of each stub are in parallel. The dual circuit to that in Figure 10-37(a) is shown in Figure 10-37(d) together with its schematic representation in Figure 10-37(e). This circuit has a short-circuited stub that realizes a shunt inductance.

Mixing lumped capacitors with a transmission line element, as shown in Figure 10-37(f), realizes a much more space-efficient network design. There are many variations to stub-based matching network design, including the two-stub design in Figure 10-37(g).

A common situation encountered in the laboratory is the matching of circuits that are in development. Laboratory items available for matching include the **stub tuner**, shown in Figure 10-38(a), and the **double-stub tuner**, shown in Figure 10-38(b). With the double-stub tuner the length of the series transmission line is fixed, but stubs can have variable length using lengths of transmission lines with sliding short circuits. Not all impedances can be matched using a double stub tuner, however. A triple-stub tuner can match all impedances presented to it [1]. The **double-slug tuner** shown in Figure 10-38(c) has dielectric slugs each of which introduces a short section of lower impedance line. The slugs are moved up and down the line and avoid the rapid changes in impedances that occur with the stub tuners and as a result the double-slug tuner provides a broader bandwidth match than does the double stub tuner.. The **slide-screw slug tuner** shown in Figure 10-38(d) can achieve a broadband match. Here a metal slug can be lowered into the slabline changing the impedance of a section of transmission line and mostly affects the magnitude of the reflection coefficient while moving the metal slug along the line mostly affects the phase. This is the type of tuner incorporated in computer-controlled automated tuners.

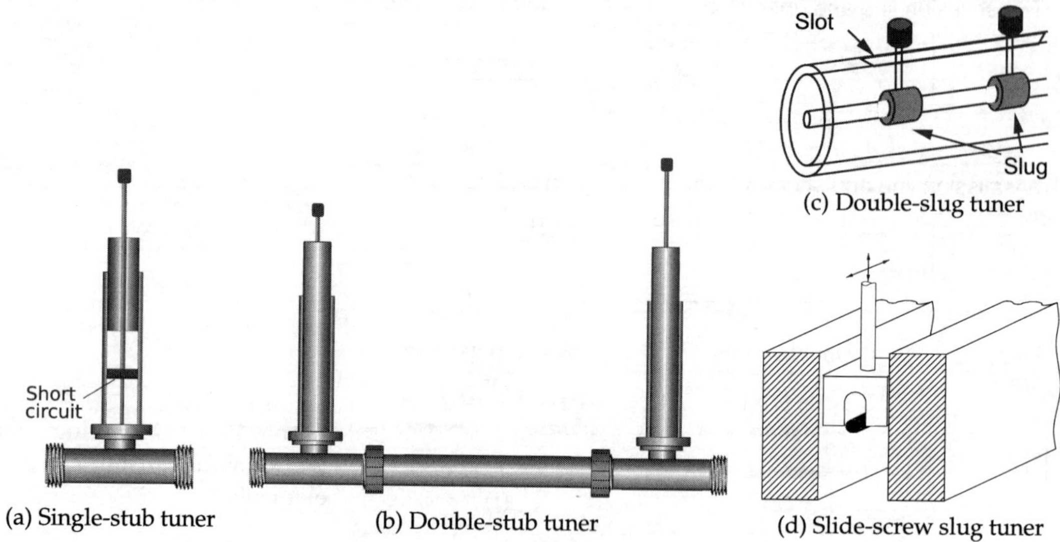

(a) Single-stub tuner (b) Double-stub tuner (c) Double-slug tuner (d) Slide-screw slug tuner

Figure 10-38: Laboratory tuners.

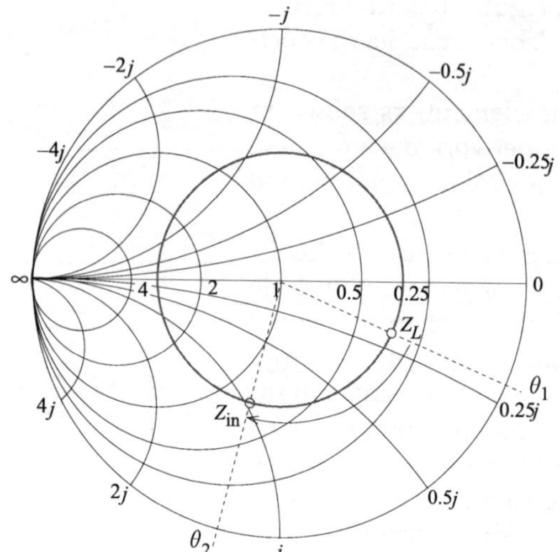

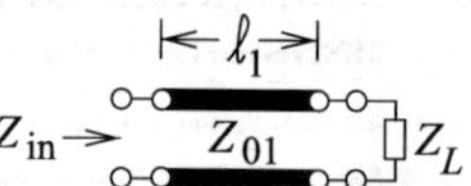

Figure 10-39: Rotation of the input impedance of a transmission line on a Smith chart normalized to Z_{01} as the line length increases.

10.8.1 Stub Matching

In this section matching using one series transmission line and one stub will be considered. This corresponds to the microstrip circuit topologies shown in Figures 10-37(a and d). First, consider the terminated transmission line shown in Figure 10-39. When the length, ℓ_1, of the line is zero, the input impedance of the line, Z_{in}, equals Z_L. How it changes is best described by considering the input reflection coefficient, Γ_{in}, of the line. If the reflection coefficient is normalized to Z_{01}, then the magnitude of Γ_{in} and its phase varies as twice the electrical length of the line. This situation is shown on the Smith chart in Figure 10-39, where Z_L is chosen arbitrarily. The input

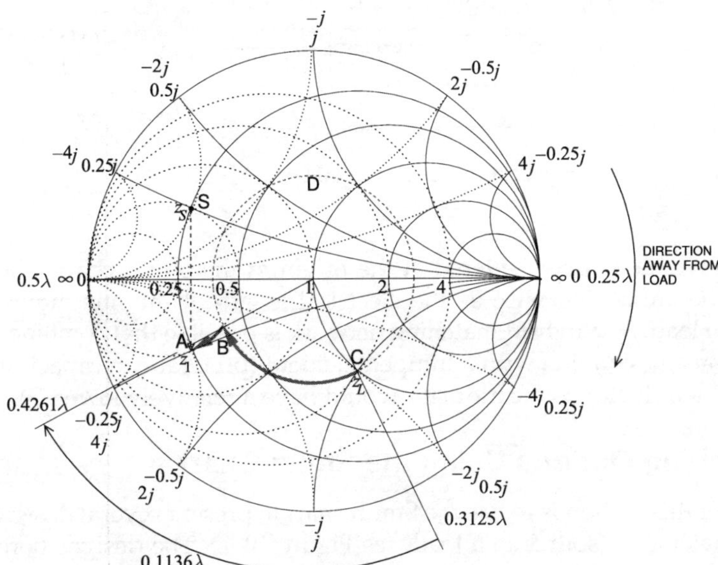

Figure 10-40: Design for Example 10.12.

reflection coefficient of the line rotates in a clockwise direction as the length of the line increases. One way of remembering this is to consider an open-circuited line. When the line length is zero, $Y_{\text{in}} = 0$ and $\Gamma_{\text{in}} = +1$. A short length of this line is capacitive so that its reflection coefficient will be in the bottom half of the Smith chart. A length of line can be used to rotate the impedance to an appropriate point to follow a line of constant conductance to the desired input impedance.

EXAMPLE 10.12 Matching Network Design With a Transmission Line and a Single Stub

Design a two-element matching network to match a source with an impedance $Z_S = 12.5 + 12.5\jmath\ \Omega$ to a load $Z_L = 50 - 50\jmath\ \Omega$, as shown in Figure 10-34. This example repeats the design in Example 10.11, but now using a transmission line.

Solution:

As in Example 10.11, choose $Z_0 = 50\ \Omega$ and the design path is from $z_L = Z_L/Z_0 = 1 - \jmath$ to z_s^*, where $z_s = 0.25 + 0.25\jmath$. One possible design solution is indicated in Figure 10-40. The line length, ℓ (taking the locus from Point C to Point B), is

$$\ell = 0.4261\lambda - 0.3125\lambda = 0.1136\lambda,$$

and the normalized shunt susceptance, b_P (taking the locus from Point B to Point C), is

$$b_P = b_A - b_B = 2 - 1 = 1.$$

Thus $X_P = (-1/b_P) \times 50\ \Omega = -50\ \Omega$. The final design is shown in Figure 10-41(a). The stub design of Figure 10-41(b) follows the procedure described in Example 6.5.

10.8.2 Hybrid Lumped-Distributed Matching

A lossless matching network can have transmission lines as well as inductors and capacitors. If the system reference or normalization impedance is the characteristic impedance of a transmission line, then the locus of the input impedance (or reflection coefficient) of the line with respect to the length of

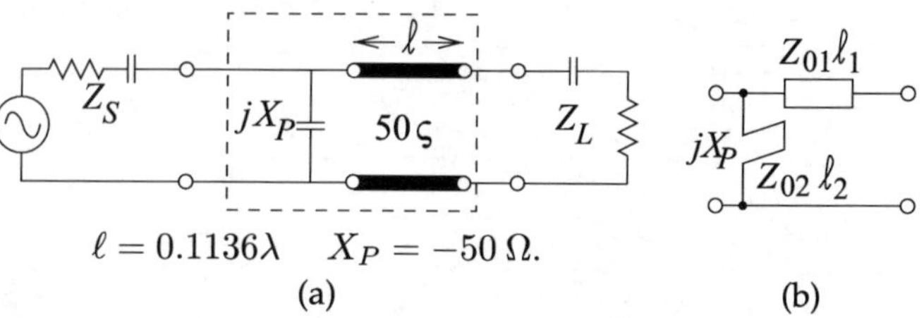

Figure 10-41: Single-stub matching network design of Example 10.12: (a) electrical design; and (b) electrical design with a shunt stub.

the line is an arc on a circle centered at the origin of the Smith chart. The direction of the arc is clockwise as the electrical length of the line moves away from the load. So a hybrid matching network is possible that combines a length of transmission line with a lumped element (preferably a capacitor rather than an a inductor as the inductor would have a relatively lower Q).

10.9 Matching Options Using the Smith Chart

The purpose of this section is to use the Smith chart to present several design options for matching a source to a load, see Figure 10-42. The designs here provide another view of design using the Smith chart.

10.9.1 Locating the Design Points

The first design choice to be made is the reference impedance to use. Here $Z_{\rm REF} = 50\ \Omega$ will be chosen largely because this is in the center of the design space for microstrip lines. Generally the characteristic impedance, Z_0, of a microstrip line needs to be between 20 Ω and 100 Ω. A microstrip line with $Z_0 < 20\ \Omega$ will be wide and there is a possibility of multimoding due to transverse resonance. Also a 20 Ω line is about six times wider than a 50 Ω line and so takes up a lot of room and there is a good chance that it could be close to other microstrip lines or perhaps the wall of an enclosure. This is based on the rule of thumb (developed in Example 3.4 of [2].) that $Z_0 \propto \sqrt{h/w}$ where h is the substrate thickness and w is the strip width. The thickness is usually fixed. If $Z_0 > 100\ \Omega$ the characteristic impedance is getting close to the wave impedance of free space or of the dielectric of the substrate. As such it is likely that field lines are not tightly constrained by the metal of the strip and the fields can more likely radiate. Then radiation loss can be high or coupling to a neighboring microstrip can be high.
The normalized source and load impedances are $z_S = Z_S/Z_{\rm REF} = [(29.36 - \jmath 12.05)\ \Omega]/(50\ \Omega) = 0.587 - \jmath 0.241$ and $z_L = Z_L/Z_{\rm REF} = [(132.7 - \jmath 148.8)\ \Omega/(50\ \Omega) = 2.655 - \jmath 2.976 = r_L + \jmath x_L$, respectively. Maximum power transfer requires that the input impedance of the matching network terminated in Z_L be $Z_1 = Z_S^*$, i.e. $z_1 = z_S^* = 0.587 + \jmath 0.241 = r_1 + \jmath x_1$. These impedances are plotted on the normalized Smith chart in Figure 10-43.
The normalized load impedance is Point L. To locate this point the arcs corresponding to the real and imaginary parts of z_L are considered

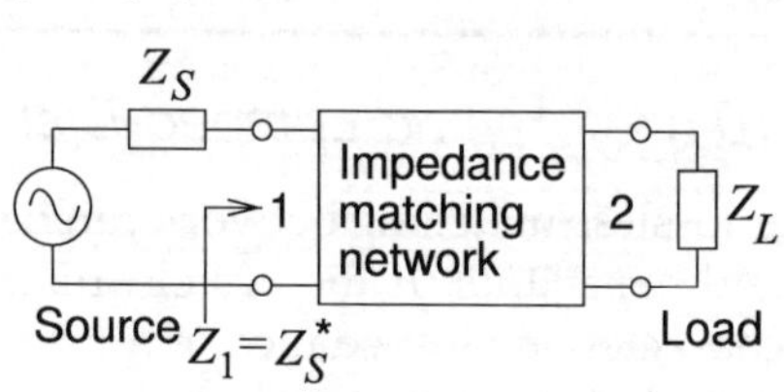

Figure 10-42: Matching problem with the matching network between the source and load designed for maxium power transfer. $Z_S = R_S + \jmath X_S = 29.36 - \jmath 12.05$, $Z_1 = R_1 + \jmath X_1 = Z_S^* = R_S - \jmath X_S = 29.36 + \jmath 12.05$, and $Z_L = R_L + \jmath X_L = 32.7 - \jmath 148.8$.

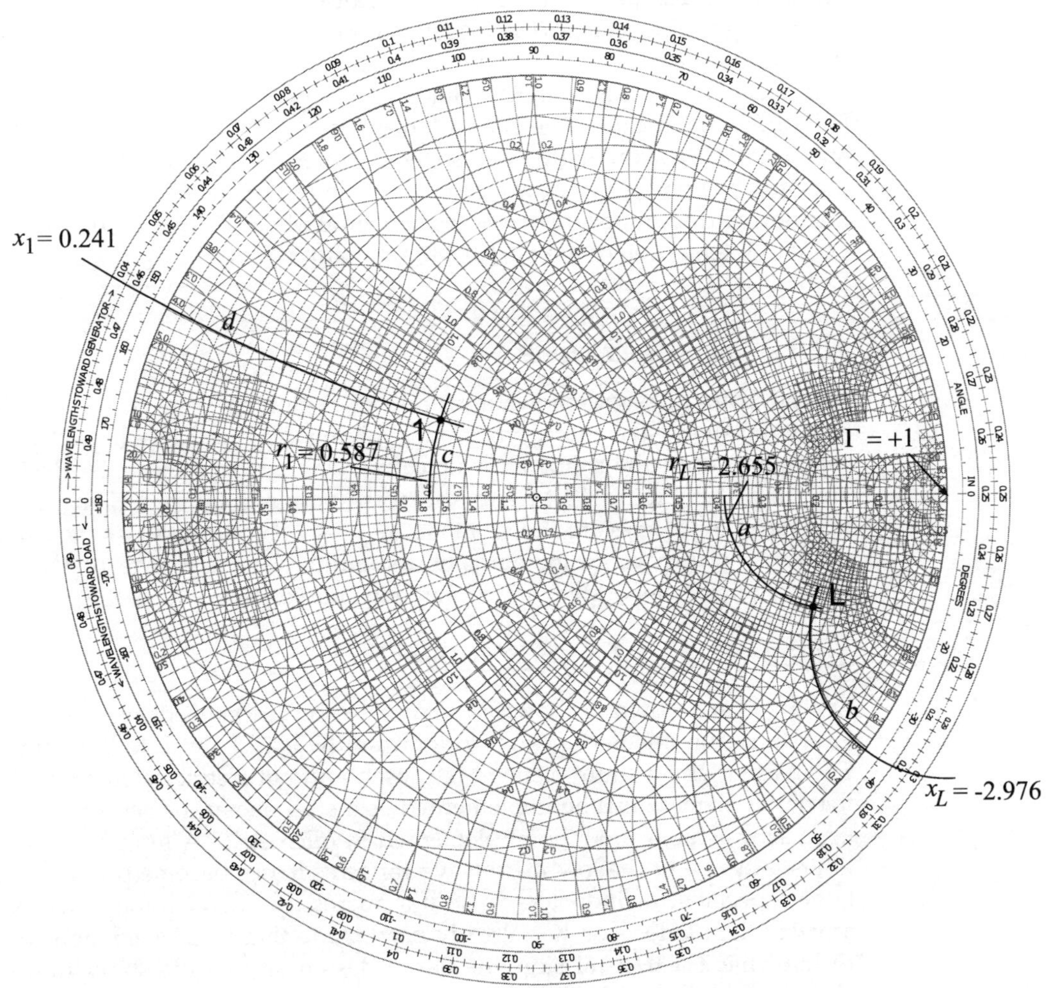

Figure 10-43: Locating Z_L at Point L and $Z_1 = Z_S^*$ at Point 1.

separately. The resistive part of z_L is $r_L = 2.655$ and the resistance labels are located on the horizontal axis (or equator) of the Smith chart. There are two sets of labels, one for normalized resistance, r (which is above the horizontal axis), and one for normalized conductance, g (which is below the horizontal axis). The way to remember which is which is to realize that the infinite impedance point is at the $\Gamma = +1$ (open circuit) location on the right of the graph. At the origin (center) of the Smith chart $r = 1 = g$ and to the right of the center the values of r should be greater than one. The closest r labels to $r_L = 2.655$ are $r = 2.0$ and $r = 3.0$. There are five divisions so the unlabeled curves correspond to 2.2, 2.4, ... The arc corresponding to $r = 2.655$ must be interpolated and this interpolation is shown as the Path 'a'.

The imaginary part of z_L is $x_L = -2.976$. The labels for the arcs of constant reactance are given adjacent to the unit circle. There are two sets of labels, one for reactance and one for susceptance. To recall which is which, the point of infinite impedance can be used and the required reactance labels should increase towards the $\Gamma = +1$ point. Recall that the Smith chart does not include signs of reactances (there is not enough room) so note must be made that positive reactances are in the top half of the Smith chart and negative reactances are in the bottom half. Since x_L is negative it will be in the bottom half of the Smith chart. The closest labels are $x = 2.0$ (this is actually -2.0) and $x = 3.0$ (this is actually -3.0) so the arc for $x = -2.976$ is interpolated as the Path 'b'. Point L, i.e. z_L, is located at the intersection of Paths 'a' and'b'. The normalized impedance z_1 is located similarly at Point 1 by finding the point of intersection of the $r_1 = 0.587$ arc, Path 'c', and the $x_1 = +0.241$ arc, Path 'd'.

10.9.2 Design Options

By convention design follows a process of beginning with z_L and adding series and shunt elements in front of it evolving the impedance (or reflection coefficient) until the input impedance is $z_1 = z_s^*$. Two electrical designs are shown in Figure 10-44 and the corresponding lumped-element and microstrip topologies are shown in Figure 10-45. The subscript on the circuit elements correspond to the paths on the Smith chart in Figure 10-44 . The designs will be elaborated in the following subsections.

10.9.3 Design 1, Hybrid Design

Design 1 on its own is shown in Figure 10-46. The concept here is to use a transmission line and a shunt to go from the load Point L to the Point 1. The reason why a shunt element is chosen and not a series element is that the shunt element can be implemented as a stub line and a series element, i.e. a series stub, cannot be implemented in microstrip. A lumped element limits a design to the low microwave range as losses become prohibitively large especially for inductors. Also if a microstrip line is going to be used anyway then a decision has already been made that there is enough room to implement a transmission-line based design and so the shunt lumped element can reasonably bereplaced by a stub.

Design follows trial and error. The first attempt, and the one that works here, is to draw a circle through L centered on the origin at Point O. This circle describes a transmission line whose characteristic impedance is the same as the reference impedance of the Smith chart, here 50 Ω. The next step is to draw a circle of constant conductance through Point 1. The combination path from L to 1 needs to lie on these circles and the intermediate point will be where these circles intersect. One other constraint is that with a transmission line the locus (as the line length increases) of the input reflection coefficient of the line must rotate in the clockwise direction. It is seen that there are two points of intersection and the first of these, at A, is chosen in design. So the electrical design is defined by the directed Paths 'g' and 'h'. Path 'g' defines the properties of the transmission line and Path 'h' defines the properties of the shunt element. As 'h' is directed towards the infinite inductive susceptance point, Path 'h' defines an inductor. The topology of this design is shown in Figure 10-45(a).

The characteristic impedance of the transmission line (defined by Path 'g') is $Z_{0g} = 50\ \Omega$ and the electrical length of the line is defined by the angle subtended by the arc 'g'. The electrical length of the line is determined from

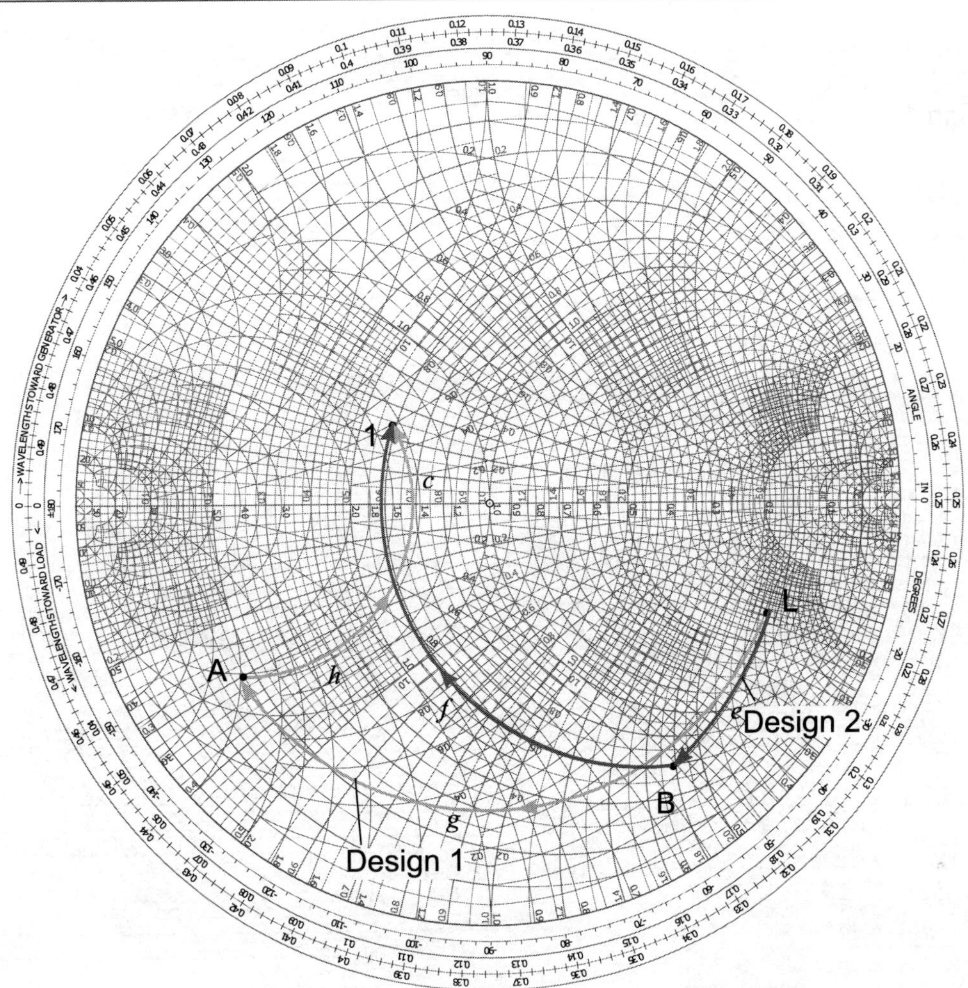

Figure 10-44: Two matching network electrical designs matching a load impedance Z_L at Point L to a source Z_S showing $Z_1 = Z_S^*$ at Point 1.

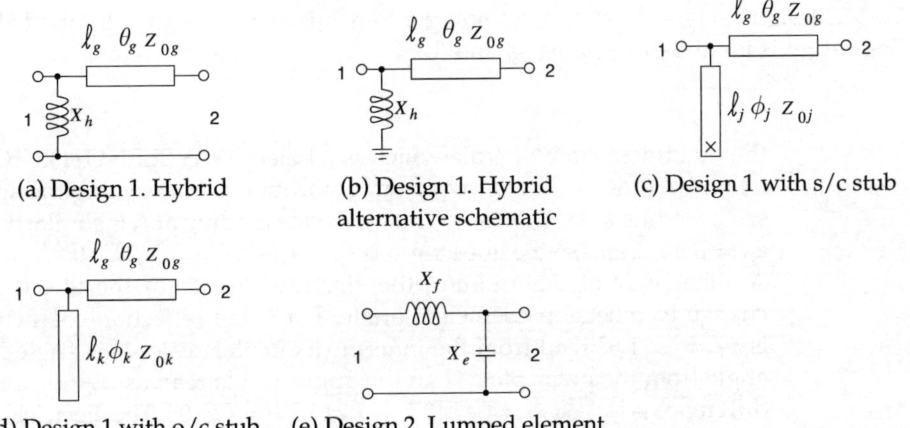

Figure 10-45: Matching network topologies using lumped elements and microstrip lines. In the stub layouts x is a via to the ground plane implementing a short circuit (s/c) and an open circuit o/c simply does not show a connection to the microstrip ground plane.

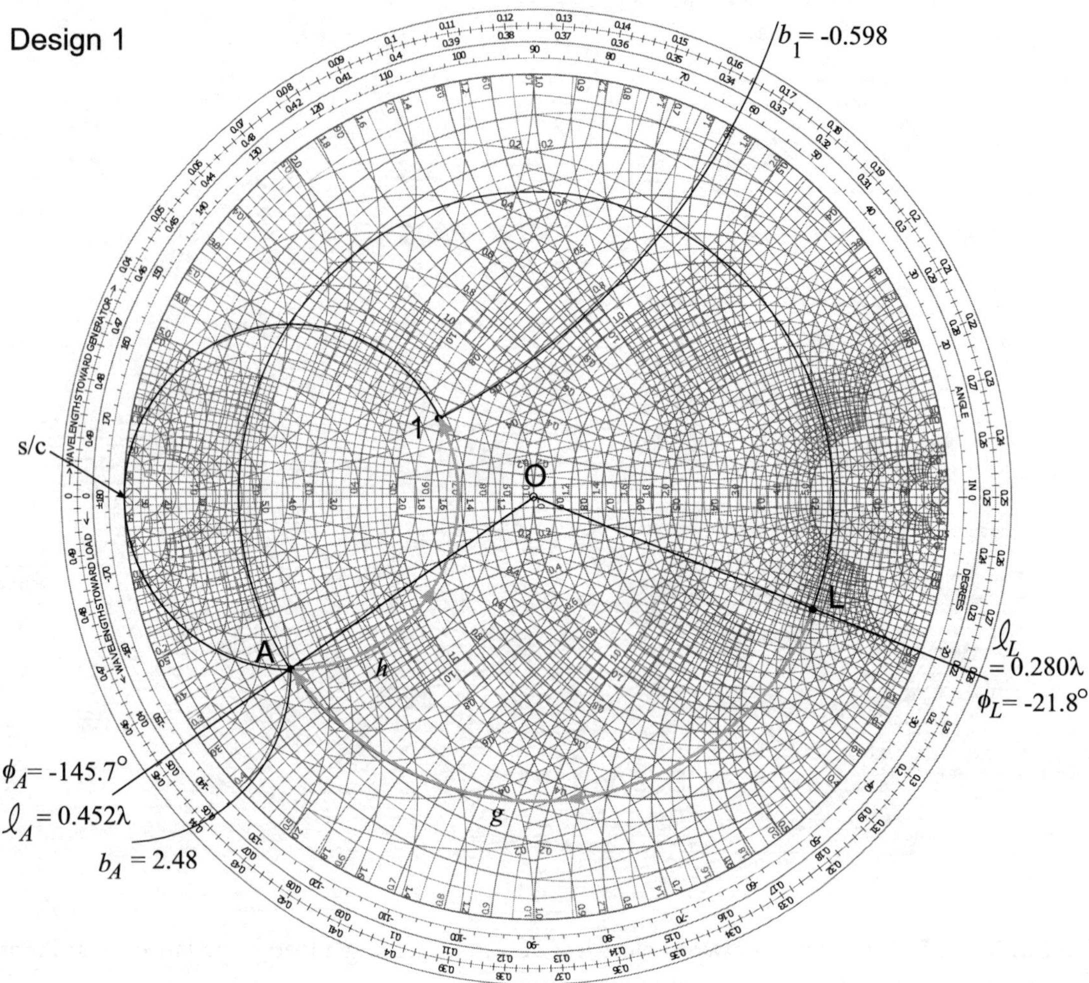

Figure 10-46: Design 1. Hybrid design combining a transmission line with a lumped element in shunt. The design is identified by paths 'g' and 'h'.

the outermost circular scale which is labeled 'WAVELENGTHS TOWARDS GENERATOR.' A line drawn from O through L intersecting the scale has a scale reading of $\ell_L = 0.280\lambda$. Then the scale reading at A is similarly found as $\ell_A = 0.452\lambda$ and so the line length is $\ell_g = \ell_A - \ell_L = 0.452\lambda - 0.280\lambda = 0.172\lambda$. Another way of determining the electrical length of the line is from the change in reflection coefficient angle. For L the reflection coefficient angle is $\phi_L = -21.8°$ read from the innermost circular scale. This angle is just the angle from the polar plot. Then the angle at A is read as $\phi_A = -145.7°$. The difference is $|\phi_A - \phi_L| = |-145.7 - (-21.8°)| = 124.9°$. The electrical length of the line is half the change in reflection coefficient angle and so the electrical length of the line is $\theta_g = \frac{1}{2}124.9° = 62.5°$. Now λ corresponds to an electrical length of 360° so θ_g corresponds to $62.5/360\lambda = 0.174\lambda$ corresponding to the previously determined length of 0.172λ which is very good agreement given that these were derived from graphical readings.

Path 'h' defines a shunt inductor and a circle of constant conductance is followed with only the susceptance changing. The susceptance indicated by Path 'h' is $b_h = b_1 - b_A$. To obtain b_A extend the circle of constant susceptance through A out to the unit circle. The extended circle crosses the unit circle between the susceptance labels 2.0 and 3.0. A check is that susceptance is positive in the bottom half of the Smith chart so the signs of the labels do not need to be adjusted. There are two scales adjacent to the unit circle, one for normalized susceptance and one for normalized reactance. The intersection is close to the infinite susceptance point at the s/c (short-circuit) so the values that are becoming very large towards s/c are used. Interpolation results in the reading $b_A = 2.48$. A similar process applied to Point 1 results in $b_1 = -0.598$ where the negative sign has been applied to the scale reading since Point 1 is in the top half of the Smith chart. Thus $b_h = b_1 - b_A = -0.598 - 2.48 = -3.08$ and so the normalized reactance of the shunt element is $x_h = -1/b_h = 0.325$. The un-normalized reactance of the shunt element is $X_h = x_h Z_{\text{REF}} = 16.2\ \Omega$.
The final Design 1 hybrid layout is shown in Figure 10-45(a) with $X_h = 16.2\ \Omega$, $Z_{0g} = 50\ \Omega$, and $\ell_g = 0.172\lambda$. That is all that is needed to define the electrical design, providing the electrical length in degrees, $\theta_g = 62.5°$ is redundant but provided anyway. The transmission line in Figure 10-45(a) is shown as as the top view of the strip of a microstrip line as is commonly done. A more common way of representing this schematic is shown in Figure 10-45(b) where the ground connections at Ports 1 and 2 have been removed and the ground connection of the inductor shown separately.

10.9.4 Design 1 with an Open-Circuited Stub

In the previous section Design 1 was left as a hybrid design with a transmission line and a lumped-element inductor. In this section the lumped-element inductor is implemented as an open-circuited stub, see Figure 10-45(d). Recall that the 50 Ω-normalized susceptance of the inductor is $b_h = -3.08$. If the stub is also implemented as a 50 Ω line then b_h can be used unchanged. Point C in Figure 10-47 corresponds to the normalized admittance $0 - \jmath 3.08$. The unit circle is the zero conductance circle (and is also the zero resistance circle) and the susceptance is read from the scale adjacent to the unit circle again noting that susceptances in the top half of the Smith chart need to incorporate a negative sign, and the susceptance scale is identified by the susceptance values becoming larger approaching the s/c point. Point C also corresponds to $x_h = -1/b_h = 3.25$ and indeed this is the value read from the normalized reactance scale.
A transmission line needs to be designed to have a normalized input susceptance of $b_h = -3.08$. Choosing an open circuit, o/c, termination the point corresponding to o/c is as identified in the figure. At the o/c point the length scale reads $\ell_{\text{o/c}} = 0.250\lambda$. The locus rotates in the clockwise direction up to Point C where the direct electrical reading reading is $\ell_C = 0.050\lambda$. Using this directly to determine the line length $\ell_k = \ell_C - \ell_{\text{o/c}} = 0.050\lambda - 0.250\lambda = -0.20\lambda$ which indicates that the stub has a negative length. Clearly an erroneous result. This apparent discrepancy comes about because the length scale resets at the short circuit point where the length scale abruptly goes from 0.5λ to 0λ. Thus the corrected ℓ_C reading needs to have an additional 0.5λ. Thus the corrected value of $\ell_C = (0.5 + 0.050)\lambda = 0.550\lambda$

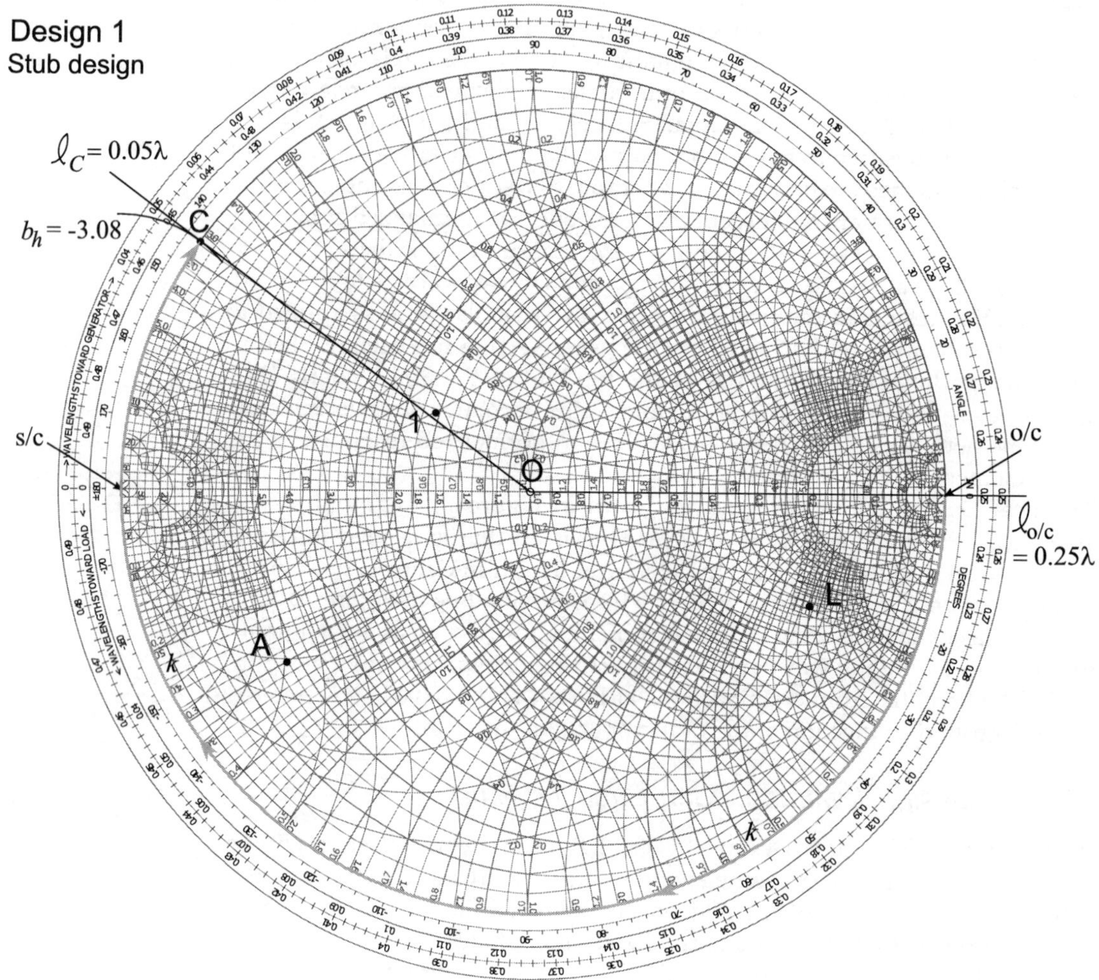

Figure 10-47: Design 1. Design of an open-circuit stub having normalized input susceptance b_h.

and $\ell_k = \ell_C - \ell_{o/c} = 0.550\lambda - 0.250\lambda = 0.300\lambda$.

Thus the final design is as shown in Figure 10-45(d) with $Z_{0k} = 50\ \Omega$, and $\ell_g = 0.300\lambda$, $Z_{0g} = 50\ \Omega$, and $\ell_g = 0.172\lambda$.

The stub could also have been implemented as a short-circuit stub as shown in Figure 10-45(c). Now the beginning of the line would be at the s/c point and the line length would be 0.050λ

10.9.5 Design 2, Lumped-Element Design

Design 2 is a lumped-element design and the Smith-chart-based electrical design is shown in Figure 10-48 resulting in the schematic shown in Figure 10-45(e). Design proceeds by identifying where circles of constant conductance and constant resistance passing through the Points L and 1 intersect. One solution is shown in Figure 10-48. A circle of constant conductance passes through L and part of a circle of constant resistance passes through 1. If the circle had continued there would have been a

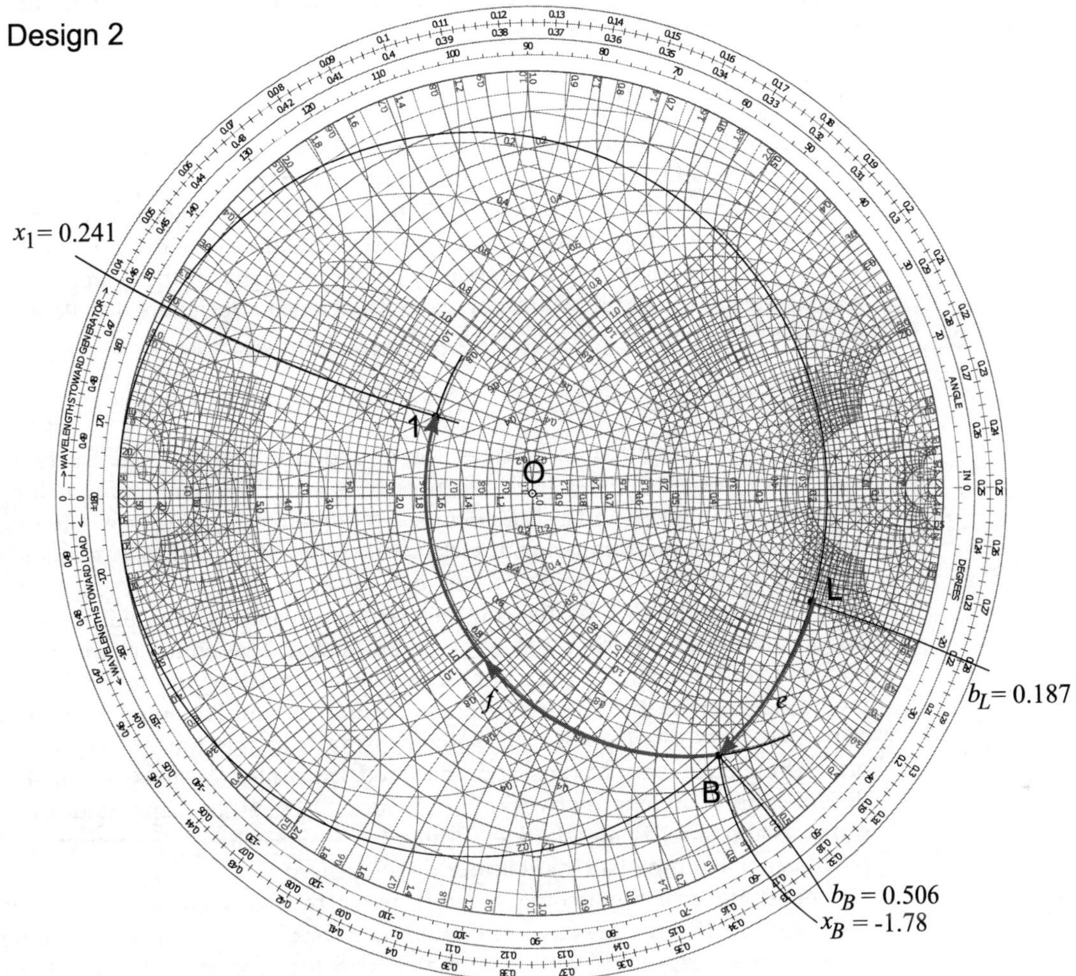

Figure 10-48: Design 2.

second intersection with the circle through L. Both of these intersections mean that there is a shunt element adjacent to the load and series element adjacent to the source. Recall that in a lumped-element design that under no circumstances can the locus a lumped element pass through the short circuit or open-circuit points (the susceptance and reactance infinity points respectively).

Returning to the actual design shown in Figure 10-48. The first intersection of the two circles is Point B so that the design is specified by the Paths 'e' and 'f'. Design has largely been completed by identifying these paths and the next stage is determining the circuit elements that correspond to these paths. Path 'e' follows a circle of constant conductance and so indicates a shunt susceptance and the direction of the locus indicates a capacitance. The value of this normalized susceptance is $b_e = b_B - b_L = 0.506 - 0.187 = 0.319$. (Remember to check the signs of the readings.) Path 'f' identifies a series inductor with a reactance $x_f = x_1 - x_B = 0.241 - (-1.78) = 2.02$. The final design is shown in Figure 10-45(e) with $X_e = Z_0/b_e = 50/0.319\ \Omega = 158\ \Omega$ and $X_f = Z_0 x_f = 50 \cdot 2.02\ \Omega = 101\ \Omega$.

10.9.6 Summary

This section presented two designs for a matching network. One of the particular benefits of using the Smith chart is identifying topologies and initial design values. Design can then transfer to a microwave circuit simulator. The Smith chart enables back-of-the-envelope design studies. While with experience it is possible to complete many of these steps with a computer-based Smith chart tool, even experienced designers doodle with a printed Smith chart when exploring design options.

10.10 Summary

This chapter presented techniques for impedance matching that achieve maximum power transfer from a source to a load. The simplest matching network uses a series and a shunt element, a two-element matching network, to realize a single-frequency match. This type of impedance matching network uses lumped elements and can be used up to a few gigahertz. Performance is limited by the self-resonant frequency of lumped elements and by their loss, particularly that of inductors. The shunt element can be replaced by a shunt stub, but in most transmission line technologies, including microstrip, the series element cannot be implemented as a stub. Matching networks can also be realized using transmission line segments only, principally shunt stubs and cascaded transmission lines. A tunable double-stub matching network, which uses two stubs separated by a transmission line, is standard equipment in microwave laboratories and facilitates matching of a circuit under development.

The bandwidth of a matching network is set by the maximum allowable reflection coefficient of the terminated network. Two-element matching nearly always results in a narrow match and for typical communications applications often achieves acceptable matching over bandwidths of only 1%–3%. The most significant determinant of the quality of the match that can be achieved is the ratio of the source and load resistances, as well as the reactive energy storage of the source and load.

An important concept in matching network design is a technique for controlling bandwidth. The concept is based on matching to an intermediate resistance, typically designated as R_v. Compared to a two-element network, increased bandwidth is obtained if R_v is the geometric mean of the source and load resistances. This new network consists of two two-element matching networks. If R_v is greater or less than both the source and load resistances, then the bandwidth of the matching network is reduced. The matching network synthesis problem can also be addressed using filter design techniques, and this enables simultaneous control over the quality and bandwidth of the match. It is always a good idea to have no more bandwidth in the system than is needed, as this minimizes the propagation of noise.

10.11 References

[1] R. Collins, *Foundations for Microwave Engineering*. McGraw Hill, 1966.

[2] M. Steer, *Microwave and RF Design, Transmission Lines*, 3rd ed. North Carolina State University, 2019.

10.12 Exercises

1. Consider the design of a magnetic transformer that will match the 3 Ω output resistance of a power amplifier (this is the source) to a 50 Ω load. The secondary of the transformer is on the load side.
   (a) What is the ratio of the number of primary turns to the number of secondary turns for ideal matching?
   (b) If the transformer ratio could be implemented exactly (the ideal situation), what is the reflection coefficient normalized to 3 Ω looking into the primary of the transformer with the 50 Ω load?
   (c) What is the ideal return loss of the loaded transformer (looking into the primary)? Express your answer in dB.
   (d) If there are 100 secondary windings, how many primary windings are there in your design? Note that the number of windings must be an integer. (This practical situation will be considered in the rest of the problem.)
   (e) What is the input resistance of the transformer looking into the primary?
   (f) What is the reflection coefficient normalized to 3 Ω looking into the primary of the transformer with the 50 Ω load?
   (g) What is the actual return loss (in dB) of the loaded transformer (looking into the primary)?
   (h) If the maximum available power from the amplifier is 20 dBm, how much power (in dBm) is reflected at the input of the transformer?
   (i) Thus, how much power (in dBm) is delivered to the load ignoring loss in the transformer?

2. Consider the design of a magnetic transformer that will match a 50 Ω output resistance to the 100 Ω load presented by an amplifier. The secondary of the transformer is on the load (amplifier) side.
   (a) What is the ratio of the number of primary turns to the number of secondary turns for ideal matching?
   (b) If the transformer ratio could be implemented exactly (the ideal situation), what is the reflection coefficient normalized to 50 Ω looking into the primary of the transformer with the load?
   (c) What is the ideal return loss of the loaded transformer (looking into the primary)? Express your answer in dB.
   (d) If there are 20 secondary windings, how many primary windings are there in your design? Note that the number of windings must be an integer? (This situation will be considered in the rest of the problem.)
   (e) What is the input resistance of the transformer looking into the primary?
   (f) What is the reflection coefficient normalized to 50 Ω looking into the primary of the loaded transformer?
   (g) What is the actual return loss (in dB) of the loaded transformer (looking into the primary)?
   (h) If the maximum available power from the source is −10 dBm, how much power (in dBm) is reflected from the input of the transformer?
   (i) Thus, how much power (in dBm) is delivered to the amplifier ignoring loss in the transformer?

3. Consider the design of an L-matching network centered at 1 GHz that will match the 2 Ω output resistance of a power amplifier (this is the source) to a 50 Ω load. [Parallels Example 10.3 but note the DC blocking requirement below.]
   (a) What is the Q of the matching network?
   (b) The matching network must block DC current. Draw the topology of the matching network.
   (c) What is the reactance of the series element in the matching network?
   (d) What is the reactance of the shunt element in the matching network?
   (e) What is the value of the series element in the matching network?
   (f) What is the value of the shunt element in the matching network?
   (g) Draw and label the final design of your matching network including the source and load resistances.
   (h) Approximately, what is the 3 dB bandwidth

of the matching network?

4. Consider the design of an L-matching network centered at 100 GHz that will match a source with a Thevenin resistance of 50 Ω to the input of an amplifier presenting a load resistance of 100 Ω to the matching network. [Parallels Example 10.4 but note the DC blocking requirement below.]
   (a) What is the Q of the matching network?
   (b) The matching network must block DC current. Draw the topology of the matching network.
   (c) What is the reactance of the series element in the matching network?
   (d) What is the reactance of the shunt element in the matching network?
   (e) What is the value of the series element in the matching network?
   (f) What is the value of the shunt element in the matching network?
   (g) Draw and label the final design of your matching network including the source and load resistance.
   (h) Approximately, what is the 3 dB bandwidth of the matching network?
5. Design a Pi network to match the source configuration to the load configuration below. The design frequency is 900 MHz and the desired Q is 10. [Parallels Example 10.8]

50 Ω
2 pF 10 nH 100 Ω
Source Load

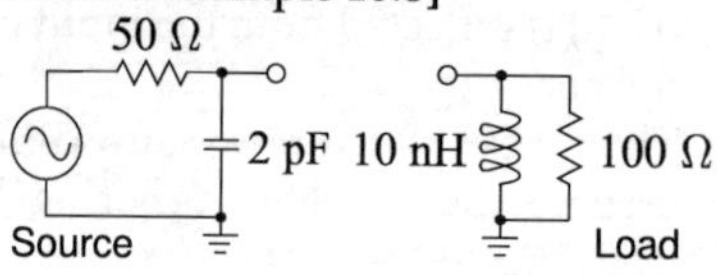

6. Design a Pi network to match the source configuration to the load configuration below. The design frequency is 900 MHz and the desired Q is 10. [Parallels Example 10.8]

50 Ω
10 nH 1 pF 100 Ω
Source Load

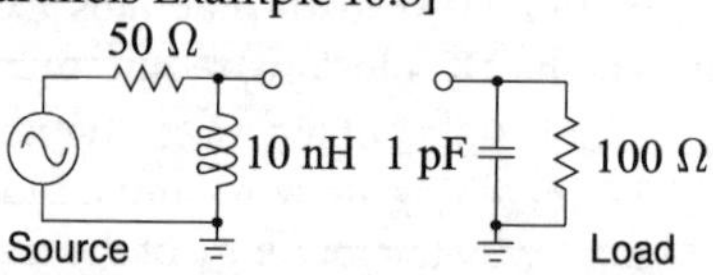

7. Develop the electrical design of an L-matching network to match the source to the load below.

50 Ω
$-\jmath 50$ Ω $\jmath 40$ Ω 100 Ω
Source Load

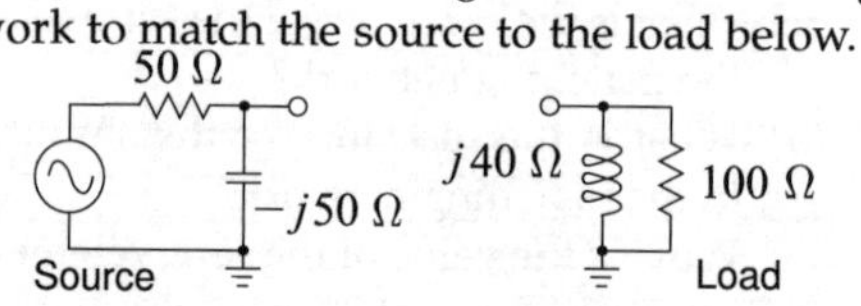

8. Develop the electrical design of an L-matching network to match the source to the load below.

50 Ω
$-\jmath 150$ Ω 100 Ω
Source Load

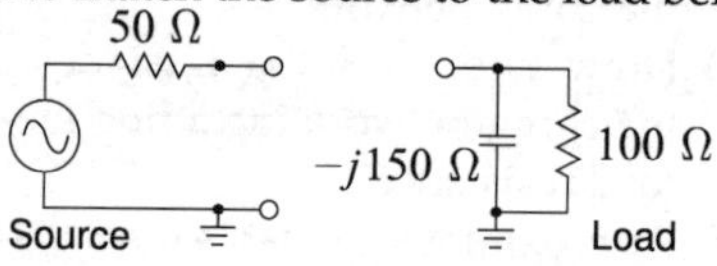

9. Design a lowpass lumped-element matching network to match the source and load shown below. The design frequency is 1 GHz. You must use a Smith Chart and clearly show your working and derivations. You must develop the final values of the elements.

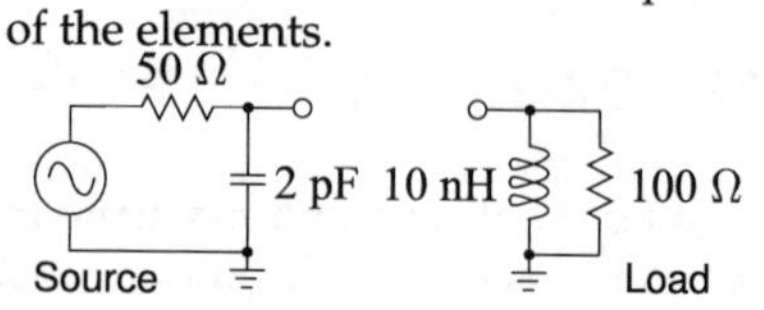

50 Ω
2 pF 10 nH 100 Ω
Source Load

10. Consider the design of an L-matching network centered at 100 GHz that will match a source with a Thevenin resistance of 50 Ω to the input of an amplifier presenting a load resistance of 200 Ω to the matching network. [Parallels Example 10.4 but note the DC blocking requirement below.]
    (a) What is the Q of the matching network?
    (b) The matching network must block DC current. Draw the topology of the matching network.
    (c) What is the reactance of the series element in the matching network?
    (d) What is the reactance of the shunt element in the matching network?
    (e) What is the value of the series element in the matching network?
    (f) What is the value of the shunt element in the matching network?
    (g) Draw and label the final design of your matching network including the source and load resistance.
    (h) Approximately, what is the 3 dB bandwidth of the matching network?
11. Design a two-element matching network to interface a source with a 25 Ω Thevenin equivalent impedance to a load consisting of a capacitor in parallel with a resistor so that the load admittance is $Y_L = 0.02 + \jmath 0.02$ S. Use the absorption method to handle the reactive load.
12. Design a matching network to interface a source with a 25 Ω Thevenin equivalent impedance to a load consisting of a capacitor in parallel with a resistor so that the load admittance is $Y_L = 0.01 + \jmath 0.01$ S.
    (a) If the complexity of the matching network is not limited, what is the minimum Q that could possibly be achieved in the complete network consisting of the matching network and the source and load impedances?
    (b) Outline the procedure for designing the matching network for maximum bandwidth if only four elements can be used in the network. You do not need to design the net-

work.

13. Design a Pi network to match the source configuration to the load configuration below. The design frequency is 900 MHz and the desired Q is 10.

50 Ω
2 pF 1 pF 500 Ω
Source Load

14. Design a passive matching network that will achieve maximum bandwidth matching from a source with an impedance of 2 Ω (typical of the output impedance of a power amplifier) to a load with an impedance of 50 Ω. The matching network can have a maximum of three reactive elements. You need only calculate reactances and not the capacitor and inductor values.

15. Design a passive matching network that will achieve maximum bandwidth matching from a source with an impedance of 20 Ω to a load with an impedance of 125 Ω. The matching network can have a maximum of four reactive elements. You need only calculate reactances and not the capacitor and inductor values.
    (a) Will you use two, three, or four elements in your matching network?
    (b) With a diagram, and perhaps equations, indicate the design procedure.
    (c) Design the matching network. It is sufficient to use reactance values.

16. Design a passive matching network that will achieve maximum bandwidth matching from a source with an impedance of 60 Ω to a load with an impedance of 5 Ω. The matching network can have a maximum of four reactive elements. You need only calculate reactances and not the capacitor and inductor values.
    (a) Will you use two, three, or four elements in your matching network?
    (b) With a diagram and perhaps equations, indicate the design procedure.
    (c) Design the matching network. It is sufficient to use reactance values.

17. Design a T network to match a 50 Ω source to a 1000 Ω load. The desired loaded Q is 15.

18. Repeat Example 10.2 with an inductor in series with the load. Show that the inductance can be adjusted to obtain any positive shunt resistance value.

19. Design a three-lumped-element matching network that interfaces a source with an impedance of 5 Ω to a load with an impedance consisting of a resistor with an impedance of 10 Ω. The network must have a Q of 6.

20. A two-port matching network is shown below with a generator and a load. The generator impedance is 40 Ω and the load impedance is $Z_L = 50 - \jmath 20$ Ω. Use a Smith chart to design the matching network.

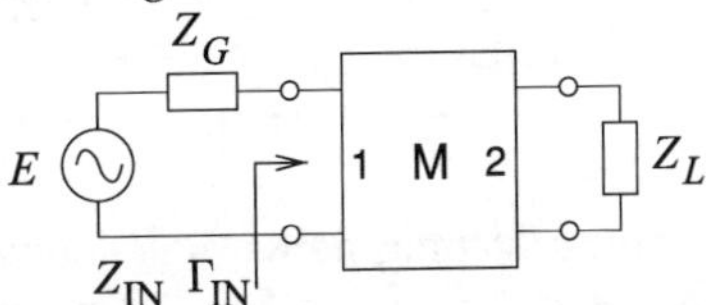

    (a) What is the condition for maximum power transfer from the generator? Express your answer using impedances.
    (b) What is the condition for maximum power transfer from the generator? Express your answer using reflection coefficients.
    (c) What system reference impedance are you going to use to solve the problem?
    (d) Plot Z_L on the Smith chart and label the point. (Remember to use impedance normalization if required.)
    (e) Plot Z_G on the Smith chart and label the point.
    (f) Design a matching network using only transmission lines. Show your work on the Smith chart. You must express the lengths of the lines in terms of electrical length (either degrees or wavelengths). Characteristic impedances of the lines are required. (You will therefore have a design that consists of one stub and one other length of transmission line.)

21. Use a lossless transmission line and a series reactive element to match a source with a Thevenin equivalent impedance of $25 + \jmath 50$ Ω to a load of 100 Ω. (That is, use one transmission line and one series reactance only.)
    (a) Draw the matching network with the source and load.
    (b) What is the value of the series reactance in the matching network (you can leave this in ohms)?
    (c) What is the length and characteristic impedance of the transmission line?

22. Consider a load $Z_L = 80 + \jmath 40$ Ω. Use the Smith chart to design a matching network consisting of only two transmission lines that will match the load to a generator of 40 Ω.
    (a) Draw the matching network with transmission lines. If you use a stub, it should be a short-circuited stub.

(b) Indicate your choice of characteristic impedance for your transmission lines. What is the normalized load impedance? What is the normalized source impedance?
(c) Briefly outline the design procedure you will use. You will need to use Smith chart sketches.
(d) Plot the load and source on a Smith chart.
(e) Complete the design of the matching network, providing the lengths of the transmission lines.

23. A two-port matching network is shown below with a generator and a load. The generator impedance is 40 Ω and the load impedance is $Z_L = 20 - \jmath 50$ Ω. Use a Smith chart to design the matching network.

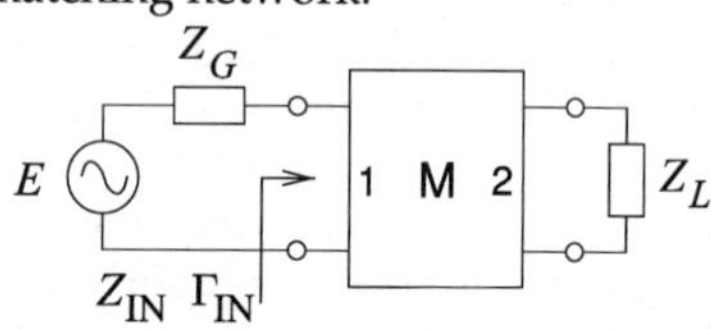

(a) What is the condition for maximum power transfer from the generator? Express your answer using impedances.
(b) What is the condition for maximum power transfer from the generator? Express your answer using reflection coefficients.
(c) What system reference impedance are you going to use to solve the problem?
(d) Plot Z_L on a Smith chart and label the point. (Remember to use impedance normalization if required.)
(e) Plot Z_G on a Smith chart and label the point.
(f) Design a matching network using only transmission lines and show your work on a Smith chart. You must express the lengths of the lines in terms of electrical length (either degrees or wavelengths long). Characteristic impedances of the lines are required. (You will therefore have a design that consists of one stub and one other length of transmission line.)

24. Use a Smith chart to design a microstrip network to match a load $Z_L = 100 - \jmath 100$ Ω to a source $Z_S = 34 - \jmath 40$ Ω). Use transmission lines only and do not use short-circuited stubs. Use a reference impedance of 40 Ω.
(a) Draw the matching network problem labeling impedances and the impedance looking into the matching network from the source as Z_1.
(b) What is the condition for maximum power transfer in terms of impedances?
(c) What is the condition for maximum power transfer in terms of reflection coefficients?
(d) Identify, i.e. draw, at least two suitable microstrip matching networks.
(e) Develop the electrical design of the matching network using the Smith chart using 40 Ω lines only. You only need do one design.
(f) Draw the microstrip layout of the matching network identify critical parameters such characteristic impedances and electrical length. Ensure that you identify which is the source side and which is the load side. You do not need to determine the widths of the lies or their physical lengths.

25. Repeat exercise 36 but now with $Z_L = 10 - \jmath 40$ Ω and $Z_S = 28 - \jmath 28$ Ω).

26. Use a Smith chart to design a two-element lumped-element lossless matching network to interface a source with an admittance $Y_S = 6 - \jmath 12$ mS to a load with admittance $Y_L = 70 - \jmath 50$ mS.

27. Use a Smith chart to design a two-element lumped-element lossless matching network to interface a load $Z_L = 50 + \jmath 50$ Ω to a source $Z_S = 10$ Ω.

10.12.1 Exercises by Section

†challenging

§10.3 1, 2
§10.4 3, 4
§10.5 5, 6, 7, 8, 9, 10, 11†, 12†, 13†
§10.6 14†, 15†, 16†, 17†, 18†, 19†
§10.7 20†, 21†, 22†, 23†
§10.9 24, 25, 26, 27

10.12.2 Answers to Selected Exercises

15(c) $Q = 1.22467$
18 $C = 1/(\omega_d^2 L_P)$
20(d) $1.25 - \jmath 0.5$
21(b) $-\jmath 50$ Ω
23(f) 40 Ω, 0.085 λ-long line before load, 40 Ω, 0.076 λ-long shorted stub

CHAPTER 11

RF and Microwave Modules

11.1 Introduction to Microwave Modules

Most microwave design uses modules such amplifiers, integrated circuits (ICs), filters, frequency multipliers, and passive components to create systems. Economics necessitate that since modules are expensive to design that they be developed for multiple applications. In system design modules are chosen for their dynamic range, noise performance, DC power consumption, and cost. Foremost the system designer must have knowledge of available modules and be prepared to design a module itself if this results in competitive performance or better manages cost. Modules are interconnected by transmission lines, and bias settings and matching networks must be designed. The system frequency plan must be developed that trades-off cost and performance while minimizing interference. This chapter begins with examples of module usage and then develops metrics that characterize microwave modules. These include characterizations of nonlinear distortion, noise, and dynamic range.

11.2 RF System as a Cascade of Modules

Most RF and microwave engineers work at the circuit board level and begin system design using modules. Some companies develop some of their own proprietary modules, thus providing a competitive advantage, but still use many modules developed by others. Examples of commercially available modules are shown in Figure 11-1. Modules can range in complexity from the simple surface mount resistor of Figure 11-1(a) and the transformer of Figure 11-1(d), up to the mixer and synthesizer modules shown in Figure 11-1(b and c). Modules can have very good performance as it is cost effective to put considerable design effort into a module that can be used in many applications and thus design costs shared.

Modules comprising a receiver are shown in Figure 11-2. Beginning with the bandpass filter after the antenna, each module contributes noise and nonlinear distortion. The system design objectives are generally to maximize dynamic range, the region between the signal being sufficiently above the noise level to be detected but before nonlinear distortion introduces spurious signals that limit the detectability of signals. At the same time power consumption and time to market must be minimized.

11.2.1 A 15 GHz Receiver Subsystem

An example of a microwave subsystem is the 15 GHz receiver shown in Figure 11-3 with details of the frequency conversion section shown in Figure 11-4. This subsystem is itself a module used in a point-to-point microwave link. The amplifier, frequency multiplier, mixer, circulator, and waveguide adaptor modules are available as off-the-shelf components from companies that specialize in particular types of modules.

11.3 Amplifiers

Amplifier modules must be optimized for low noise, moderate to high gain, high efficiency, stability, low distortion, and particular output power. Usually

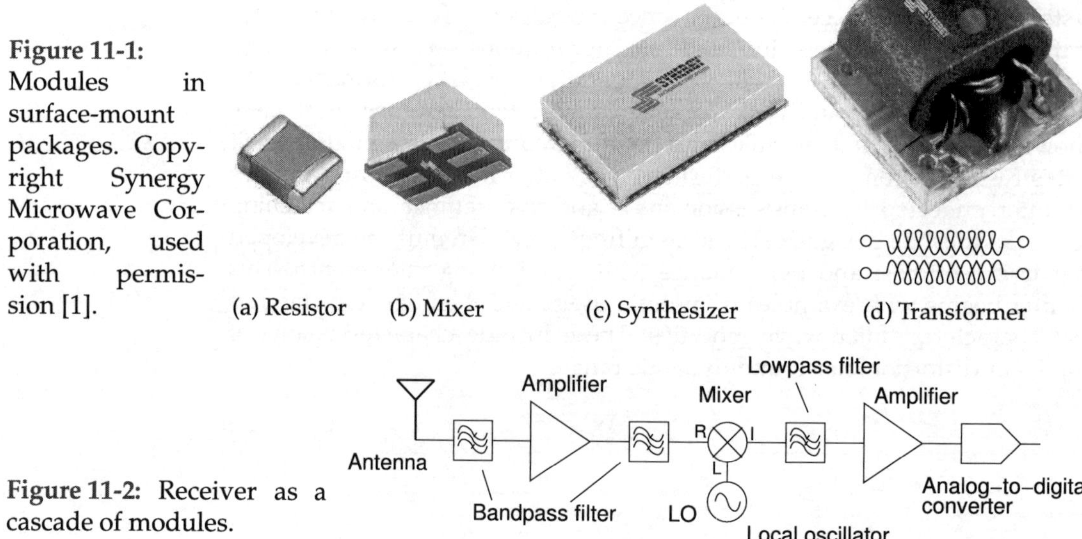

Figure 11-1: Modules in surface-mount packages. Copyright Synergy Microwave Corporation, used with permission [1].

Figure 11-2: Receiver as a cascade of modules.

an application note provided by the amplifier module vendor indicates how to do this for their module. Amplifiers require input and output matching networks as shown in Figure 11-5. The DC bias control circuit is fairly standard and the lowpass filters (in the bias circuits) are often incorporated into the input and output matching networks. Manufacturers of amplifier modules provide substantial information, including S parameters and, in many cases, reference designs.

Some amplifier modules come with input and output matching networks embodied in the module package. Since matching networks have limited bandwidth, incorporating them in the module sets the operating frequency range. Alternatively the designer can design the input and output matching networks and thus control the operating frequency.

Usually the module manufacturer provides a reference design and often

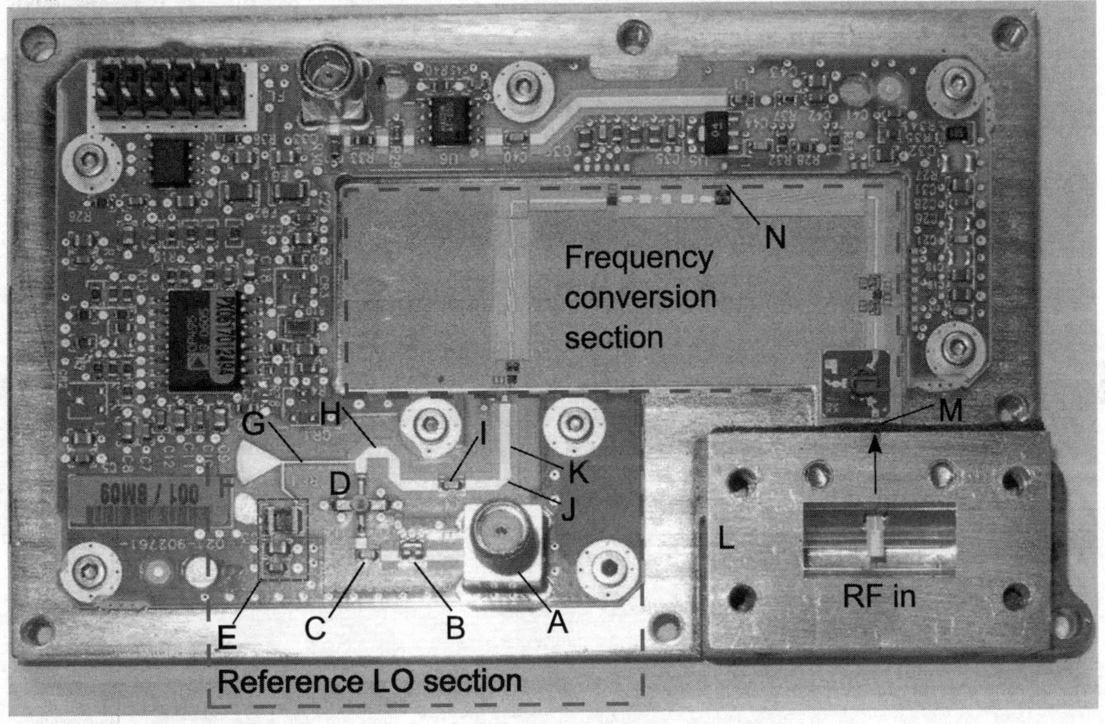

A	SMA connector, Reference LO in	**H**	50 Ω transmission line
B	Attenuator	**I**	DC blocking capacitor
C	DC blocking capacitor	**J**	Mitered bend
D	Reference LO amplifier	**K**	50 Ω transmission line
E	Bias line	**L**	Waveguide-to-microstrip adaptor, RF in
F	Radial stub	**M**	Interface to frequency conversion section
G	High-impedance transmission line	**N**	Interface to IF section

Figure 11-3: A 14.4–15.35 GHz receiver consisting of cascaded modules interconnected by microstrip lines. Surrounding the microwave circuit are DC conditioning and control circuitry. RF in is 14.4 GHz to 15.35 GHz, LO in is 1600.625 MHz to 1741.875 MHz. The frequency of the IF is 70–1595 MHz. Detail of the frequency conversion section is shown in Figure 11-4.

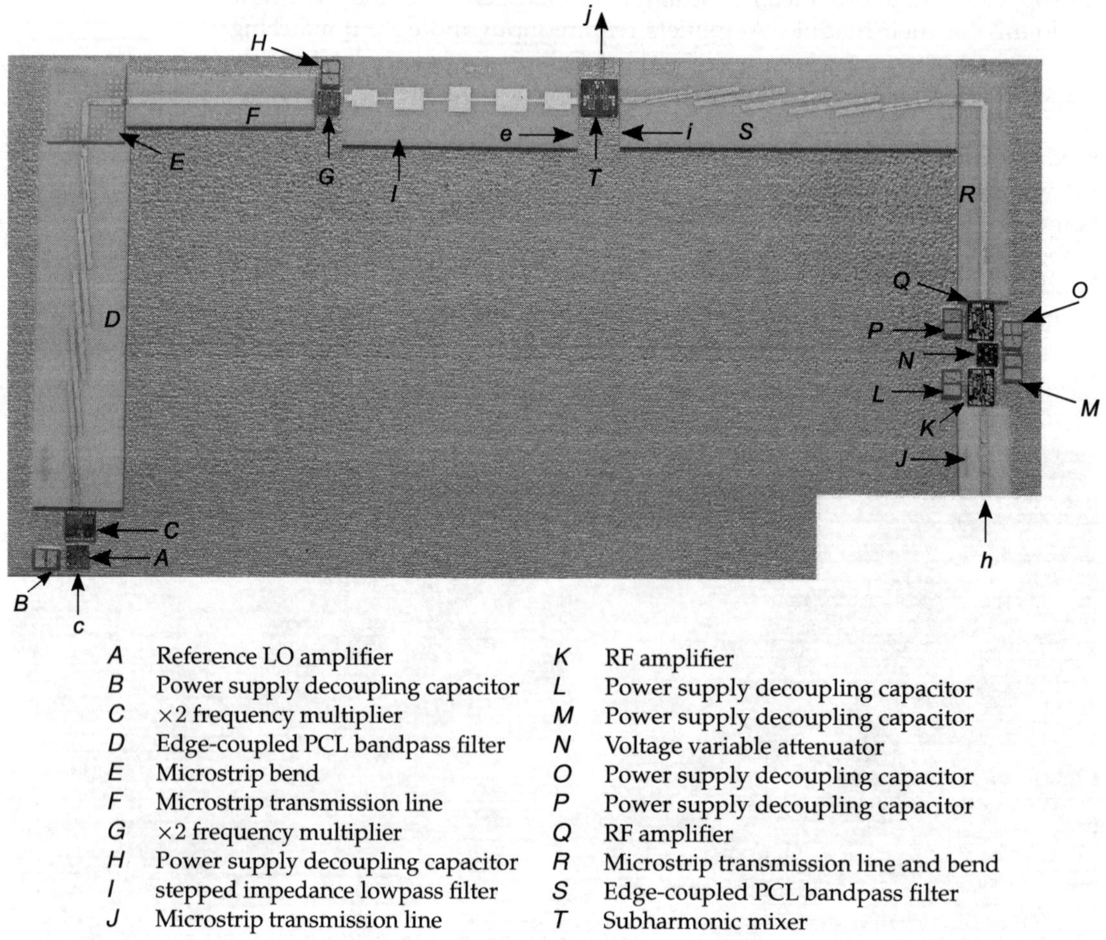

Figure 11-4: Frequency conversion section of the receiver module. The reference LO is applied at *c*, the RF is applied at *h* following the isolator. The IF is output at *j*.

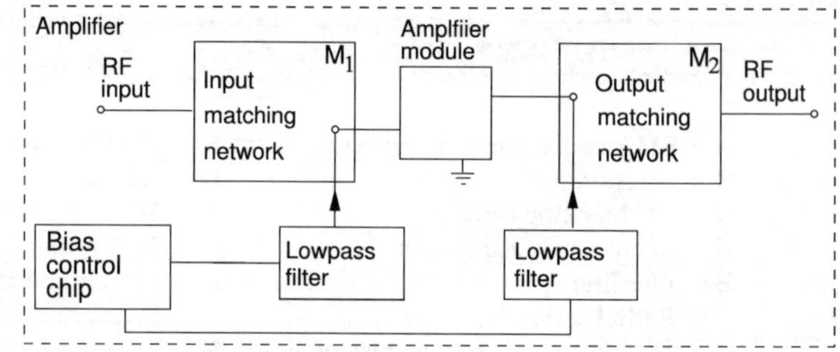

Figure 11-5: Block diagram of an RF amplifier including biasing networks.

an evaluation board, e.g. see the evaluation board in Figure 11-6 for a power amplifier. This module does not include input and output matching networks but these are provided on the evaluation board.

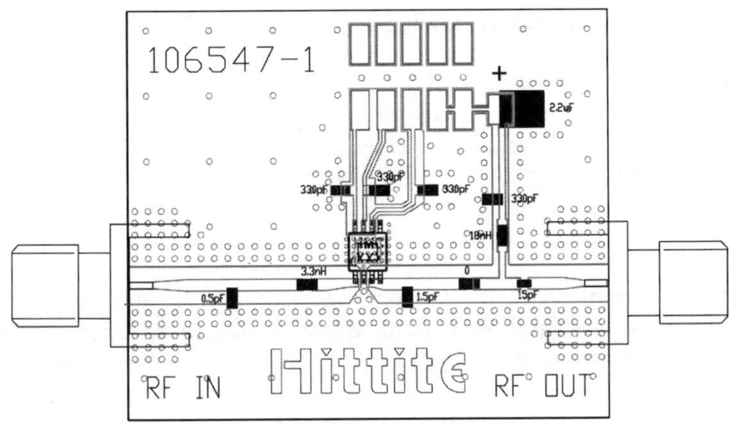

(a) Evaluation board

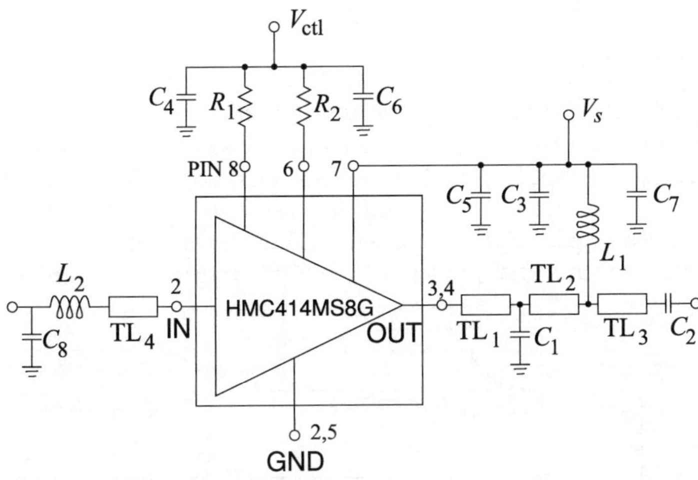

(b) Schematic of the evaluation board

Figure 11-6: Evaluation board for the HMC414MS8G GaAs InGaP HBT MMIC power amplifier module operating between 2.2 and 2.8 GHz. The amplifier provides 20 dB of gain and +30 dBm of saturated power at 32% PAE from a +5V supply voltage. The amplifier can also operate with a 3.6 V supply selectable by the resistors R_1 and R_2. Copyright Hittite Microwave Corporation, used with permission [2].

11.4 Filters

A microwave filter can consist solely of lumped elements, solely of distributed elements, or a mix. Loss of lumped elements, particularly above a few gigahertz, means that the performance of distributed filters nearly always exceeds that of lumped-element filters. However, since the basic component of a distributed filter is a one-quarter wavelength long transmission line, distributed filters can be prohibitively large below a few gigahertz. The basic types of responses required at RF as follows:

(a) Lowpass—providing maximum power transfer at frequencies below the corner frequency, f_0. See Figure 11-7(a).
(b) Highpass—passing signals at frequencies above f_0. Below f_0, transmission is blocked. See Figure 11-7(b).
(c) Bandpass—passing signals at frequencies between lower and upper corner frequencies (defining the passband) and blocking transmission outside the band. This is the most common type. See Figure 11-7(c).
(d) Bandstop (or notch)—which blocks signals between lower and upper corner frequencies (defining the stopband). See Figure 11-7(d).
(e) Allpass—which equalizes a signal by adjusting the phase generally to

correct for phase distortion elsewhere. See Figure 11-7(e).

In the passband, filters can be treated as though they are attenuators with typical losses, or attenuation, of 0.1 dB for a large basestation filter to 3 dB for a miniature filter in a handset. Out-of-band losses are typically at least 40 dB for a handset filter to 120 dB or more for a basestation filter.

11.5 Noise

Amplifiers, filters, and mixers in an RF front end process (e.g., amplify, filter, and mix) input noise the same way as an input signal. In addition, these modules contribute **excess noise** of their own. Without loss of generality, the following discussion considers noise with respect to the amplifier shown in Figure 11-8(a), where v_s is the input signal. The noise signal, with source designated by v_n, is uncorrelated and random, and described as an RMS voltage or by its noise power.

11.5.1 Noise Figure

The most important noise-related metric is the SNR. Denoting the noise power input to the amplifier as N_i, and denoting the signal power input to the amplifier as S_i, the input signal-to-noise power ratio is $\text{SNR}_i = S_i/N_i$. If the amplifier is noise free, then the input noise and signal powers are amplified by the power gain of the amplifier, G. Thus the output noise power is $N_o = GN_i$, the output signal power is $S_o = GS_i$, and the output SNR is

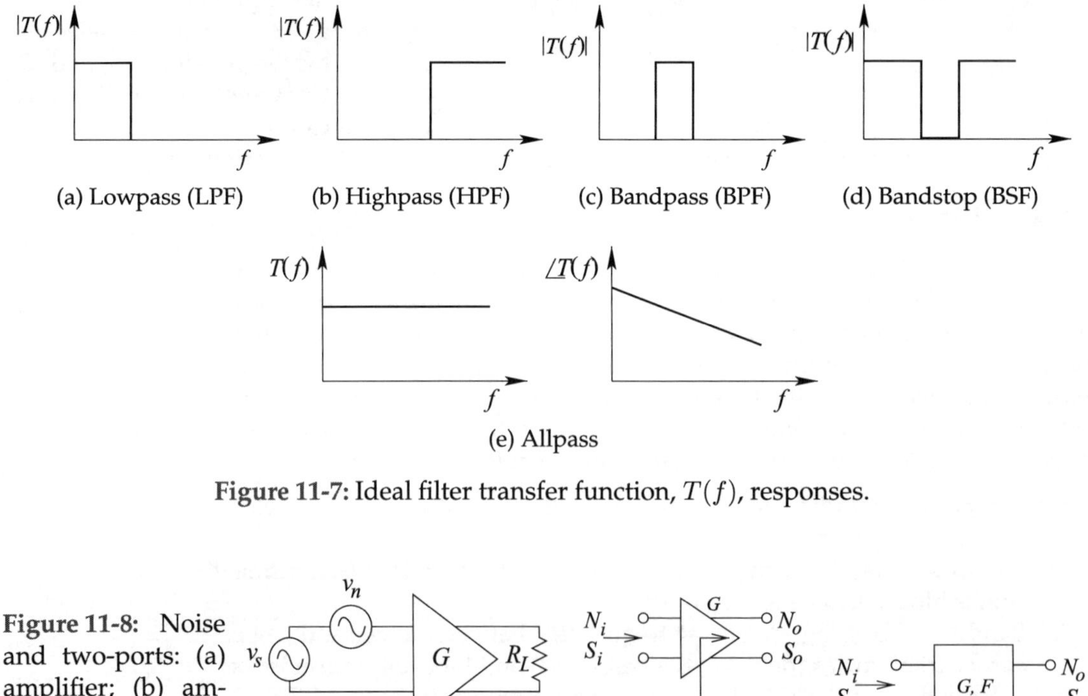

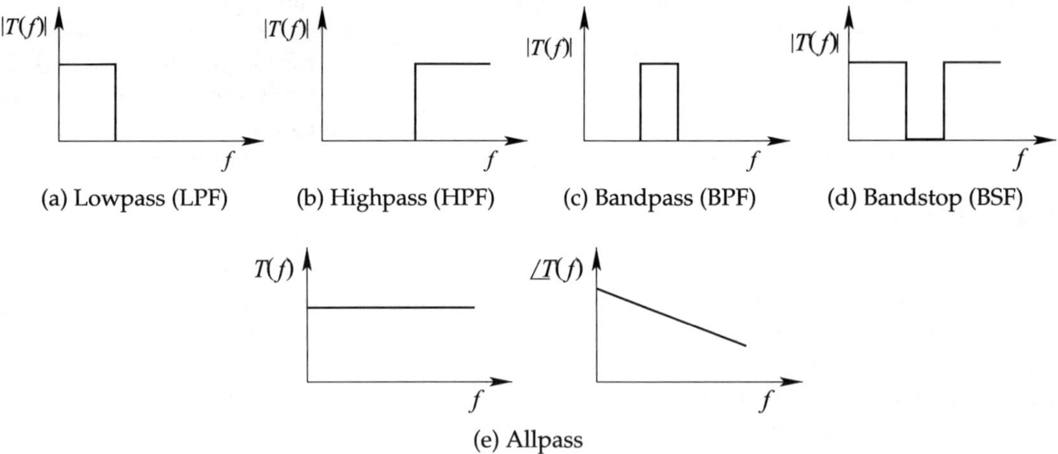

(a) Lowpass (LPF) (b) Highpass (HPF) (c) Bandpass (BPF) (d) Bandstop (BSF)

(e) Allpass

Figure 11-7: Ideal filter transfer function, $T(f)$, responses.

Figure 11-8: Noise and two-ports: (a) amplifier; (b) amplifier with excess noise; and (c) noisy two-port network.

$\text{SNR}_o = S_o/N_o = \text{SNR}_i$.

In practice, an amplifier is noisy, with the addition of excess noise, N_e, indicated in Figure 11-8(b). The excess noise originates in different components in the amplifier and is either referenced to the input or to the output of the amplifier. Most commonly it is referenced to the output so that the total output noise power is $N_o = GN_i + N_e$. In the absence of a qualifier, the **excess noise** should be assumed to be referred to the output. N_e is not measured directly. Instead, the ratio of the SNR at the input to that at the output is measured and called the **noise factor**, F:

$$F = \frac{\text{SNR}_i}{\text{SNR}_o}, \tag{11.1}$$

If the circuit is noise free, then $\text{SNR}_o = \text{SNR}_i$ and $F = 1$. If the circuit is not noise free, then $\text{SNR}_o < \text{SNR}_i$ and $F > 1$. With the excess noise, referred to the output,

$$F = \frac{\text{SNR}_i}{\text{SNR}_o} = \frac{\text{SNR}_i}{1}\frac{1}{\text{SNR}_o} = \frac{S_i}{N_i}\frac{N_o}{S_o} = \frac{S_i}{N_i}\frac{GN_i + N_e}{GS_i} = 1 + \frac{N_e}{GN_i}. \tag{11.2}$$

One of the conclusions that can be drawn from this is that F depends on the available noise power at the input of the circuit. As reference, the available noise power, N_R, from a resistor at **standard temperature,** T_0 (290 K), and over bandwidth, B (in Hz), is used,

$$N_i = N_R = kT_0B, \tag{11.3}$$

where k ($= 1.381 \times 10^{-23}$ J/K) is **the Boltzmann constant**. If the input of an amplifier is connected to this resistor and all of the available noise power is delivered to the amplifier, then

$$F = 1 + \frac{N_e}{GN_i} = 1 + \frac{N_e}{GkT_0B}. \tag{11.4}$$

When expressed in decibels, the **noise figure (NF)** is used:

$$\text{NF} = 10\log_{10} F = \text{SNR}_i|_{\text{dB}} - \text{SNR}_o|_{\text{dB}}\,.$$

where the SNR is expressed in decibels. Rearranging Equation (11.4), the output-referred excess noise power is

$$N_e = (F - 1)GkT_0B. \tag{11.5}$$

Also from Equation (11.4), the output noise is

$$N_o = GN_i + N_e = FGN_i. \tag{11.6}$$

That is, $$N_o|_{\text{dBm}} = \text{NF} + G|_{\text{dB}} + N_i|_{\text{dBm}}\,. \tag{11.7}$$

EXAMPLE 11.1 Noise Figure of an Attenuator

What is the noise figure of a 20 dB attenuator in a 50 Ω system?

Solution:

The appropriate circuit model to use in the analysis consists of the attenuator driven by a generator with a 50 Ω source impedance, and the attenuator drives a 50 Ω load. Also, the input impedance of the terminated attenuator is 50 Ω, as is the impedance looking into the output of the attenuator when it is connected to the source. The key point is that the noise coming from the source is the noise thermally generated in the 50 Ω source impedance, and this noise is equal to the noise that is delivered to the load. So the input noise, N_i, is equal to the output noise:

$$N_o = N_i. \tag{11.8}$$

The input signal is attenuated by 20 dB (= 100), so $S_o = S_i/100$, (11.9)

and thus the noise factor is $F = \frac{SNR_i}{SNR_o} = \frac{S_i}{N_i}\frac{N_o}{S_o} = \frac{S_i}{N_i}\frac{N_i}{S_i/100} = 100$ (11.10)

and the noise figure is NF = 20 dB. (11.11)

THat is, the noise figure of an attenuator (or a filter) is just the loss of the component. This is not true for amplifiers of course, as there are other sources of noise, and the output impedance of a transistor is not a thermal resistance.

11.5.2 Noise of a Cascaded System

Section 11.5.1 developed the noise factor and noise figure measures for a two-port. This can be generalized for a system. Considering the second stage of the cascade in Figure 11-9, the excess noise at the output of the second stage, due solely to the noise generated internally in the second stage, is

$$N_{2e} = (F_2 - 1)kT_0BG_2. \tag{11.12}$$

Then the total noise power at the output of the two-stage cascade is

$$\begin{aligned} N_{2o} &= (F_2 - 1)kT_0BG_2 + N_{o,1}G_2 \\ &= (F_2 - 1)kT_0BG_2 + F_1kT_0BG_1G_2. \end{aligned} \tag{11.13}$$

This assumes (correctly) that the excess noise added in one stage is uncorrelated to the other sources of noise. Thus noise powers can be added. Generalizing this result the total system noise factor:

$$F^{\mathrm{T}} = F_1 + \frac{F_2 - 1}{G_1} + \frac{F_3 - 1}{G_1G_2} + \frac{F_4 - 1}{G_1G_2G_3} + \cdots. \tag{11.14}$$

This equation is known as **Friis's formula** [3].

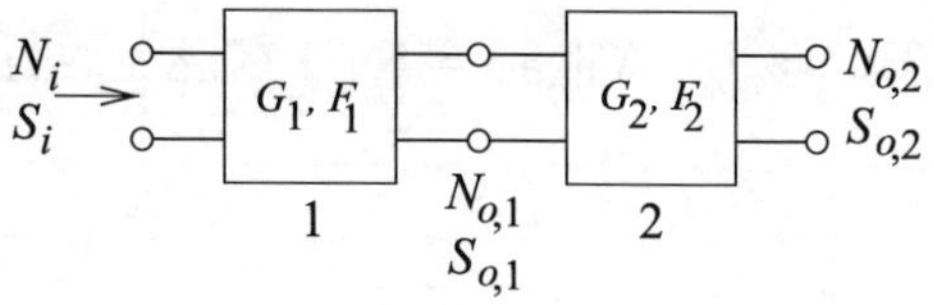

N_i S_i G_1, F_1 1 $N_{o,1}$ $S_{o,1}$ G_2, F_2 2 $N_{o,2}$ $S_{o,2}$

Figure 11-9: Cascaded noisy two-ports.

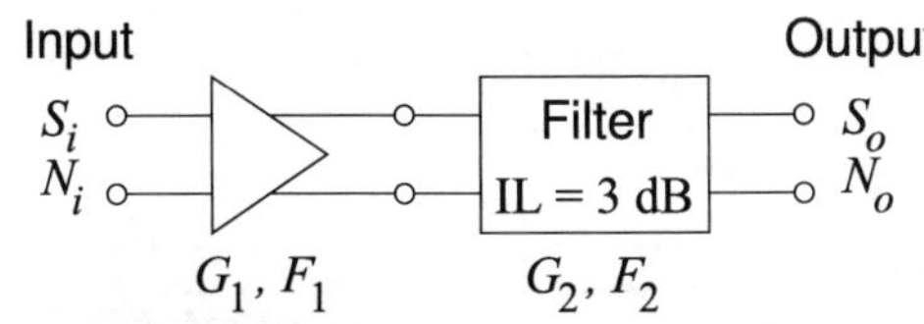

Figure 11-10: Differential amplifier followed by a differential filter.

EXAMPLE 11.2 Noise Figure of Cascaded Stages

Consider the cascade of a differential amplifier and a filter shown in Figure 11-10.

(a) What is the midband gain of the filter in decibels? Note that IL is insertion loss.
(b) What is the midband noise figure of the filter?
(c) The amplifier has a gain $G_1 = 20$ dB and a noise figure of 2 dB. What is the overall gain of the cascade system in the middle of the band?
(d) What is the noise factor of the cascade system?
(e) What is the noise figure of the cascade system?

Solution:

(a) $G_2 = 1/\text{IL}$, thus $G_2 = -3$ dB.
(b) For a passive element, $\text{NF}_2 = \text{IL} = 3$ dB.
(c) $G_1 = 20$ dB and $G_2 = -3$ dB, so $G_{\text{TOTAL}} = G_1|_{\text{dB}} + G_2|_{\text{dB}} = 17$ dB.
(d) $F_1 = 10^{\text{NF}_1/10} = 10^{2/10} = 1.585$, $F_2 = 10^{\text{NF}_2/10} = 10^{3/10} = 1.995$, $G_1 = 10^{20/10} = 100$, and $G_2 = 10^{-3/10} = 0.5$. Using Friis's formula

$$F_{\text{TOTAL}} = F_1 + \frac{F_2 - 1}{G_1} = 1.585 + \frac{1.995 - 1}{100} = 1.594. \tag{11.15}$$

(e) $\text{NF}_{\text{TOTAL}} = 10\log_{10}(\text{F}_{\text{TOTAL}}) = 10\log_{10}(1.594) = 2.03$ dB.

EXAMPLE 11.3 Noise Figure of a Two-Stage Amplifier

Consider a room-temperature (20°C) two-stage amplifier where the first stage has a gain of 10 dB and the second stage has a gain of 20 dB. The noise figure of the first stage is 3 dB and the second stage is 6 dB. The amplifier has a bandwidth of 10 MHz.

(a) What is the noise power presented to the amplifier in 10 MHz?
(b) What is the total gain of the amplifier?
(c) What is the total noise factor of the amplifier?
(d) What is the total noise figure of the amplifier?
(e) What is the noise power at the output of the amplifier in 10 MHz?

Solution:

(a) Noise power of a resistor at room temperature is -174 dBm/Hz (or more precisely -173.86 dBm/Hz at 293 K). In 10 MHz the input noise power is $N_i = -173.86\ \text{dBm} + 10\log(10^7) = -173.86 + 70\ \text{dBm} = -103.86$ dBm.
(b) Total gain $G^{\text{T}} = G_1G_2 = 10\ \text{dB} + 20\ \text{dB} = 30\ \text{dB} = 1000$.
(c) $F_1 = 10^{\text{NF}_1/10} = 10^{3/10} = 1.995$, $F_2 = 10^{\text{NF}_2/10} = 10^{6/10} = 3.981$. Using Friis's formula, the total noise figure is $F^{\text{T}} = F_1 + \dfrac{F_2 - 1}{G_1} = 1.995 + \dfrac{3.981 - 1}{10} = 2.393$.
(d) The total noise figure is $\text{NF}^{\text{T}} = 10\log_{10}(\text{F}^{\text{T}}) = 10\log_{10}(2.393) = 3.79$ dB.
(e) Output noise power in 10 MHz bandwidth is $N_o = F^{\text{T}}kT_0BG^{\text{T}} = (2.393)\cdot(1.3807\cdot 10^{-23}\cdot \text{J}\cdot\text{K}^{-1})\cdot(293\ \text{K})\cdot(10^7\cdot\text{s}^{-1})(1000) = 9.846\cdot 10^{-11}\ \text{W} = -70.07$ dBm. Alternatively, $N_o|_{\text{dBm}} = N_i|_{\text{dBm}} + G^{\text{T}}|_{\text{dB}} + \text{NF}^{\text{T}} = -103.86\ \text{dBm} + 30\ \text{dB} + 3.79\ \text{dB} = -70.07$ dBm.

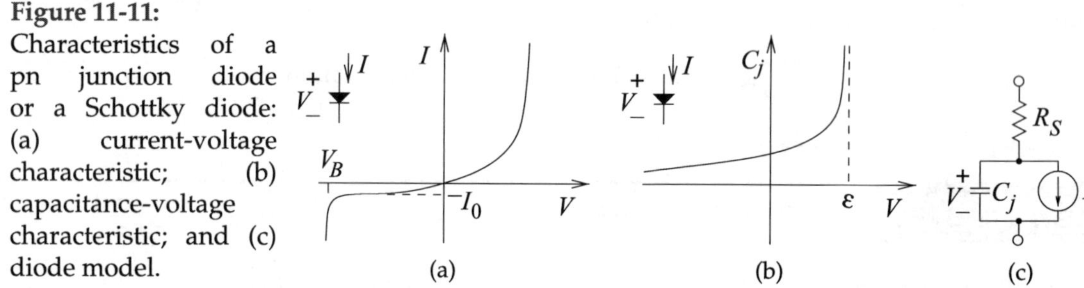

Figure 11-11: Characteristics of a pn junction diode or a Schottky diode: (a) current-voltage characteristic; (b) capacitance-voltage characteristic; and (c) diode model.

11.6 Diodes

Diodes are two-terminal devices that have nonlinear current-voltage characteristics. The most common diodes are listed in Table 11-1.

Junction and Schottky Diodes: A junction diode has an asymmetric current-voltage characteristic, see Figure 11-11(a),

$$I = I_0 \left[\exp\left(\frac{qV}{nkT}\right) - 1 \right], \tag{11.16}$$

where $q(= -e)$ is the absolute value of the charge of an electron, k is the Boltzmann constant ($1.37 \cdot 10^{-23}$ J/K), and T is the absolute temperature (in kelvin). I_0 is the reverse saturation current and is small, with values ranging from 1 pA to 1 nA. The quantity n is the diode ideality factor, with $n = 2$ for a graded junction pn junction diode, and $n = 1.0$ when the interface between p-type and n-type semiconductor materials is abrupt. The abrupt junction is most closely realized by a Schottky diode, where a metal forms one side of the interface. When the applied voltage is sufficiently positive to cause a large current to flow, the diode is said to be forward biased. When the voltage is negative, the current flow is negligible and the diode is said to be reverse biased. In a diode, charge is separated over distance and so a diode has junction capacitance modeled as

$$C_j(V) = \frac{C_{j0}}{(1 - (V/\phi))^{\gamma}}, \tag{11.17}$$

Table 11-1: IEEE standard symbols for diodes and a rectifier [4]. ([1]In the direction of anode (A) to cathode (K). [2]Use symbol for general diode unless it is essential to show the intrinsic region.)

Component	Symbol
Diode, general (including Schottky)[1]	
IMPATT diode[1]	
Gunn diode	
PIN diode[1,2]	
Light emitting diode (LED)[1]	
Rectifier[1]	
Tunnel diode[1]	
Varactor diode[1]	or
Zener diode[1]	

where ϕ is the built-in potential difference across the diode. This capacitance profile is shown in Figure 11-11(b). The built-in potential is typically 0.6 V for silicon diodes and 0.75 V for GaAs diodes. The doping profile can be adjusted so that γ can be less than the ideal $\frac{1}{2}$ of an abrupt junction diode. Current must flow through bulk semiconductor before reaching the active region of the semiconductor diode, and so there will be a resistive voltage drop. Combining effects leads to the equivalent circuit of a pn junction or Schottky diode, shown in Figure 11-11(c).

Varactor Diode: A varactor diode is a pn junction diode operated in reverse bias and optimized for good performance as a tunable capacitor. A common application of a varactor diode is as the tunable element in a voltage-controlled oscillator (VCO) where the varactor, with voltage-dependent capacitance, C, is part of a resonant circuit.

PIN Diode: A PIN diode is a variation on a pn junction diode with a region of intrinsic semiconductor (the I in PIN) between the p-type and n-type semiconductor regions. The properties of the PIN diode depend on whether there are carriers in the intrinsic region. The PIN diode has the current-voltage characteristics of a pn junction diode at low frequencies; however, at high frequencies it looks like a linear resistor, as carriers in the intrinsic region move slowly. When a forward DC voltage is applied to the PIN diode, the intrinsic region floods with carriers, and at microwave frequencies the PIN diode is then modeled as a low-value resistor. At high frequencies there is not enough time to remove the carriers in the intrinsic region, so even if the total voltage (DC plus RF) across the PIN diode is negative, there are carriers in the intrinsic region throughout the RF cycle. If the DC voltage is negative, carriers are removed from the intrinsic region and the diode looks like a large-value resistor at RF. The PIN diode is used as a microwave switch controlled by a DC voltage.

Zener Diode: Zener diodes are pn junction or Schottky diodes that have been specially designed to have sharp reverse breakdown characteristics. They can be used to establish a voltage reference or, used as a limiter diode, to provide protection of more sensitive circuitry. As a limiter, they are found in communication devices in a back-to-back configuration to limit the voltages that can be applied to sensitive RF circuitry.

11.7 Switch

In many cellular systems a phone does not transmit and receive simultaneously. Consequently a switch can be used to alternately connect a transmitter and a receiver to an antenna. An ideal switch is shown in Figure 11-12(a), where an input port, $\mathrm{RF_{IN}}$, and an output port, $\mathrm{RF_{OUT}}$, are shown. At microwave frequencies, realistic switches must be modeled with parasitics and with finite on and off resistances. A realistic model applicable to many switch types is shown in Figure 11-12(b). The capacitive parasitics, the C_{PS}, limit the frequency of operation and the on resistance, R_{ON}, causes loss. Ideally the off resistance, R_{OFF}, is very large, however, the parasitic shunt capacitance, C_{OFF}, is nearly always more significant. The result is that

Table 11-2: Typical properties of small microwave switches. (Sources: [1]Radant MEMS, [2]RF Micro Devices, and [3]Tyco Electronics.)

Switch type	Power handling	Insertion loss	Operating frequency	Actuation voltage	Response time
MEMS[1]	0.5 W	0.5 dB	to 10 GHz	90 V	10 μs
MEMS[1]	4 W	0.8 dB	to 35 GHz	110 V	10 μs
pHEMT[2]	10 W	0.3 dB	to 6.5 GHz	5 V	0.5 μs
pHEMT[2]	0.3 W	1.1 dB	to 25 GHz	5 V	0.5 μs
PIN[3]	13 W	0.35 dB	to 2 GHz	12 V	0.5 μs
PIN[3]	10 W	0.4 dB	to 6 GHz	12 V	0.5 μs

at high frequencies there is an alternative capacitive connection between the input and output through C_{OFF}. The on resistance of the switch introduces voltage division that can be seen by comparing the ideal connection shown in Figure 11-12(c) and the more realistic connection shown in Figure 11-12(d). There are four main types of microwave switches: mechanical (rarely used except in instrumentation), PIN diode, FET, and **microelectromechanical system (MEMS)**, see Figures 11-12(e–g) and Table 11-2.

A MEMS switch is fabricated using photolithographic techniques similar to those used in semiconductor manufacturing. They are essentially miniature mechanical switches with a voltage used to control the position of a shorting arm, see Figure 11-13.

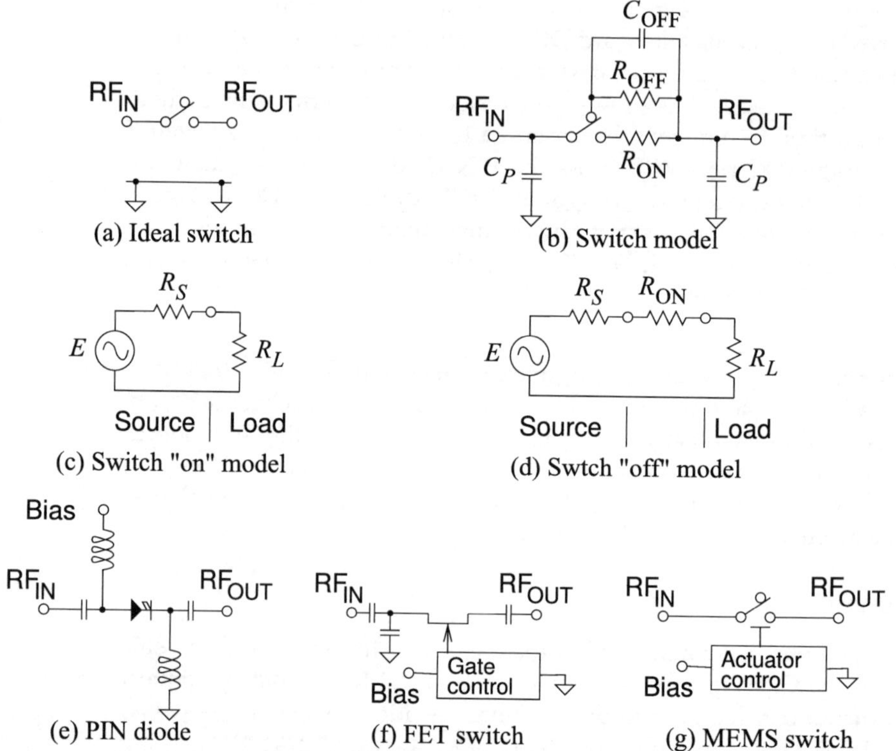

Figure 11-12: Microwave switches: (a) ideal switch connecting RF_{IN} and RF_{OUT} ports; (b) model of a microwave switch; (c) ideal circuit model with switch on and with source and load; (d) realistic low-frequency circuit model with switch on; (e) switch realized using a PIN diode; (f) switch realized using an FET; and (g) switch realized using a MEMS switch.

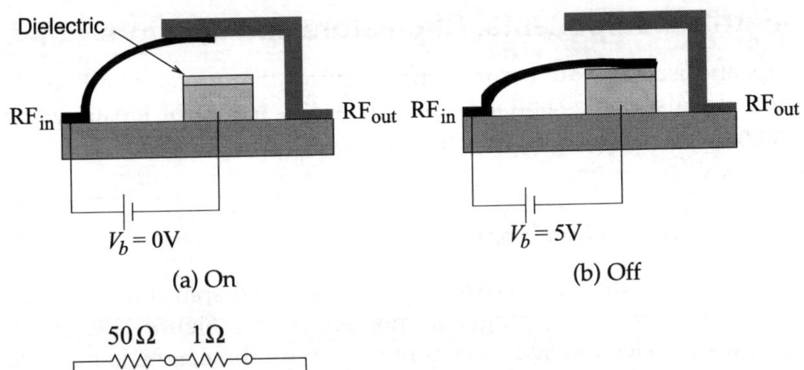

Figure 11-13: RF MEMS switch: (a) the RF_{in} line in contact with the RF_{out} line; and (b) the cantlever beam electrostatically attracted to the pedestal and there is no RF connection.

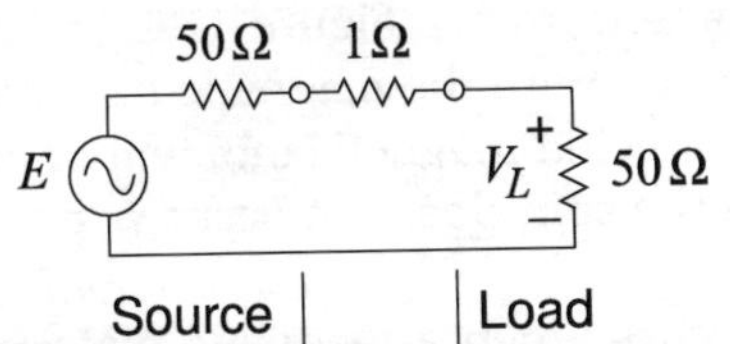

Figure 11-14: Model used in calculating the loss of a switch in a 50 Ω system.

EXAMPLE 11.4 Insertion Loss of a Switch

What is the insertion loss of a switch with a 1 Ω on resistance when it used in a 50 Ω system?

Solution:

The model to be used for evaluating the insertion loss of the switch is shown in Figure 11-14. The insertion loss is found by first determining the available power from the source and then the actual power delivered to the load. The available input power is calculated by first ignoring the 1 Ω switch resistance. Then there is maximum power transfer from the source to the load. The available input power is

$$P_{Ai} = \frac{1}{2}\frac{\left(\frac{1}{2}E\right)^2}{50} = \frac{E^2}{400}, \tag{11.18}$$

where E is the peak RF voltage at the Thevenin equivalent source generator. The power delivered to the 50 Ω load is found after first determining the peak load voltage:

$$V_L = \frac{50}{50+1+50}E = \frac{50}{101}E. \tag{11.19}$$

Thus the power delivered to the load is

$$P_D = \frac{1}{2}\frac{V_L^2}{50} = \frac{1}{100}\left(\frac{50}{101}\right)^2 E^2. \tag{11.20}$$

The insertion loss is

$$\text{IL} = \frac{P_{Ai}}{P_D} = \frac{E^2}{400}\frac{100}{E^2}\left(\frac{101}{50}\right)^2 = 1.020 = 0.086\text{ dB}. \tag{11.21}$$

11.8 Ferrite Components: Circulators and Isolators

Circulators and isolators are nonreciprocal devices that preferentially route microwave signals. The essential element is a disc (puck) of ferrite which when magnetized supports a preferred direction of propagation due to the gyromagnetic effect.

11.8.1 Gyromagnetic Effect

An electron has a quantum mechanical property called spin (there is not a spinning electron) creating a magnetic moment m, see Figure 11-15(a). In most materials the electron spin occurs in pairs with the magnetic moment of one electron canceled by the oppositely-directed magnetic moment of an other electron. However, in some materials the spin does not occur in pairs and there is a net magnetic moment.

A most interesting microwave property occurs when a magnetic material is biased by a DC magnetic field resulting in the gyromagnetic effect. This affects the way an RF field propagates and this is described by a nine element permeability called a tensor. The permeability of a magnetically biased magnetic material is:

$$[\mu] = \begin{bmatrix} \mu_{xx} & \mu_{xy} & \mu_{xz} \\ \mu_{yx} & \mu_{yy} & \mu_{yz} \\ \mu_{zx} & \mu_{zy} & \mu_{zz} \end{bmatrix} = \begin{bmatrix} \mu_0 & 0 & 0 \\ 0 & \mu & \jmath\kappa \\ 0 & -\jmath\kappa & \mu \end{bmatrix}. \quad (11.22)$$

That is,

$$\begin{bmatrix} B_x \\ B_y \\ B_z \end{bmatrix} = \begin{bmatrix} \mu_0 & 0 & 0 \\ 0 & \mu & \jmath\kappa \\ 0 & -\jmath\kappa & \mu \end{bmatrix} \begin{bmatrix} H_x \\ H_y \\ H_z \end{bmatrix}. \quad (11.23)$$

Thus a z-directed H field produces a y-directed B field. The effect on propagation of an EM field is shown in Figure 11-15(b). The EM wave does not travel in a straight line and instead curves, in this case, to the right. Thus forward- and backward-traveling waves diverge from each other and propagation is not reciprocal. This separates forward- and backward-traveling waves.

11.8.2 Circulator

A circulator exploits the gyromagnetic effect. Referring to the schematic of a circulator shown in Figure 11-16(a), where the arrows indicate that the signal that enters Port 1 of the circulator leaves the circulator at Port 2 and not at Port 3. Similarly power that enters at Port 2 is routed to Port 3, and

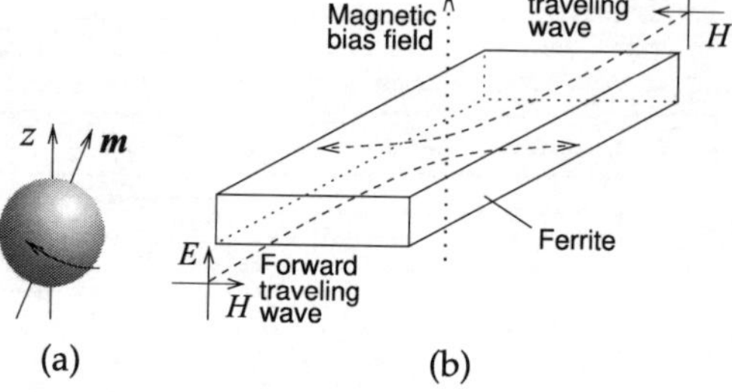

Figure 11-15: Magnetic moments: (a) electron spin; and (b) Gyromagnetic effect impact on the propagation of EM waves in a magnetic material with an externally applied magnetic bias field.

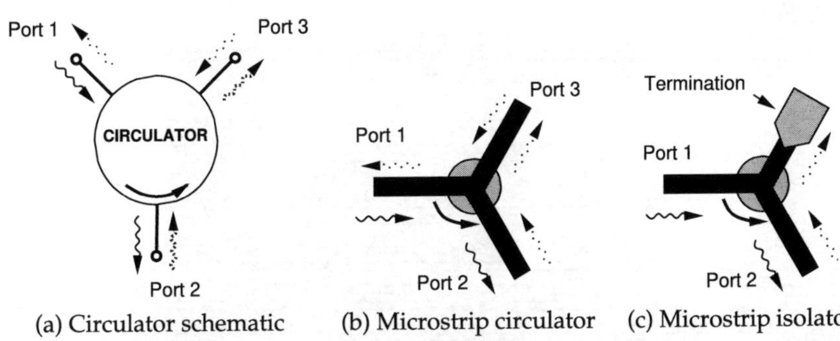

(a) Circulator schematic (b) Microstrip circulator (c) Microstrip isolator

Figure 11-16: Ferrite components with a magnetized ferrite puck in the center in (b) and (c).

power entering at Port 3 is routed to Port 1. In terms of S parameters, an ideal circulator has the scattering matrix

$$\mathbf{S} = \begin{bmatrix} 0 & 0 & S_{13} \\ S_{21} & 0 & 0 \\ 0 & S_{32} & 0 \end{bmatrix} = \begin{bmatrix} 0 & 0 & 1 \\ 1 & 0 & 0 \\ 0 & 1 & 0 \end{bmatrix}. \tag{11.24}$$

A microstrip circulator is shown in Figure 11-16(b), where a disc of magnetized ferrite can be placed on top of a microstrip Y junction to realize a preferential direction of propagation of the EM fields. In the absence of the biasing magnetic field, the circulation function does not occur.

11.8.3 Circulator Isolation

The isolation of a circulator is the insertion loss from what is the output port to the input port, i.e. in the reverse direction. Referring to the circulator in Figure 11-16(a), if port 1 is the input port there are two output ports and so there are two isolations equal to the return loss from port 3 to port 2, and from port 2 to port 1. The smaller of these is the quoted isolation. If this is an ideal circulator and port 2 is perfectly matched, then the isolation would be infinite. Then the S parameters of the circulator are

$$\mathbf{S} = \begin{bmatrix} \Gamma & \alpha & T \\ T & \Gamma & \alpha \\ \alpha & T & \Gamma \end{bmatrix} \approx \begin{bmatrix} 0 & 0 & 1 \\ 1 & 0 & 0 \\ 0 & 1 & 0 \end{bmatrix} \tag{11.25}$$

where T is the transmission factor, Γ is the reflection coefficient at the ports, and α is the leakage.

11.8.4 Isolator

Isolators are devices that allow power flow in only one direction. Figures 11-16(c) and 11-17 show microstrip isolators based on three-port circulators. Power entering Port 1 as a traveling wave is transferred to the ferrite and emerges at Port 2. Virtually none of the power emerges at Port 3. A traveling wave signal applied at Port 2 appears at Port 3, where it is absorbed in a termination created by resistive material placed on top of the microstrip. The resistive material forms a lossy transmission line and no power is reflected. Thus power can travel from Port 1 to Port 2, but not in the reverse direction.

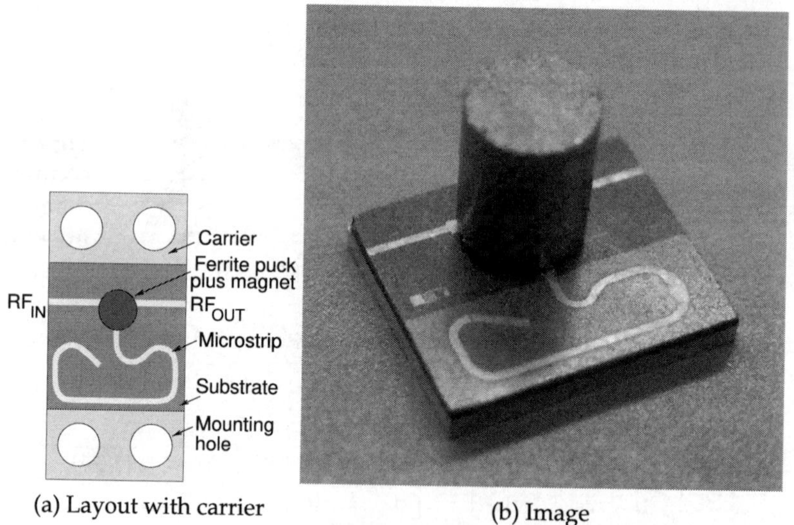

(a) Layout with carrier (b) Image

Figure 11-17: A microstrip isolator operating from 29 to 31.5 GHz. Isolator in (b) has the dimensions 5 mm × 6 mm and is 6 mm high. The isolator supports 2 W of forward and reverse power with an isolation of 18 dB and insertion loss of 1 dB. Renaissance 2W9 series, copyright Renaissance Electronics Corporation, used with permission.

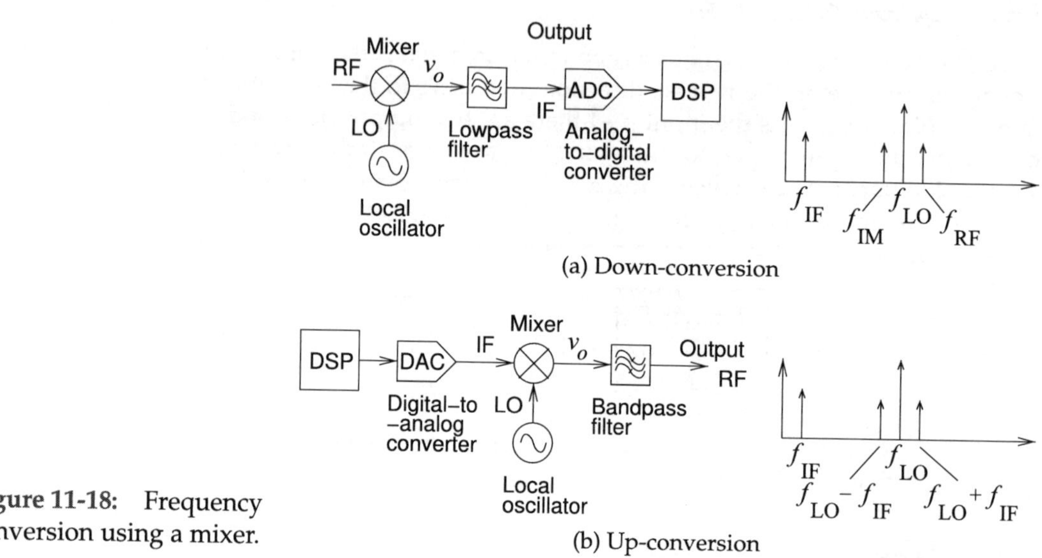

Figure 11-18: Frequency conversion using a mixer.

11.9 Mixer

Frequency conversion or mixing is the process of converting information centered at one frequency (present in the form of a modulated carrier) to another frequency. The second frequency is either higher, in the case of frequency **up-conversion**, where it is more easily transmitted; or lower when mixing is called frequency **down-conversion**, where it is more easily captured, see Figure 11-18.

Conversion loss, L_C: This is the ratio of the available power of the input signal, $P_{\text{in}}(\text{RF})$, to that of the output signal after mixing, $P_{\text{out}}(\text{IF})$:

$$L_C = \frac{P_{\text{in}}(\text{RF})}{P_{\text{out}}(\text{IF})}. \tag{11.26}$$

EXAMPLE 11.5 Mixer Calculations

A mixer has an LO of 10 GHz. The mixer is used to down-convert a signal at 10.1 GHz and has a conversion loss, L_c of 3 dB and an image rejection of 20 dB. A 100 nW signal is presented to the mixer at 10.1 GHz. What is the frequency and output power of the down-converted signal at the IF?

Solution:

The IF is at 100 MHz. L_c = 3 dB = 2 and from Equation (11.26) the output power at IF of the intended signal is

$$P_{\text{out}} = P_{\text{in}}(\text{RF})/L_c = 100\ \text{nW}/2 = 50\ \text{nW} = -43\ \text{dBm}. \tag{11.27}$$

11.10 Local Oscillator

In an oscillator noise close to the oscillation center frequency is called flicker noise or $1/f$ noise and in offsets below a few tens of megahertzis much larger than thermal noise and so a big concern in microwave systems. The noise manifests itself as random fluctuations of amplitude and phase of the carrier. The amplitude fluctuations are quenched by saturation in the oscillator and so are not of concern. Thus the close-in noise of concern is just phase noise.
The phase noise of an oscillator with a low Q feedback loop is shown in Figure 11-19(a) and in Figure 11-19(b) for a high Q loop. The physical origin of the straight line phase regions is nt understood.
Phase noise is expressed as the ratio of the phase noise power in a 1 Hz bandwidth of a single sideband (SSB) to the total signal power. This is measured at a frequency f_m offset from the carrier and denoted $\mathcal{L}(f_m)$ with the units of dBc/Hz (i.e., decibels relative to the carrier power per hertz).
The phase noise that is important in RF and microwave oscillators (having relatively low Q) is usually dominated by a $1/f_m^2$ shape. Then the phase noise

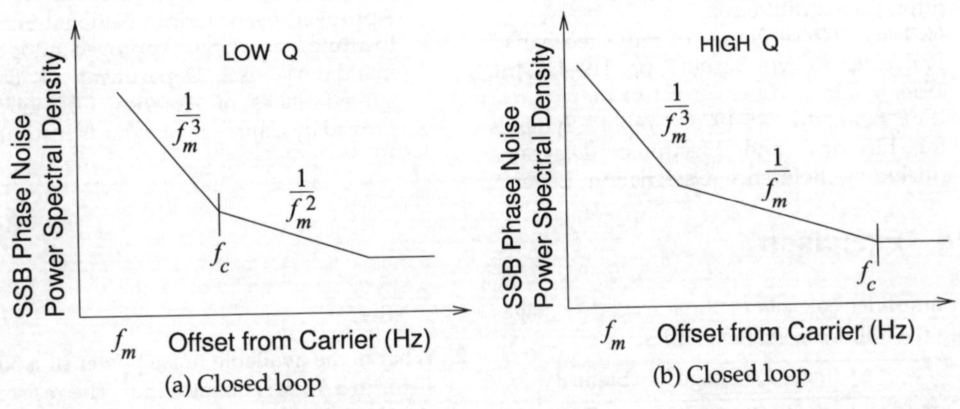

(a) Closed loop (b) Closed loop

Figure 11-19: Log-log plot of oscillator noise spectra: (a) closed-loop noise with low-Q loop; and (b) closed-loop noise with high-Q loop.

at 1 MHz (a common frequency for comparing the phase noise performance of different oscillators) is related to the phase noise measured at f_m by

$$\mathcal{L}(1\text{ MHz}) = \mathcal{L}(f_m) - 10\log\left(\frac{1\text{ MHz}}{f_m}\right)^2. \quad (11.28)$$

11.11 Frequency Multiplier

A microwave frequency multiplier uses a nonlinear element to generate harmonics and a bandpass filter selects the appropriate harmonic for the output, see Figure 11-20. If the input signal is

$$x(t) = A\cos(\omega_t + \phi) \quad (11.29)$$

then the output at the n harmonic is

$$y(t) = A_n \cos(n\omega_t + n\phi). \quad (11.30)$$

11.12 Summary

The use of modules has become increasingly important in microwave engineering. A wide variety of passive and active modules are available and high-performance systems can be realized enabling many microwave systems of low to medium volumes to be realized cost effectively and with stellar performance. Module vendors are encouraged by the market to develop competitive modules that can be used in a wide variety of applications.

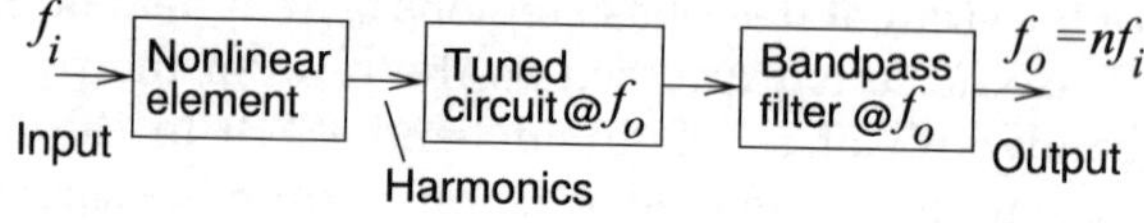

Figure 11-20: Frequency multiplier using a nonlinear tuned circuit.

11.13 References

[1] http://www.synergymwave.com.
[2] http://www.hittite.com.
[3] H. Friis, "Noise figures of radio receivers," *Proc. of the IRE*, vol. 32, no. 7, pp. 419–422, Jul. 1944.
[4] IEEE Standard 315-1975, Graphic Symbols for Electrical and Electronics Diagrams (Including Reference Designation Letters), Adopted Sept. 1975, Reaffirmed Dec. 1993. Approved by American National Standards Institute, Jan. 1989. Approved adopted for mandatory use, Department of Defense, United States of America, Oct. 1975. Approved by Canadian Standards Institute, Oct. 1975.

11.14 Exercises

1. An amplifier consists of three cascaded stages with the following characteristics:

	Stage 1	Stage 2	Stage 3
Gain (dB)	−3	15	5
NF (dB)	3	2	2

   (a) What is the overall gain of the amplifier?
   (b) What is the overall noise figure of the amplifier?

2. What is the available noise power of a 50 Ω resistor in a 10 MHz bandwidth. The resistor is at standard temperature.

3. A 50 Ω resistor a 20 Ω resistor are in shunt. If both resistors have a temperature of 300 K, what is the total available noise power spectral den-

sity of the shunt resistors?

4. A 2 GHz amplifier in a 50 Ω system has a bandwidth of 10 MHz, a gain of 40 dB, and a noise figure of 3 dB. The amplifier is driven by a circuit with a Thevenin equivalent resistance of 50 Ω held at 290 K (standard temperature). What is the available noise power at the output of the amplifier?

5. A 30 dB attenuator is terminated at Port 2 in a matched resistor and both are at 290 K. What is the noise temperature at Port 1 of the attenuator?

6. A receive amplifier with a gain of 30 dB, a noise figure of 2 dB, and bandwidth of 5 MHz is connected to an antenna which has a noise temperature of 20 K. [Parallels Example 11.2]
   (a) What is the available noise power presented to the input of the amplifier in the 5 MHz bandwidth (recall that the antenna noise temperature is 20 K?
   (b) If instead the input of the amplifier is connected to a resistor held at standard temperature, what is the available noise power presented to the input of the amplifier in the 5 MHz bandwidth?
   (c) What is the noise factor of the amplifier?
   (d) What is the excess noise power of the amplifier referred to the its output?
   (e) What is the effective noise temperature of the amplifier when the amplifier is connected to the antenna with a noise temperature of 20 K. That is, what is the effective noise temperature of the resistor in the Thevenin equivalent circuit of the amplifier output?

7. A receive amplifier has a bandwidth of 5 MHz, a 1 dB noise figure, a linear gain of 20 dB. The minimum acceptable SNR is 10 dB.
   (a) What is the output noise power in dBm?
   (b) What is the minimum detectable output signal in dBm?
   (c) What is the minimum detectable input signal in dBm?

8. The system shown below is a receiver with bandpass filters, amplifiers, and a mixer. [Parallels Example 11.2]

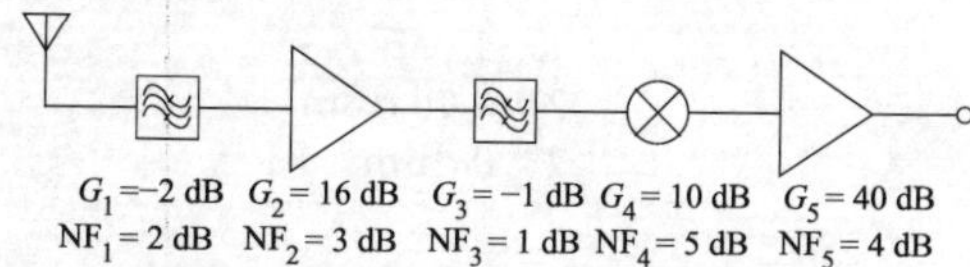

   (a) What is the total gain of the system?
   (b) What is the noise factor of the first filter?
   (c) What is the system noise factor?
   (d) What is the system noise figure?

9. An amplifier consists of three cascaded stages with the following characteristics:

	Stage 1	Stage 2	Stage 3
Gain	10 dB	15 dB	30 dB
NF	0.8 dB	2 dB	2 dB

What is the noise figure (NF) and gain of the cascade amplifier?

10. The first stage of a two-stage amplifier has a linear gain of 40 dB and a noise figure if 3 dB. The second stage has a gain of 10 dB and a noise figure of 5 dB.
   (a) What is the overall gain of the amplifier?
   (b) What is the overall noise figure of the amplifier?

11. A subsystem consists of a matched filter with an insertion loss of 2 dB then an amplifier with a gain of 20 dB and a noise figure, NF, of 3 dB.
   (a) What is the overall gain of the subsystem?
   (b) What is NF of the filter?
   (c) What is NF of the subsystem?

12. A subsystem consists of a matched amplifier with a gain of 20 dB and a noise figure of 2 dB, followed by a 2 dB attenuator, and then another amplifier with a gain of 10 dB and NF of 3 dB.
   (a) What is the overall gain of the subsystem?
   (b) What is NF of the attenuator?
   (c) What is NF of the subsystem?

13. A connector used in a 50 Ω system introduces a series resistance of 0.5 Ω. What is the insertion loss of the connector?

14. A microwave switch is used in a 75 Ω system and has a 5 Ω on resistance. The reactive parasitics of the switch are negligible.
   (a) What is the insertion loss of the switch in the on state?
   (b) What is the return loss of the switch in the on state?

15. A microwave switch is used in a 50 Ω system and has a 5 Ω on resistance. The reactive parasitics of the switch are negligible.
   (a) What is the insertion loss of the switch in the on state?
   (b) If the available power of the source is 50 W, what is the power dissipated by the switch?

16. A microwave switch is used at 1 GHz in a 50 Ω system and it has a 2 Ω on resistance and a 2 kΩ off resistance. The reactive parasitics of the switch are negligible.
   (a) What is the insertion loss of the switch?

(b) What is the isolation of the switch (i.e., what is the insertion loss of the switch when it is in the off state)?

17. The RF front end of a communications unit consists of a switch, then an amplifier, and then a mixer. The switch has a loss of 0.5 dB, the amplifier has a gain of 20 dB, and the mixer has a conversion gain of 3 dB. What is the overall gain of the cascade?

18. A three-port circulator has the S parameters
$$\begin{bmatrix} 0 & 0 & 0.5 \\ 20.5 & 0 & 0 \\ 0 & 0.5 & 0 \end{bmatrix}.$$
Port 3 is terminated in a matched load creating a two-port network.
   (a) Find the S parameters of the two-port.
   (b) What is the return loss in dB at Port 1 if Port 2 is terminated in a matched load?
   (c) What is the insertion loss in dB for a signal applied at Port 1 and leaving at Port 2 with matched source and load impedances?
   (d) What is the insertion loss in dB for a signal applied at Port 2 and leaving at Port 1 with matched source and load impedances?

19. Two isolators are used in cascade. Each isolator has an isolation of 20 dB. The isolators are matched so that their input and output reflection coefficients are zero. Determine the isolation of the cascaded isolator system?

20. A three-port circulator in a 50-Ω system has the S parameters
$$\begin{bmatrix} 0.1 & 0.01 & 0.5 \\ 0.5 & 0.1 & 0.01 \\ 0.01 & 0.5 & 0.1 \end{bmatrix}.$$
If port 3 is terminated in a matched load to create a two-port network
   (a) Find the S parameters of the two-port.
   (b) What is the return loss in dB at Port 1 if Port 2 is terminated in 50-Ω?
   (c) What is the insertion loss in dB for a signal applied at Port 2 and leaving at Port 1 with 50-Ω source and load impedances?
   (d) What is the insertion loss in dB for a signal applied at Port 1 and leaving at Port 2 with 50-Ω source and load impedances?
   (e) What is is the name of this network?

21. A mixer in a receiver has a conversion loss of 16 dB. If the applied RF signal has an available power of 100 μW, what is the available power of the IF at the output of the mixer?

22. The RF signal applied to the input of a mixer has a power of 1 nW and the output of the mixer at the IF has a power level of 100 pW. What is the conversion loss of the mixer in decibels?

23. A mixer in a receiver has a conversion gain of 10 dB. If the applied RF signal has a power of 100 μW, what is the available power of the IF at the output of the mixer?

24. A mixer in a receiver has a conversion loss of 6 dB. If the applied RF signal has a power of 1 μW, what is the available power of the IF at the output of the mixer?

25. The phase noise of an oscillator was measured as −125 dBc/Hz at 100 kHz offset. What is the normalized phase noise at 1 MHz offset, assuming that the phase noise power varies as the square of the inverse of frequency?

26. The phase noise of an oscillator was measured as −125 dBc/Hz at 100 kHz offset. What is the normalized phase noise at 1 MHz offset, assuming that the phase noise power varies inversely with frequency offset?

11.14.1 Exercises by Section

†challenging,

§11.5 1, 2, 3, 4, 5, 6, 7, 8, 9, 10, 11, 12 §11.8 18, 19, 20 §11.10 25†, 26
§11.7 13, 14†, 15†, 16†, 17† §11.9 21, 22, 23, 24

11.14.2 Answers to Selected Exercises

8(d) 5.17 dB
10(b) 3 dB
12(b) 60 dB
14(b) 29.8 dB
15(a) 0.424 dB
17 22.5 dB
19 40 dB
21 −26 dBm
23 0 dBm

Index